WHY METALS FAIL

WHY METALS FAIL

R. D. Barer
Head, Materials Engineering

B. F. Peters
Head, Metallurgy and Non-destructive Testing

Defence Research Establishment Pacific
Victoria B.C.

GORDON AND BREACH SCIENCE PUBLISHERS
New York London Paris

Gordon and Breach Science, Publishers, Inc.
One Park Avenue,
New York, NY 10016

Gordon and Breach, Science Publishers Ltd.
42 William IV Street
London WC2N 4DE

Gordon & Breach
58, rue Lhomond
75005 Paris

First Published November 1970
Second Printing August 1971
Third Printing November 1974
Fourth Printing February 1982

Library of Congress catalog card number 72 107155. ISBN 0 677 02630 7. Printed in the United States of America.

PREFACE

THE MAIN PURPOSE of this book is to acquaint engineering students and practising engineers with the mechanisms by which failures occur in metallic structures and components. Also, I believe that the book has an educational function; by the use of case histories the authors demonstrate the type of scientific reasoning or diagnostic ability which is essential in solving such problems.

All engineers regardless of their field of specialization are concerned with the behaviours of metallic materials in service and should be concerned with the mechanisms by which failures occur. Such failures rarely occur because knowledge of the fundamental properties of the materials or of the principles of design is not available. They occur, in most cases, in metallic materials which have been used widely for a long time and in components and environments which are not new or unusual.

The authors have attempted to present a broad coverage of failure mechanisms and many well-documented case histories. Although the examples are drawn from their own extensive experience, the principles are obviously applicable to many other fields.

The book is one of very few written on this subject in over ten years and will be invaluable as a reference text in a field where information is widely scattered if, indeed, it is published at all.

W. M. Armstrong
Professor of Metallurgy
Former Dean, Faculty of Applied Science
Deputy President, University of British Columbia

CONTENTS

PART 1

FAILURE MECHANISMS AND EXAMINATION TECHNIQUES

I

INTRODUCTION

MANY ARTICLES and publications are available today on metallurgical advances in "frontiers" such as space, deep ocean, ultra high strength metals, high temperature, solid state and so on. This activity is all necessary and it engages many capable people—but there is a large area involving many practising physical metallurgists and other engineers who have failings of "ordinary" metals to contend with. There are also many students in undergraduate courses who are not always exposed sufficiently to the mechanisms by which "ordinary" (and also exotic) metals may fail. It is in recognition of these two large groups for whom not as much is conveniently available—some material there certainly is but often it is widely scattered in the technical literature—that the authors believed a volume such as the present one would help to meet a need.

On the various occasions when one of us has presented a paper illustrating metal failures with case histories, the reaction has been most favorable—not because of any particular gift for erudition, but because the examples seemed to meet a universal need to see metal failures and to understand how one might proceed to find an answer. At such times someone has invariably remarked how useful these and similar examples would be if available in book form. To those who by this type of comment sparked the idea of this volume, the authors express their appreciation.

Another reason for believing this type of approach could fill a need in the midst of advancing metallurgy, is the type of plea first raised forcefully by Admiral Rickover*. It was a plea for "painstaking care" and "attention to detail". It was a plea aimed at preventing failure of common everyday components and *not* failure of the advanced or the exotic. This "attention to detail", this need for reliability of everyday components, is related to the understanding of how and why *ordinary* metals behave as they do and fail as they do. Possibly these case histories can contribute to this understanding.

The examples chosen for this display of case histories are not intended to be definitive—they simply were the best we had available to illustrate a type of problem. They are not intended to be exhaustive—rather to help point the way—but enough of them were included to give the feeling that similarities are common even in entirely different metals or applications. While it soon

* Hyman G. Rickover, "The never-ending challenge", *Metals Engineering Quarterly*, February 1963, Vol. **3**, No. 1, pp. 1–16.

becomes obvious that the examples were from a marine environment—largely because the authors' experience is currently in this area—it should be borne in mind that such an apparently limited selection is not necessarily detrimental for the diagnosis of metal failures. One needn't be an expert in railroad machinery or aircraft or submarines to diagnose metal failure. In fact, in these diverse fields, some of the most serious of metal failures are almost predictable; i.e. austentitic stainless steel will suffer crevice corrosion whether used in a train or a seaplane or a submarine.

We believe that the failures illustrated and others that individuals experience in their own plants, are of considerable value to designers, engineers, and draughtsmen. It should be possible for all involved in design and manufacture of a product to profit from these failures, to ensure that "attention to detail" includes the avoidance of so many of these failings whether in design, selection of material, or protective measures which were omitted.

The many case histories are not intended for reading straight through but for examples to be used in an educational sense—to provide understanding of *approaches* in solving this type of metallurgical problem. The book is also aimed at final year undergraduate metallurgical, chemical, mechanical and materials engineers who are usually exposed to much fundamental and theoretical material in their studies but often very little, unless they're fortunate in summer employment, of the type of problem exemplified in these case histories.

Some of the case histories may appear trivial to metallurgists with even minimal experience—they were included again for their educational value as examples of what to look for and of the value of certain techniques in solving them. Some of the case histories are considerably more complex than they have been made to appear—often representing studies and developments extending over a considerable period and including comparison with work of others in the extensive literature—nevertheless the key ideas or clues are valuable and in a book of this size had to be condensed. It is hoped that with the supplementary reading suggested, plus further searching in current metallurgical journals, the interested student or engineer will be able to "dig" as deeply in any specific area as he wishes.

In our experience, there is a hunger on the part of advanced undergraduates to see, feel, and explore actual failures. In somewhat similar fashion engineers who are not by training metallurgists, are called upon to evaluate some failures in the absence of any one else with remotely relevant experience, and these engineers also hunger for expositions of case histories for their educational value and to serve for comparison with their own diagnosis of comparable problems. These people will also gain much insight (we hope) from an approach which permits more than just judging from surface appearance which can be misleading.

While it may not be readily apparent to those who have not worked in a materials group—the interdependency of the various skills, techniques and background are very much a part of the way these problems were diagnosed. For example; the availability of excellent spectrochemical analysis is vital in assessing correct choice of material; the understanding of how metals behave when exposed to corrosive environments is an insight of major importance; and the establishment of exact stress levels sometimes provides the definitive clue. There are many other illustrations of how the various disciplines work together effectively in the broader materials engineering function and these will become apparent in the case histories.

On a number of occasions in our experience, metal failures have been considered normal and simply a matter for replacement: the sort of "it bust" type of approach, but fortunately, on enough occasions some other person was sufficiently curious or wanted technical verification so that pieces became available for diagnosis. And in many of these instances uncovering of an important and somewhat widespread mechanism resulted therefrom. This applied particularly to the martensitic stainless steel valve parts, case history 5, B-3(a), some of which might readily have been attributed to failure from a design fault or to failure from fatigue. It is one of the authors' hopes that study of these case histories will lead to some sense of what are the best approaches in first attempting to diagnose a problem—some sense of judging the best move or moves to give reasonable economy of effort and still not miss important clues. It's true, of course, that any real judgement of this nature must come from practical experience—preferably from working in a group constantly exposed to a variety of such problems.

One diagnostic tool that probably stands out as most necessary from scanning these case histories, is the metallurgical microscope. With few exceptions, important clues can be derived from such an examination, and in some instances, the complete answer. Again, as with any method, experience is not easily by-passed. Study of these examples and others, should however, provide guidelines of value.

In the chapters to follow a number of case histories have been set out. These case histories are sufficiently varied to serve, we believe, for instructional purposes. They may even provide instructors with ideas on the type of defect or failure they might try to pick up from a local metallurgical laboratory for first hand instructional purposes—in the same way geologists handle mineral samples. They have been set out in functional categories because it was believed this could prove useful to someone interested, for example, specifically in boilers. The other useful categorization, i.e. by failure mechanism has not been overlooked—it is in fact the way each functional category is broken down. For example if one is interested in stress corrosion cracking one need only examine case histories of interest under B3 throughout. This mechanism index is tabulated on page 39.

Some amplification of several common failure mechanisms is given on pages 9 to 16. These mechanisms include brittle failure, fatigue, corrosion fatigue, stress corrosion cracking, crevice corrosion, and galvanic corrosion. Brief though these outlines are, it was believed they could be useful in an introductory sense and by including them in this fashion it would save some unnecessary repetition under the relevant case histories. For those interested in additional understanding the excellent sources given in the References and Supplementary Sources (page 335) should prove most useful.

In a similar sense, we have included a brief look at the tools commonly of value in diagnosis of failures such as those in this text. Again the descriptions of the tools are certainly not intended to be comprehensive, but of an appropriate level to those not yet specialized, and to some extent, to illustrate the interdependency of one specialist on others. This is the pattern in many materials engineering groups. An example of the interdependency, already mentioned is the support provided by spectrochemical analysis, because metallurgists are rarely able to provide their own chemical analyses, and even more rarely would they ever have the time. Similarly, support from stress analysts and radiography specialists is often provided in larger groups.

Some even more specialized tools may be called upon for very specific problems—an example is electron microprobe analysis which was required in case history 1.C 2(a). It is beyond the scope of the present text to deal with such specialized tools, which are certainly outside what one might expect in the usual materials engineering group.

The first chapter on case histories deals with boiler tube failures, not because there is something extra valuable about boiler tubes which have failed (!) but because this is an area to which the authors have been well exposed and from which some useful approaches may be gleaned. These examples have been explored rather more fully than most others in the book partly because they are more complex and partly to illustrate how one assesses the factors involved and the information available in attempting to reach an answer. The second chapter deals with a variety of boiler-related problems. These were chosen because in each there was some aspect that was interesting and worthwhile as a demonstration of metallurgical diagnosis. The third chapter concerns itself with failures in heat exchangers, which, with a couple of exceptions, are the result of some variety of corrosive action. These are examined and evaluated, and remedial measures suggested.

In chapter four, a different assortment of failure mechanisms are explored in a number of pressure vessels. In this group the troubles largely relate back to errors in manufacturing. Chapter five deals with valves and valve components in which problems have been experienced. These troubles range from a variety of corrosion effects to original manufacturing failings such as an inferior casting and an inferior weld.

In chapter six, several failures of piping systems are explored for their interesting characteristics. Two failures from excessive pressure demonstrate that no fault attaches to the metals involved, whereas other failures occurred through poor choice of metal, abuse of metal in fabrication, or where the metal was not resistant to corrosion by the particular environment.

In chapter seven, a variety of bolts and fittings are examined for causes of failure. Many of these are vital parts because human life may depend on their reliability: among these were the rupture of a bolt in a ship's steering system which came very close to causing loss of the ship, and failure of a scaffold clamp which fortunately fell apart only when a high scaffold was being dismantled. Correct diagnosis here is obviously critical. Chapter eight is a short one dealing with three examples of weakening in wire ropes. One is by kinking, another by crevice corrosion of aluminium wire rope, and the third demonstrates the danger in braze faults propagating into sound wires.

Chapter nine deals with what we have labelled, for want of a better category, structural and fittings. Included are three corrosion fatigue failures of steel in sea water, and failures of aluminum alloys prone to corrosion in a marine environment.

In chapter ten, failures of pumps and fittings cover those initiated by fatigue, those by stress-corrosion cracking and other types of corrosion attack. Two of the problems were traced back to manufacturing errors. Chapter eleven reviews a few problems associated with vehicles—a particularly important one involving human life is the unexpected fatigue rupture in axles or steering assemblies.

Chapter 12 deals with an assortment of failings in internal combustion engines. As might be expected, fatigue failures and their prevention are significant here. Among other examples, the role of strain gauges in establishing a corrosion fatigue mechanism is particularly interesting. Chapter thirteen explores only a few of the many bearing problems, both ball and babbitt type, that one might encounter. Again the microscope provides major clues, while chemical analysis and ultrasonic examination also contribute.

Chapter fourteen on springs brings out the need for care in these highly stressed units. Some of the troubles encountered involve corrosion and also defects, whether originally in the steel or introduced by grinding. Chapter fifteen explores propeller failures by impact and by corrosion, and the toughness needed in anchor shanks.

Chapter sixteen is a "clean-up" chapter to examine failures and problems which could not readily fit into earlier chapter groupings. Included are discussions on the importance of correct dynamic balance in helicopter engine components, liquid metal penetration of another (stressed) metal, and the importance of adequate and correct heat treatment in steel to avoid failure in diverse uses.

Acknowledgements

The authors wish to express their appreciation to their associates in the Materials Engineering Section at the Defence Research Establishment Pacific (formerly Pacific Naval Laboratory), in particular the assistance provided by Mr. C. A. Waggoner and Mr. J. A. H. Carson in their respective fields. Mr. Waggoner was responsible for the spectrochemical analyses which are an integral part of many diagnoses in this text. Mr. Carson kindly provided from his files a number of examples in which corrosion mechanisms were involved; in particular we are indebted to him for case histories 12.B-7 (cavitation) and 6.B-8 (aluminum clad piping).

We also wish to acknowledge the very considerable help provided in preparing the manuscript by Mr. R. H. Temple, Mr. D. S. G. Tweedhope, and Mrs. A. A. James.

The authors acknowledge the permission of the Defence Research Board of Canada to publish the material contained in this book. The problems and investigations discussed have arisen in the course of the authors' work at the Defence Research Establishment Pacific. Any opinions and decisions expressed are the responsibility of the authors.

II

SEVERAL COMMON FAILURE MECHANISMS

THE FOLLOWING PAGES provide a brief explanation of several common failure mechanisms. These include

a) Brittle Failure
b) Fatigue
c) Corrosion Fatigue
d) Stress Corrosion Cracking
e) Crevice Corrosion
f) Galvanic Corrosion.

In this chapter, these failure mechanisms are introduced sufficiently to provide a reasonable working understanding. For those who need a deeper understanding the references suggested (p 335) will generally provide the needed depth. For more current work the technical literature may be consulted. Often there are texts covering symposia on a single one of these mechanisms, as for example, those on Stress Corrosion Cracking.

By presenting these mechanisms in this fashion, considerable repetition was avoided when dealing with the specific failures in the remainder of the text. Conversely when studying a particular problem it will be useful to refer back to this chapter on failure mechanisms.

a) **Brittle Failure**

For a long time the most puzzling of metal failures to metals engineers have been those which can be classified as brittle failure. This includes those failures which give no evidence of ductility or stretching, but appear to separate or fall apart, often with little warning. One might also include here those failures which on further diagnosis prove to be the result of fatigue or corrosion fatigue or stress corrosion cracking. These three types will be discussed separately.

It was not until the great number of brittle ship failures during and following World War II that several intensive investigations led to some well-defined landmarks in an understanding of factors involved.* It is actually interesting when one delves into brittle failure historically that such failures have been plaguing engineers for some time. Some 80 years ago it was recognized that there was something unusual about failure in a brittle manner of steel which met the usual tests for strength and ductility. Approximately 50 years ago a French investigator established the significance of the herringbone pattern which occurs on plate fractures and a few years later, an independent investigation in the Boston Navy Yard reproduced brittle fractures in notched plates (also with the herringbone pattern). This work and developments which might have arisen therefrom were not at all well known, and occasional brittle failures continued to occur of storage tanks, pressure vessels and even bridges, almost always during winter temperatures.

The wartime failures previously mentioned were related to the rather massive increase in the use of welding without much of a backlog of experience to draw upon. From studies of these failures certain common factors were eventually established as being significant. These included the fact that failure occured in cold weather and heavy seas, from poor workmanship tending to leave defects in welds, and that the steels were inferior in their ability to withstand sudden loading in the presence of a defect (notch) at service temperature. There proved, in fact, to be an excellent correlation between toughness as measured by the Charpy V-notch test and the tendency for these steels to rupture in brittle fashion at temperatures experienced by the ships. This information was condensed by a later requirement that steels had to absorb at least 15 ft.lbs. at a specified "transition temperature"† set somewhat below the lowest anticipated service temperature. The test used to evaluate "transition temperature" of failed ship plates was the Charpy V-notch but because of discrepancies (including a size effect) many other tests have been used, including a "tear test", explosion bulge test, a crack arrestor test, etc. More recent work by Pellini and his associates extends these principles more widely to more complex steels than used for ship plate. This work and the previous tests are all fully described in the related literature, but take us beyond our intended scope in this book. For those interested in the growing field of "fracture mechanics", there are several introductory texts[48,49] and articles on its application.‡

One may well ask at this point "what is the relevance of all this to the present text?". The answer is that some brittle fractures are described which are best understood if the reader has been exposed to the elementary concept that

* E. R. Parker, *Brittle Behavior of Engineering Structures,* John Wiley and Sons Inc., 1957, especially Chapters 11 and 12.

† Transition between tough and brittle fracture. In some tests appearance of fracture is also used.

‡ F. M. Burdekin, "The practical application of fracture tests to prevent service failures", *Welding Journal,* Vol. **47**, No. 3, March 1968, p. 129S.

steel (and other metals) can behave in a brittle fashion under one set of conditions (i.e. temperature) and in a ductile fashion under other conditions. Steel can be made tougher by changes in its chemistry (e.g. increase in manganese relative to carbon content) and by changing its microstructure.

While it is not rigorous enough for those who faced brittle failure as a major activity for years, we found that even the relatively simple Charpy V-notch has provided us with useful and practical guidelines. We would probably horrify some fundamentalists in impact studies by indicating that Charpy V-notch helped us to demonstrate a marked difference between unsound, slightly discolored hickory (for axe handles) and normal hickory! We have found Charpy V-notch capable of differentiating tidily between improperly heat treated samples of the same steel (page 200 and page 212).

We have been unorthodox enough to use Charpy V-notch to demonstrate a difference in behaviour between brass and bronze as temperature increased (page 321).

This presentation is hardly more than a glimpse of a major field of study in metallurgy (see references 44–50, p. 337).

b) **Fatigue**

Another type of brittle failure which can be disastrous particularly because of the unexpected rupture resulting, is known as fatigue.

The term fatigue implies a tiring or weakening after a lengthy period of service. What actually happens inside the metal is still the subject of theorizing by solid state experts. Essentially, the current position is that "... fatigue is the consequence of uninhibited fine slip which gradually becomes intensified and develops into a crack. This crack then propagates until the remaining cross-section is too small to carry the load, and cataclysmic failure occurs"*. Another way of expressing it might be that repetitive stressing at below usual static rupture levels results in local overstressing and a very small rupture initiates, usually at a weakness or stress-raising point, and eventually propagates to rupture of the cross-section.

The local overstressing starts at a surface notch or defect which serves as a "stress-raiser". A discontinuity on the surface produces a local stress-concentration, which under repetitive action is beyond the capability of the metal to withstand. The term "endurance limit" is applied to a stress level which a metal can withstand with indefinitely repeated cycles of stress. Above this endurance stress, failure results.

The surface defects or "stress raisers" may be machining faults, sharp corners or inadequate fillets (see page 264), file or punch marks, corrosion

* G. R. Gohn, *Materials Research and Standards,* February 1963, p. 114.

grooves or pits. If corrosion is still active, the combined action gives corrosion fatigue cracking.

It is commonly accepted engineering practice to consider that the "endurance limit" of steels is approximately half their ultimate tensile strength. Of course this is lowered by even minor imperfection in surface finish and by other variables which may affect the surface. It is also generally accepted that there is no endurance limit when any corrosive action is present; that is, the level of stress which the sample will withstand continues to fall with increasing number of cycles. To give some feel for the magnitudes involved, a steel of 200,000 psi ultimate tensile has an approximate endurance limit of 100,000 psi (in a smooth surfaced fatigue specimen) and a value of some 8,000 to 10,000 psi after several million cycles in sea water. This compares with some 30,000 psi for monel or 60,000 psi for titanium under the same number of cycles in sea water.

The consequences of cycling stresses need to be constantly borne in mind by engineers, designers and often by the people on the shop floor doing the actual work. The best engineering or design can be negated by someone cutting too sharp a fillet or leaving a rough machining groove. Defects in stressed welds are also in the same category. Ideally all of these faults should be caught by inspecting authorities, but obviously they often are not. There is no substitute for constant attention to detail and constant educational efforts so that all who are involved in almost any aspect of material use understand the significance of the factors involved in fatigue.

c) **Corrosion Fatigue**

As already indicated in the comments under "Fatigue", the ability of a metal to withstand cyclic stressing is weakened considerably when corrosive effects can also act on the surface. It is of course, possible that a particular environment which is normally considered corrosive may not be so to a particular metal (e.g. monel in sea water) and in this instance the normal fatigue strength is unimpaired. We are concerned mostly with metals which are sensitive to the environments to which they are exposed and a number of case histories illustrate this.

There is a general belief that corrosion and cyclic stressing acting jointly are more severe than might be expected cumulatively from each. One factor which influences this could be the inability of the metal surface so affected to either develop a protective oxide film or even a partly protective corrosion product film. In fact, it seems likely that corrosion products would function in a wedging fashion to intensify stress at the advancing front. It is not our function here to enter this fundamental research field which is well covered by many texts (e.g. Evans,[52] Uhlig[51]) and articles, but only to indicate some generally accepted principles.

Corrosion fatigue is often initiated by corrosion pits as may be seen in case histories 1.B-2 and 4.B-2. Corrosion fatigue is usually transgranular, displaying thereby little concern for the microstructure. This is well demonstrated in case history 1.B-2. The transgranular behaviour demonstrates the primacy of the mechanical or stress cycling component. It is also likely that some integranular action might accompany corrosion fatigue where the fatigue component is small or absent for various periods. This interesting action is demonstrated in case history 12.B-2, specifically Figure 12–16. Along similar lines is the corrosive action proceeding at right angles to the corrosion fatigue crack. In case history 7.B-2 there obviously were periods when the most corrosion-prone areas, along the rolling direction, were involved.

The important point is that in the presence of corrosive factors, a normally respectable fatigue life can readily become a rapid and sometimes catastrophic failure. The possibility of overlooking corrosion will obviously change markedly the reliability expected.

d) **Stress Corrosion Cracking**

A type of failure occurring in many metal families under certain conditions is stress corrosion cracking. In simplest terms this is a cracking which proceeds under the joint action of a steady stress in the presence of a corrosive environment. It should be particularly noted that the corrosive environment referred to here is often a specific one, and one which is not generally harmful to the metal in the absence of stress. For example, moderate caustic concentrations will not particularly corrode mild steel when no stress is involved (see Figure 1–16). Brasses are not otherwise harmed by ammonia type products in the minor concentrations which lead to stress corrosion cracking (see Figure 6–14). Similarly, the aluminum alloys which are prone to stress corrosion cracking in a marine environment may show complete absence of other corrosive action (see Figure 7–26), or at most only the slightest evidence of corrosion products as in case history 7.B-3(b). On the other hand, there are situations where the environment is markedly corrosive to the metal and the part played by the mechanical factor, or stress, is almost negligible.

The susceptibility of certain metals to stress corrosion cracking may be alleviated by heat treatment which puts them in a softer condition or one with inherently lower residual stresses. This concept applies in the case of martensitic stainless steels (case history 5.B-3(a)). Unfortunately, in this instance, the effect results in conflict with the very reason one probably chose this metal in the first place: corrosion resistance combined with high strength. Other approaches are then required. On the other hand, the resistance of the aluminum–10% magnesium alloys to stress corrosion cracking is enhanced by careful heat treatment which leaves it in a stronger condition than furnace cooling (case history 5.B-3(d)).

Characteristically stress corrosion cracking is generally evidenced under the microscope as intergranular cracking. This is definitely so for mild steel

(Figure 1-17), martensitic stainless steel (Figure 5-10), aluminum–4% copper (Figure 7-21 and Figure 7-27) and brasses (Figure 6-15 and Figure 7-19). Stainless steel generally shows transgranular stress corrosion cracking (Figure 2-17).

The cracking produced by stress corrosion is particularly hazardous because it is often unsuspected and because it can result in easy separation of a vital component at a load which would normally have been handled readily. This is particularly true where the metal is not appreciably attacked in the absence of stress and the lack of reliability is hidden. Illustrations of this consequent hazard to personnel are case histories 7.B-3(a) and 7.B-3(b).

Specific means to lessen the hazard and adverse effect on reliability are outlined in discussion of the case histories under category B-3 in the following chapters.

A comprehensive discussion of stress corrosion cracking in several metal systems is given in Evans,[52] including the influence of heat treatment and alloying elements.

As with corrosion fatigue, the relative roles of corrosion and of the mechanical component are the subject of many articles on the fundamental aspects of stress corrosion cracking. A commonly held view is that the stress raising effect of the first bit of corrosion provides a ready means for additional propagation, and that protective films which might try to form are easily ruptured by the stress. Another interesting viewpoint among many valuable papers[58] is that the wedging action of corrosion products plays an important role where stress is otherwise minimal. For those who are not specialists but would still like to understand fundamentals, including exposition of why metals crack in a transgranular fashion, we would suggest the article by Swann.[62]

e) **Crevice Corrosion**

An interesting, and in some ways insidious type of action, is known as crevice corrosion. This is the corrosion that occurs under debris on a surface that is immersed, under washers or gaskets, or between sheets of similar metals. It is the effect which can be harnessed to cut a piece of immersed stainless steel with a rubber band. (The rubber band shields the metal beneath it, which then becomes an actively corroding area which in time crumbles and the rubber band slowly advances.)

The action is related to the difference in oxygen available in the freely aerated surrounding solution and the more limited supply in the film of liquid trapped in a crevice. The resultant effect is particularly strong with those metals such as stainless steel and aluminum which depend on oxygen availability to repair weak areas in their protective oxide films. Under a deposit or other crevice, there is insufficient oxygen and, coupled as it is metallically to the rest of the metal which has access to plentiful oxygen, there is a strong tendency for the sheltered metal to become an anode and the remainder, a large cathode. A corrosion cell is thereby set up and the action can be intense.

It is well, therefore, to bear the hazards of this type of corrosive action in mind when planning to use stainless steels, aluminum alloys, and even copper alloys. Careful design is required to avoid places where moisture can be trapped, and to use only butt welded joints (not laps), and to seal with non-absorbing, non-cracking fillers those surfaces which can't be handled any other way. When used in situations where the metal is immersed, periodic cleaning to avoid debris accumulation on surfaces is a wise precaution.

It should also be borne in mind that gaskets, washers, and other spacers are also suspect. In many situations these will tend to hold small films of moisture in contact with metal surfaces. Again, if these must be used they should be sealed with non-absorbing, non-cracking sealants. There is a tendency to believe that a rubber (elastomeric) washer is safe. This assumption leads to unsafe practices—rubber perishes and does not maintain its tight seal against metal.

Screws, bolts and other fastenings are often overlooked as providing ideal situations for crevice damage unless sealed off against moisture. This attack on fastenings which may be holding critical parts together, can be dangerous. We have observed stainless steel screws which have disintegrated when used as fastenings in small wooden boats (used in sea). Case history 7.B-5 describes crevice corrosion of aluminum bolts in large wooden boats (page 210).

Other examples of destructive crevice action are given in case histories 1.B-5 (page 72) dealing with worked boiler tube ends, 5.B-5 on attack of a chromium stainless valve seat (page 153), and 3.B-5 (page 109) on an interesting crevice action in copper–nickel–iron heat exchanger tubes.

f) Galvanic Corrosion

Galvanic corrosion is well known to anyone with even an elementary understanding of corrosion mechanisms. In its simplest aspect it refers to the intensified, usually local, attack that results from coupling a "noble" metal to one less "noble". Noble here refers to the relative behaviour of a metal in relation to another metal which is more active in that particular environment. For example, there is a galvanic series or listing of metals by their relative tendencies to corrode in sea-water. In this series zinc is less noble than iron which is less noble than copper, and so on. One would therefore expect that the coupling of zinc and a copper-based alloy where moist sea air provided the electrolyte, would result in severe corrosion of the zinc. This effect would be lessened greatly if the area of copper-based alloy were small or if there were poor metallic contact between the two, as with corrosion products.

Incidentally, the principle of having a less noble metal (e.g. zinc) corrode and thereby protect another metal (e.g. steel) is used to provide "cathodic protection". Which is another way of saying that the cathode (steel) is protected by the anode (zinc).

It is most desirable to avoid metallic contact of dissimilar metals which may somehow become moist or which may operate in an environment permitting current flow (as in a battery) between two metals. It is preferable to use similar metals, i.e. those which would develop only negligible potential difference between them; to prevent presence of an electrolyte by coating in some manner; or even by careful insulating techniques to avoid metallic contact.

In this text, examples where galvanic action has played a part in metal failure are described under category B-1. An interesting problem is described under 5.B-1 where an alloy normally considered satisfactory for service in sea water, the aluminum–6% silicon alloy, suffered severe intergranular corrosion at flanges coupled to stainless steel. Another case history where severe pitting of a shaft was the result of galvanic action between bronze and steel components is 2.B-1(c).

Again more fundamental treatment is given in Evans[52] and Uhlig.[51]

III

THE METALLURGIST'S TOOLS

Introduction

THE MODERN UNIVERSITY metallurgist has at his disposal a variety of apparatus which can help him to glean information from a metal sample. These range from hardness testers to electron probe microanalysis equipment. However, most practising metallurgists, in small plants, are somewhat limited as to the tools available to them. This chapter will cover most of the basic metallurgical and analytical types of equipment present in a well equipped small materials laboratory.

It is emphasised that these are tools and as such give relatively little real information to the inexperienced. A man of experience is far more useful than many fine pieces of equipment, but of course the experienced metallurgist can be far more valuable when suitable equipment is at his disposal.

The metallurgist, engineer or plant manager will be asking "Why did this metal fail?" and "What can we do to ensure that it does not happen again?" He may be satisfied as soon as he finds that the manufacturer has not met a specification. He may be concerned with deciding what to do with similar defective equipment still in operation and may therefore want to be able to evaluate the problem non-destructively. He may want to specify a change in manufacturing techniques which will ensure that the problem does not recur. In any event, he will try to judge, based on his experience and in order to conserve effort, which of the many tools will be most informative. If a casting leaks, a radiographic examination might be a first test. If a bolt, specified to be hardened, deforms significantly, a hardness test will be in order.

If a metal has fractured, it is of course most useful to visually examine the failure. From the fracture, it can often be established that the failure was due to fatigue, as in Figure 12-9, or torsion as in Figure 7-14. If the casting was of poor quality, the fractured surface often has a dendritic appearance as in Figure T-1. Dendrites, as in Figure T-2, are often observed in sectioned defective castings. A binocular bench microscope can be used to advantage in examining these fracture surfaces. Note also the granular appearance of the fracture in Figure 15-3. Superficial evaluations of apparently brittle failures as in Figure 5-6, are often misleading. Here, it is the authors' experience that, of all the tools, the metallurgical microscope is the most useful.

In this chapter, attention will first be given to the metallurgical microscope. The usual physical tests, their significance, and the equipment will then be

Figure T-1 Fracture surface showing dendritic appearance of porous casting. ($\times 5$)

Figure T-2 Sectioned defective casting showing dendrites in casting void. ($\times 15$)

discussed briefly. Consideration will be given to the numerous non-destructive tests which are powerful tools for specific problems and provide information not otherwise readily obtained. Finally, the analytical tools, often next in importance to the metallurgical microscope in solving failure problems, will be introduced.

The Metallurgical Microscope

The standard biological microscope, familiar to most high school students, transmits light through the sample under observation. However, because metals are opaque, light has to be reflected from the surface of a metal sample.

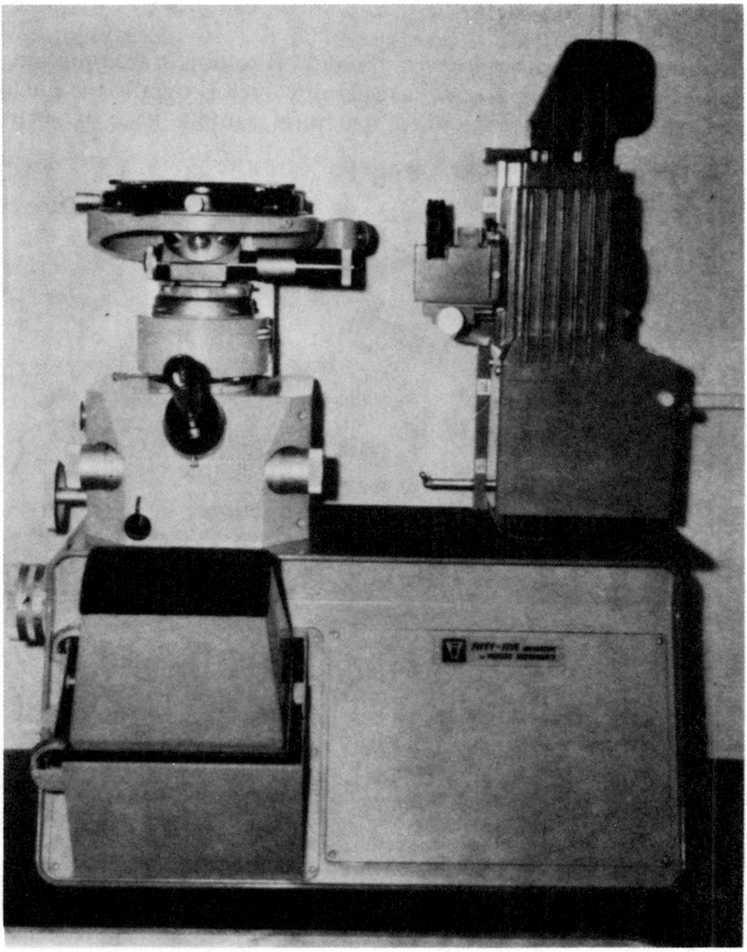

Figure T-3 Metallurgical Microscope or Metallograph.

The metallurgical microscope has, therefore, a slightly different optical arrangement from that of the familiar biological microscope. A plain glass reflector diverts a beam of light onto the specimen surface and then transmits the reflected light into the ocular lens system of the microscope. Metallurgical microscopes range from simple and inexpensive desk units to elaborate research laboratory models, complete with polarized light sources, built-in photographic systems and microhardness testing attachments. One type of metallographic unit is shown in Figure T-3.

Complete descriptions of the makeup of various metallurgical microscopes, the types of lenses, the optimum methods for illuminating samples and the proper photographic techniques are given elsewhere.[19] In any event, each microscope will have an instruction manual complete with this type of information.

The purpose of metallurgical examination is to obtain information about the grain structure of a material, a structure which is usually too fine for the unaided eye to see. To observe these structures, samples have to be cut from

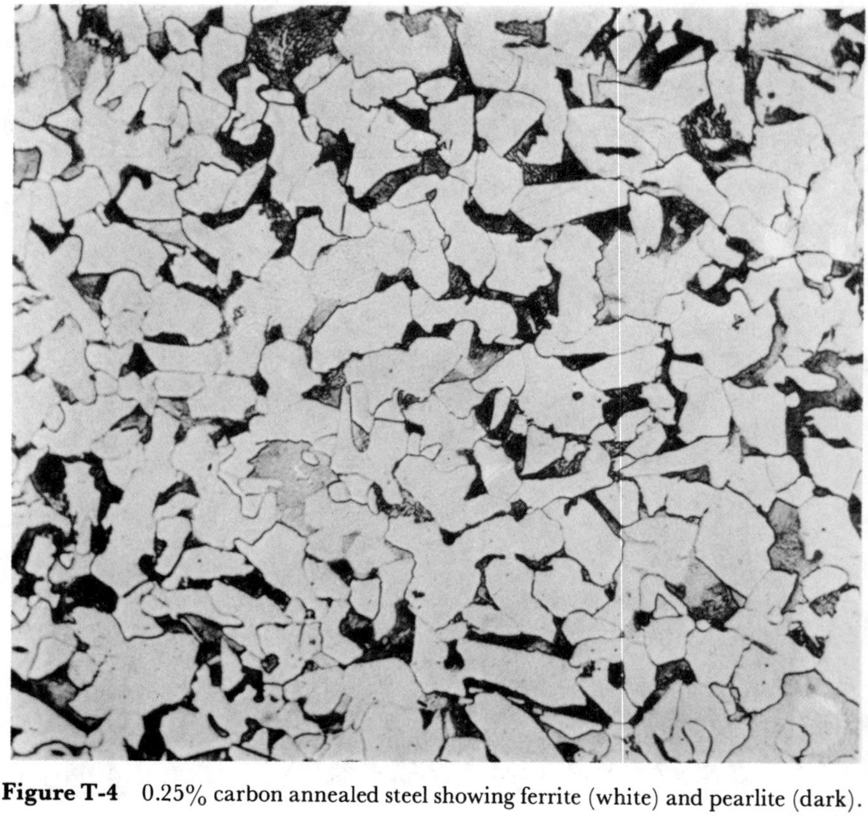

Figure T-4 0.25% carbon annealed steel showing ferrite (white) and pearlite (dark). (×500)

the components and carefully polished. The sample may, after polishing, be lightly etched by a chemical swab or immersion to outline the crystal structure. The correct selection of an etchant is a topic in itself and will not be discussed here; tables of etchants and etching techniques are outlined by Kehl,[19] The Metals Handbook[1] and elsewhere.

A considerable amount of information about composition and fabricating history can often be obtained from metallurgical examinations. For example, a common microstructure observed by metallurgists is a low carbon steel as shown in Figure T-4. The structure shows ferrite grains (white) and areas of eutectoid pearlite (dark). This structure gives an indication of both chemical analysis and fabricating history.

The relative amounts of ferrite and pearlite indicate the approximate carbon content of the steel. This information is obtained from a standard iron–carbon phase diagram.[1] Had the steel been completely a pearlite structure the metallurgist would know that the steel contained approximately 0.8% carbon, and a half and half structure represents a 0.4% carbon steel and so on. The carbon content of the steel is important because it gives an indication of the strength, ductility and potential hardness of the steel. Also, it indicates whether precautions will have to be taken while welding.

The two phase structure shows that the steel was hot worked (and not deformed by cold working) and that it was slowly cooled. The homogeneity of the structure shows that it is not an as-cast or as-welded structure. A

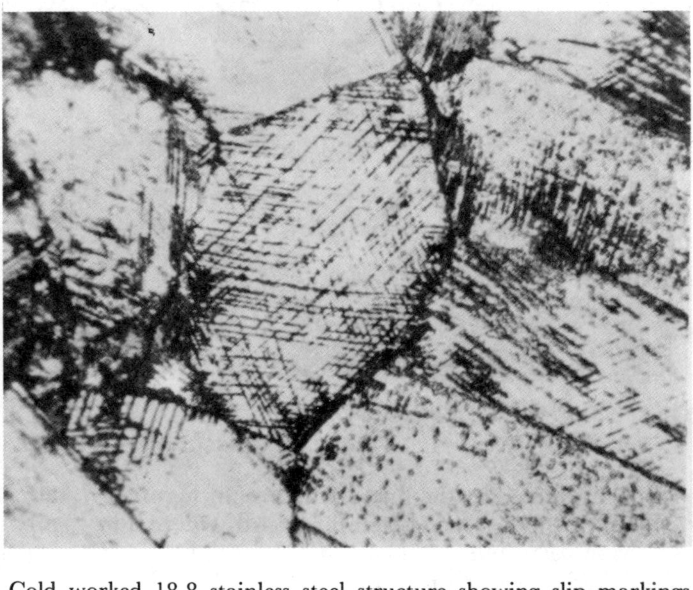

Figure T-5 Cold worked 18-8 stainless steel structure showing slip markings in grains. (×400)

normalized steel, a quenched steel (compare Figure 1-4) and a cast steel have different structures.

It should be added that increased rates of cooling from that of furnace cooling result in more pearlite than would be expected from an iron–carbon phase diagram. An example of a 0.45 carbon steel which contains relatively little ferrite because of rapid cooling is shown in Figure 7-13.

Additional information regarding the fabrication history of a metal part is often revealed by the microstructure. Slip lines associated with metal deformation at lower temperatures (e.g. room temperature) are readily observed under the optical microscope as shown in Figure T-5. The cold worked structure will be harder and more brittle than the annealed structure. At lower magnification, Figure T-6, the nature of the cold worked structure associated with a forging defect is outlined by the microscope. That a part has been exposed to sudden loading or impact, can be revealed in iron by the presence of mechanical twins, as in Figure 7-10 (page 197).

Another example of the value in using a microscope to determine fabricat-

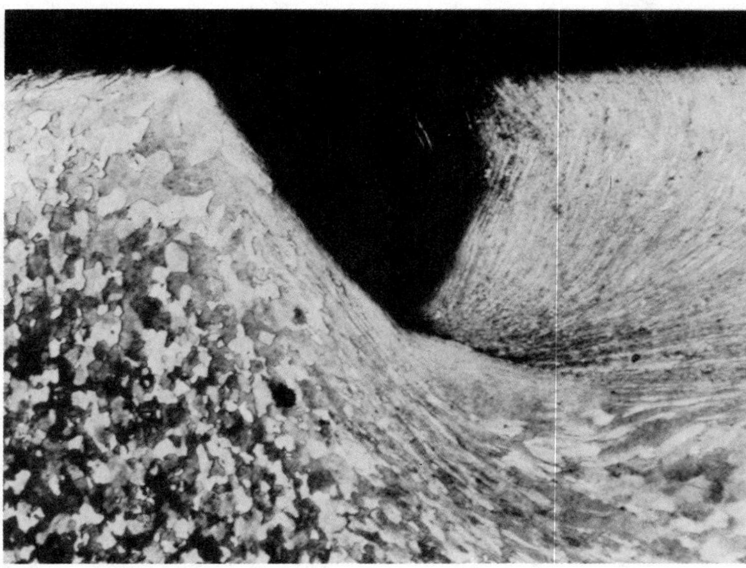

Figure T-6 Section through scored steel plate showing nature of deformation of grains. (×70)

ing history is shown by the microstructures in Figure T-7 and T-8. In Figure T-7, the bolt threads have been machined, whereas in Figure T-8, the bolt threads have been rolled. The two bolts would have different fatigue properties, the rolled bolt being much better.

The structures noted above are, of course, only a few of the many micro-

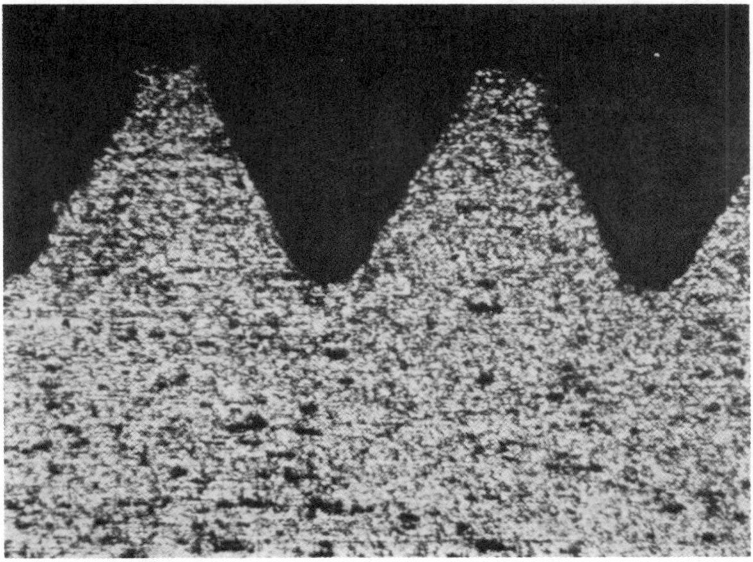

Figure T-7 Machined threads of standard steel bolt. (×80)

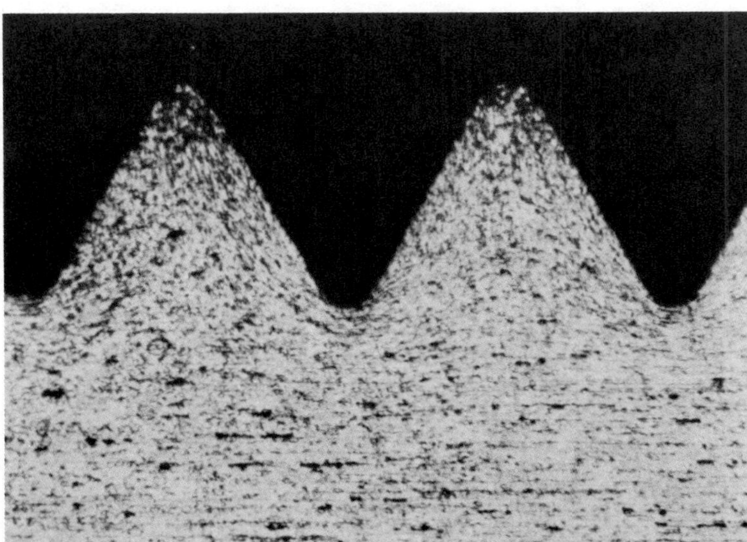

Figure T-8 Rolled threads of good quality bolt. (×80)

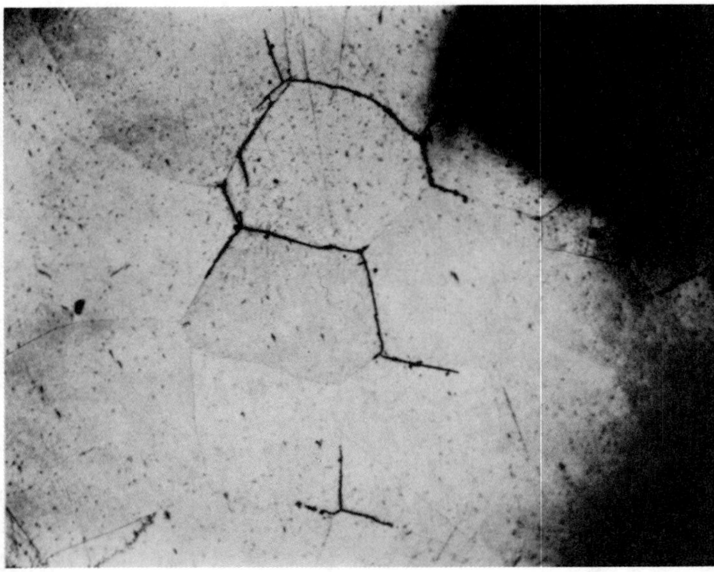

Figure T-9 Intergranular attack of high strength beta brass propeller (Note also Figure 14-4.) (×200)

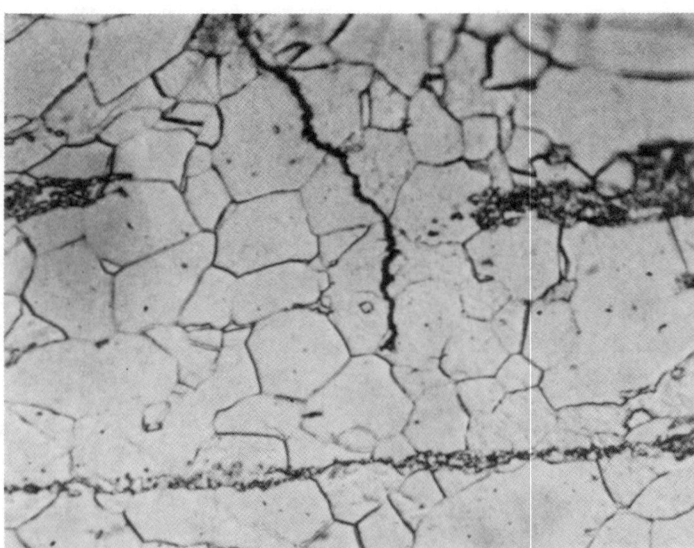

Figure T-10 Transgranular fatigure crack in low carbon steel cotter pin. (×1000)

structures which can result when various alloys are fabricated in different ways. When evaluating failures, microscopic evaluations will indicate whether cracking is intergranular (as in Figure T-9), transgranular (as in Figure T-10), branched (as in Figure 2-27) or whether attack is general. The microscope, then, is the primary tool of the trouble-shooting metallurgical engineer.

In recent years electron microscopes have taken a prominent place in metal failure analysis. While these expensive microscopes are not part of the equipment of most small materials groups, and electron microscopical results are not shown with the case history analyses in this book, it is noteworthy that very useful fracture analysis information has been obtained utilizing electron microscopical techniques.*

The electron microscopes not only have the advantage over the optical microscopes of much higher magnification and resolution (the ability to separate extremely close lines or points) but they also have the advantage

Figure T-11 Electron Microscope.

* W. Wiebe, "Fracture surface analysis by means of electron fractography", *Canadian Aeronautics and Space Journal,* **13**, 3 March 1967, p. 135.

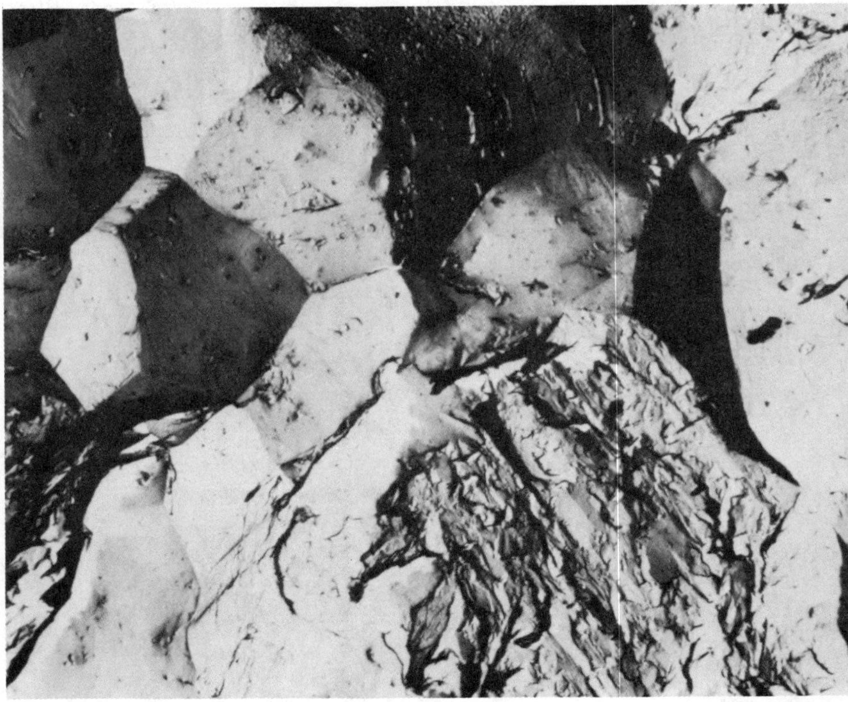

Figure T-12 Replica of quench crack in hardened steel. Crack is primarily intergranular. (×4000)

of their great depth of focus. Thus very irregular faces of fractured metals can be conveniently examined with both the "bumps" and "hollows" in focus.

Two types of electron microscopes are used in fracture studies, transmission electron microscopes of the type shown in Figure T-11, and scanning electron microscopes. In the former, a shadow picture of thin fracture replicas are observed and photographed. In the latter, the fracture face is observed directly. At present, the transmission electron microscope has been used to a much greater extent and fracture types have been catalogued*, however, it is anticipated that scanning electron microscopical evidence will become increasingly important.†

There are two basic methods of producing replicas for fracture analysis.

* M. N. Pelloux, "The analysis of fracture surfaces by electron microscopy", *Metals Engineering Quarterly,* ASM November 1965, p. 27.

† O. Jahari, "Comparison of transmission electron microscopy and scanning electron microscopy of fracture surfaces", *Journal of Metals,* **20**, 6, June 1968, p. 27.

Figure T-13 Replica showing stepped transgranular cracking in ferrite. (×8000)

In the "one step" method carbon is evaporated onto a fractured surface in vacuum and this thin carbon replica is used directly for examination. In the "two step" method, a plastic replica is made first and the carbon replica (often shaded obliquely with a heavy metal) is then made from the plastic "negative".

The electron microscopical evidence clearly indicates whether fractures have taken place by intergranular or transgranular mechanisms as outlined in Figure T-12. Here the fracture was primarily intergranular. In Figure T-13 the transgranular fracture has taken place by a stepped cleavage mechanism. A more ductile transgranular separation would show oblong dimples as in Figure T-14.

Very fine fatigue markings can also be detected using electron microscopical techniques. While these tell-tale markings can often be observed in large steel components, as in Figure 12-9, their presence cannot normally be detected in smaller non-ferrous (e.g. aluminum) components using optical means.

The utilization of electron microscopy in the evaluation of fractures is, however, a large topic and will not be dealt with further in this book.

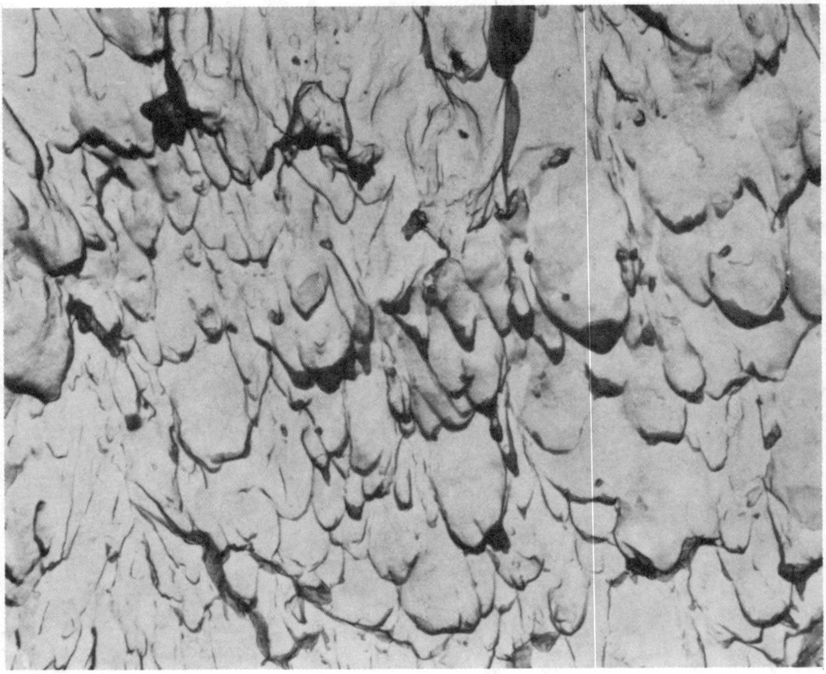

Figure T-14 Replica showing shear fracture of mild steel. (×12000)

Physical Testing

a) *Tensile Testing*

Tensile tests are applied to a large extent in metallurgical operations to ensure that the quality of products are maintained and that specifications are met. The tensile testing unit is, however, used to a much lesser degree in failure analysis. In fact, in none of the failure investigations reported in this volume did we use a tensile test as part of the diagnosis.

Few failures take place by simple overloading. Most take place because of an incorrect selection of a metal, poor design, errors in manufacture, or a lack of control of environmental factors. Because hardness tests give an indication of tensile strength properties, particularly for steel, hardness tests are usually conducted to establish whether a steel part has been adequately heat treated.

Even in the case of quality control on castings, it is often much easier and more pertinent to check chemical analysis and evaluate quality with radiography. If a casting is chemically correct, and is sound, then an expensive tensile test—the cost of machining a good sample is significant—is rarely required. It is more important to have a sound casting than that the sound portion of a casting should have its tensile strength confirmed.

It is, of course, true that a tensile test will give some important information. For example, in one metallurgical problem, the cold bending of 4″ diameter 90–10 copper–nickel piping resulted in cracking because the metal did not have adequate elongation characteristics. A suitable heat treatment, checked by tensile tests resulted in adequate elongation. Other test results (hardness, microstructure) could not be simply related to elongation properties.

b) *Hardness Testing*

A hardness testing machine is a common tool in even the more sparsely equipped metallurgical laboratories. Several types of testing units are available with variously shaped indentors but, in most cases the principle of the tests is the same: The indenter is forced into the metal specimen with a given load, and the depth of penetration is indicated. One of the commonly used testers is the Rockwell unit shown in Figure T-15. Detailed descriptions of hardness testers are given in Kehl,[19] The Metals Handbook[1] and elsewhere.

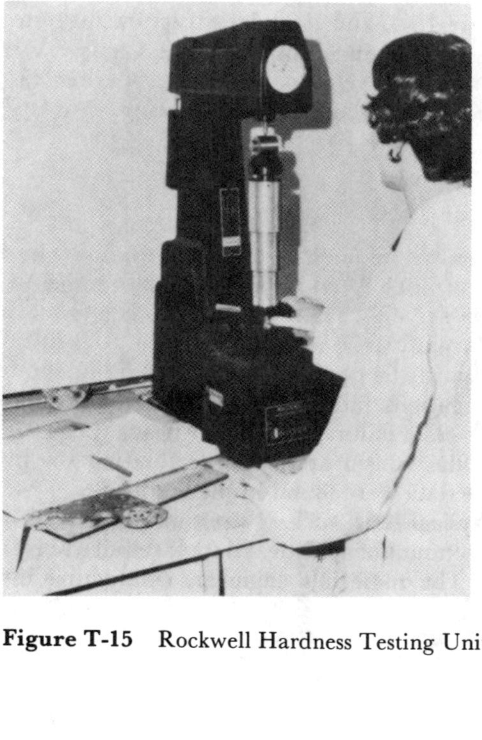

Figure T-15 Rockwell Hardness Testing Unit.

An example of the value of hardness testing is given on page 213, Martensitic stainless steel bolts were failing at the neck by a delayed fracture mechanism. Hardness tests showed that the head–neck area was very hard, whereas the shank was soft. It was readily apparent that the bolts had not been heat treated subsequent to being forged.

c) *Impact Testing*

The usefulness of ordinary impact tests has been questioned and, as a result, there are now a large number of types of tests applied to establish the resistance of metals to suddenly applied loads. The usual tests are somewhat similar: a notched bar is broken by a swinging hammer and the energy absorbed in breaking the specimen is recorded. Other more exotic tests include explosive charges or dropped weights.

For engineering applications and service failure investigations, the Charpy V-notch impact test is, in most instances, adequate for establishing whether the metal under consideration has poor impact properties related to service temperature.

Microstructures often suggest that impact properties are poor, and indeed, impact tests are often applied to confirm a microscopic examination. The impact test gives a quantitative result which can be compared with satisfactory materials or known satisfactory levels for the same material.

In the case of the anchor shank failure described on page 313 both visual examination, Figure 15-5, and the microstructure suggested poor properties and indeed, the 2 foot-pounds obtained in a Charpy V-notch impact test confirmed that a steel with very poor impact properties was being evaluated. On subsequent heat treatment the impact property was greatly improved, Figure 15-7.

d) *Fatigue and Other Tests*

Whereas a considerable amount of useful information can be obtained with one or two tensile or impact test samples—ideally suited to failure investigations—fatigue tests are only really valid when a large number can be performed. This is normally done when conducting extensive tests on particular alloys and would hardly be possible on a limited fragment from a failure.

In failure investigations, fatigue results are normally obtained from existing tables on materials. The failures of cylinder heads (page 270) resulted from a corrosion fatigue mechanism and when evaluating the problem, corrosion fatigue and fatigue data were found in the literature.

Many other physical tests, such as stress corrosion tests, fall into the same category. A large amount of statistical data is required before useful information is obtained. The materials engineer, conducting investigations, continually leans on this vast amount of data available in textbooks, handbooks. etc.

Non-destructuve Testing

a) *Radiography*

The value of radiography and other non-destructive testing procedures in quality control has been well documented. However, non-destructive testing is also used, with considerable advantage, as a metallurgical trouble-shooting tool. These procedures are of value in failure investigations to determine whether metallurgical discontinuities are present in castings, forgings, weldments, and could therefore have been responsible for the failures. Corrosion mechanisms (page 125) are also indicated by utilizing these techniques.*

Of the non-destructive testing methods, radiography has distinct advantages in failure investigations. Some of the other techniques such as ultrasonics are best applied to components for which a non-destructive testing routine has already been established. With radiography much useful information can be obtained from a part which is being evaluated for the first time.

We have found it good practice to radiographically evaluate failed metal parts before any destructive sectioning. Information obtained from the radiograph is often valuable in choosing the location for sectioning which will ultimately give the most information, or, more important, will not destroy useful information.

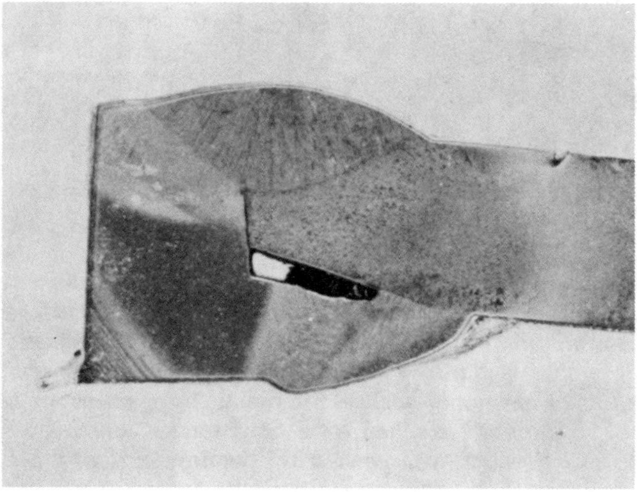

Figure T-16 Incomplete penetration and gross lack of fusion in weld. These defects are readily indicated by radiography.

* B. F. Peters, "Radiography for corrosion evaluation", *Materials Evaluation*, **V23**, 3, March 1965, p. 129.

Radiography is the most useful tool for weld evaluation whether in production or for failure investigation. The sharp notch formed by an incomplete penetration, as in Figure T-16, is readily shown by radiography. Similar defects have been responsible for more than one potential hazard in the authors' experience.

Radiography is also most useful in evaluating questionable castings. For example, a large pump housing, shown in Figure T-17, leaked because of the large cavity in the wall of the casting. Also, the core shift shown in Figure 10-17 was established with radiography.

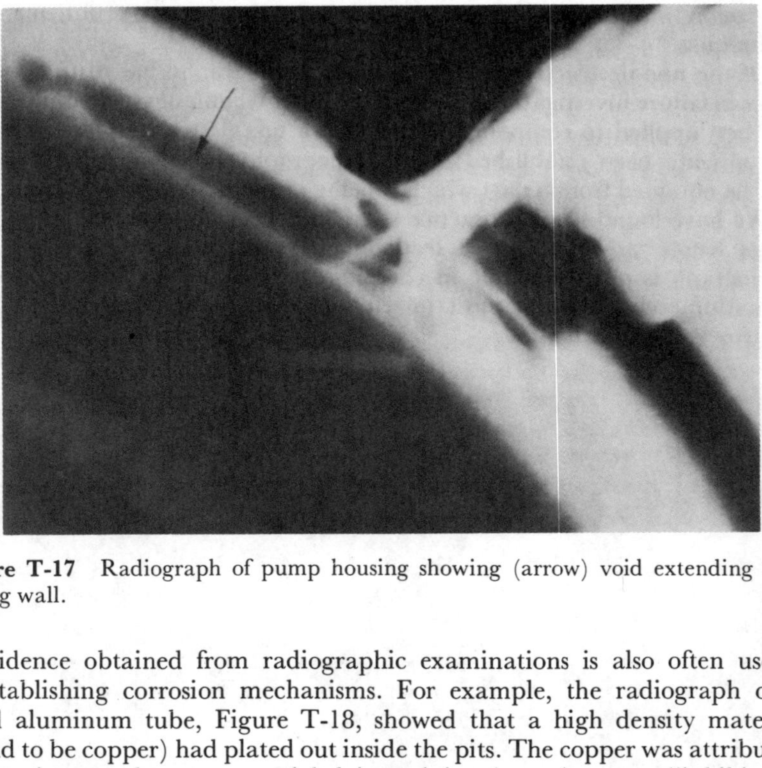

Figure T-17 Radiograph of pump housing showing (arrow) void extending into casting wall.

Evidence obtained from radiographic examinations is also often useful in establishing corrosion mechanisms. For example, the radiograph of a failed aluminum tube, Figure T-18, showed that a high density material (found to be copper) had plated out inside the pits. The copper was attributed to corrosion attack on copper–nickel–iron piping, by an incorrect "inhibitor".

The same "inhibitor" resulted in a considerable amount of corrosion damage inside the water passageways of aluminum diesel cylinder heads, as shown in Figure T-19. This attack was also evaluated non-destructively by radiography and similar copper deposits were noted.

Radiographic examinations have on occasions proved particularly useful in revealing foreign material associated with corrosion. For example, as shown in Figure T-20, mussel fouling in copper–nickel–iron tubing was shown to be associated with the pitting which was being investigated. Since

Figure T-18 Radiograph of aluminum tube showing copper (white ring) deposited in pit.

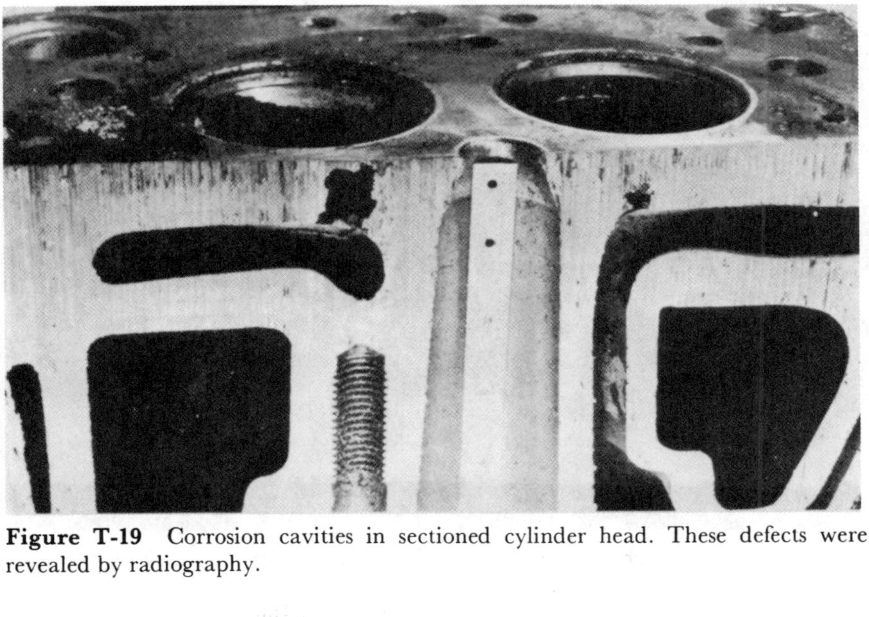

Figure T-19 Corrosion cavities in sectioned cylinder head. These defects were revealed by radiography.

copper–nickel–iron was expected to be relatively free from marine growth because of a small release of copper by the alloy, an investigation was conducted. It was established that the pipe filler materials used in the bending of the tube were not adequately removed, and the fouling took place on the residue which shielded the growth from the copper. In fact, the growth was incidental: the corrosion attack resulted from the presence of the cathodic pipe filler residue.

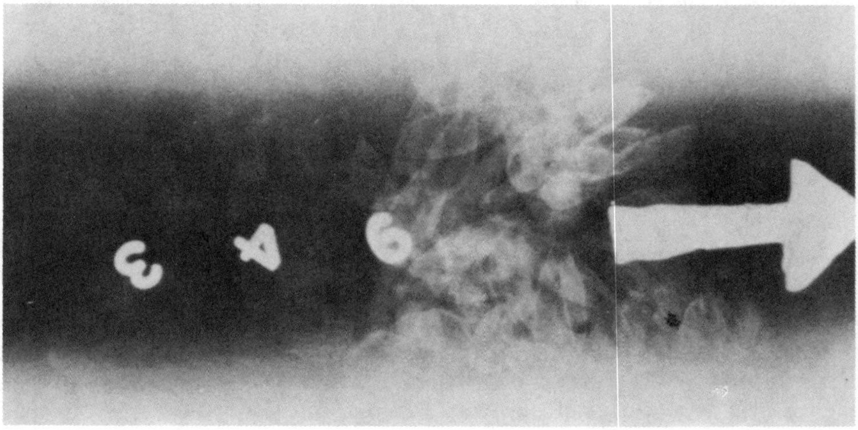

Figure T-20 Radiograph of copper-nickel tube showing mussel fouling in the tube.

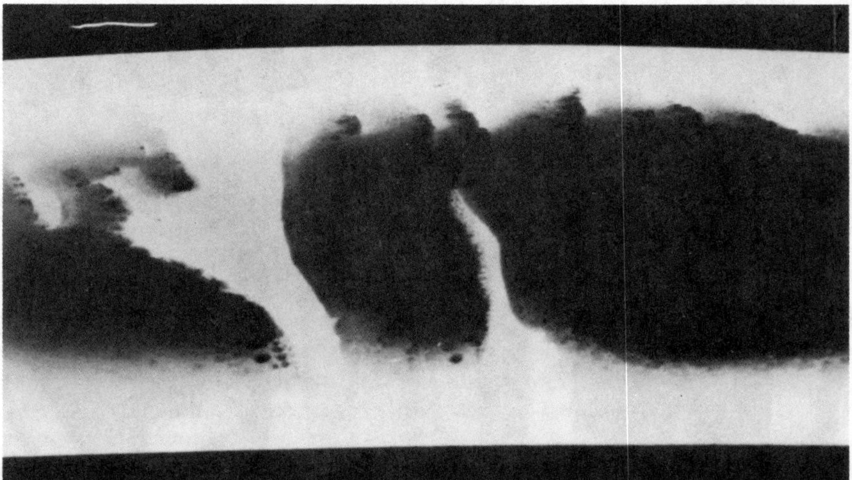

Figure T-21 Radiograph of steel tubing which failed by impingement.

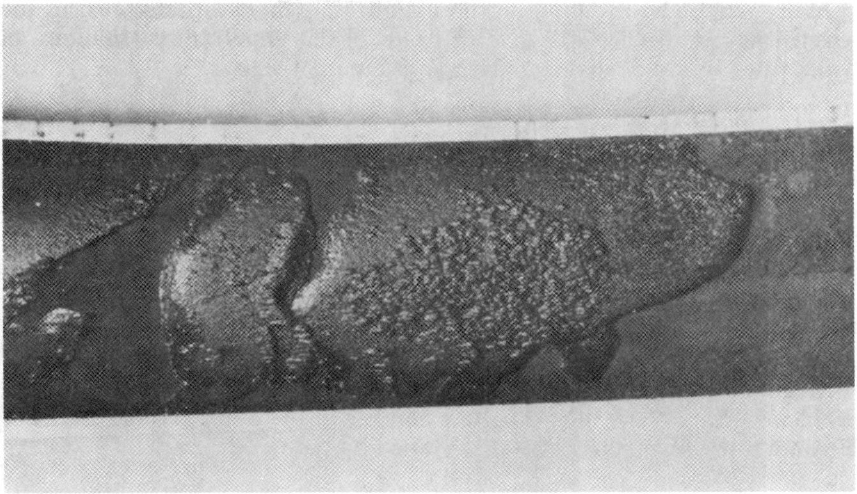

Figure T-22 Steel tube (Figure T-21) showing impingement attack.

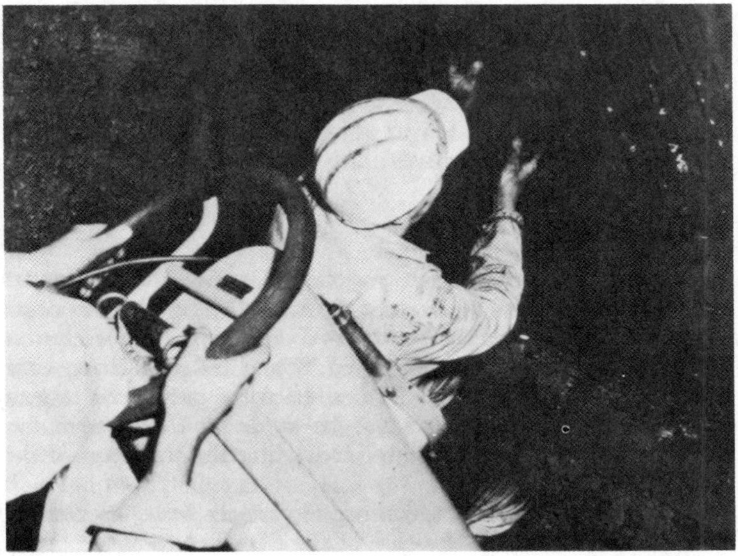

Figure T-23 Preparing for radiographic examination of boiler tubes in a marine boiler.

The leakage of the steel steam-heating coil in a large fuel oil tank was also evaluated with radiography. The radiograph Figure T-21 showed quite clearly that the corrosion damage was due to corrosion–erosion, Figure T-22.

Many additional examples of the use of radiography are outlined in the selected failure problems in this text. One of the important evaluations, of boiler tubes in situ, is shown underway in Figure T-23.

b) *Magnetic Particle Inspection*

A major use of magnetic particle inspection is to find flaws in ferrous products and machinery components, and also to detect the fine cracks which could result in fatigue failures. Magnetic particle inspections are also commonly made on ferrous parts which have failed in service. This procedure consists of magnetising the part in question, usually by induced magnetism from a coil, and dusting iron particles on it. The iron particles cling to any discontinuity in the ferrous part.

Magnetic particle inspection procedures were utilized in many of the problems described in this text. In a number of the photographs, cracks or discontinuities in ferrous parts are outlined by the powder.

c) *Dye Penetrant Examinations*

Cracks and discontinuities in both ferrous and non-ferrous parts are often outlined by dye penetrant methods. In short, a dye is applied to the metal in question, and given time for it to make its way into any crack or defect. The part is then washed and developed. The developer draws the dye out of the cracks outlining the discontinuity.

This method has been applied to non-ferrous parts in a number of the investigations described. (Note particularly Figure 14-2.)

d) *Ultrasonic Testing*

Ultrasonic testing is increasing in popularity among the non-destructive testing procedures applied to evaluating metals. The process consists of applying, with a suitable couplant such as oil or water, a piezoelectric source to the surface of the material being tested. Sound reflected from within the metal is indicated on a cathode ray screen when the pulse-echo technique is used. Thickness measurements can also be made by a resonance method: here the resonant frequency is shown, indicating the thickness of the part being tested.

In only a few instances have ultrasonic techniques been utilized by us to evaluate failure problems. In the case of the edge defect in the thick plate in Figure 16-7, ultrasonic techniques showed that a lamination was present. An ultrasonic inspection of ships plate for laminations is shown being conducted in Figure T-24.

Ultrasonic techniques are most useful when evaluating a part which has previously been tested in large numbers so that accurate correlations can be made between ultrasonic signals and defect types. Many service failures are

Figure T-24 Ultrasonic examination of ship's plate for laminations.

one-of-a-kind and ultrasonic testing is not of as great a value in failure investigations as it is in quality control.

e) *Strain Gauges*

Strain gauges have been used to advantage in evaluating service failure problems. We have found that a large number of service problems result from residual rather than service stresses, and strain gauges can be used to evaluate these residual stresses.

The electrical resistance of common strain gauges increases as the gauge is strained. If the gauge is glued onto the surface of an object, it will indicate by its change in resistance how much the object is strained.

To determine residual stresses, strain gauges are applied to the part in question and then the metal is removed from the neighbouring areas, leaving only a piece of metal slightly larger than the strain gauge. The change in strain indicating an approximation of the residual stresses, is recorded.

The evaluations of the stress corrosion cracking failure of superheater tubes (page 70) and the corrosion fatigue failure of cylinder heads (page 272) were performed using this technique. In both instances, the residual stresses were surprisingly high and indicated that other contributing effects were of relatively minor importance.

An excellent reference on strain gauges is the book edited by M. Hetényi.[66]

f) *Chemical Analysis*

A number of types of analytical equipment are used to aid in metallurgical investigations. These include:

a) Wet chemistry for alloy analysis.
b) Spectrographic equipment for alloy analysis.
c) Carbon determinators.
d) X-ray diffraction for crystallographic determinations.
e) X-ray fluorescence equipment for alloy analysis.
f) Electron probe microanalysis for determination of small inclusions and unknown phases (used in case history 1.C-2).

Apart from the metallurgical microscope, often the most important of the failure investigation tools is the analytical equipment, notably the spectrograph.

Spectrographic analysis can quickly reveal the cause when failure has resulted from the use of inferior or improper materials. The failure of a diesel bearing (page 295) resulted because the babbitt material was only solder; many of the premature stress corrosion cracking failures of aluminum alloys resulted because the alloys were copper-containing, known to be susceptible. These are only examples of the many failure problems encountered which have been composition problems.

As soon as the metallurgist is aware of the composition of a failed part, the microstructures often become more meaningful. The microstructures of aluminum bronzes and brasses are not always easily differentiated, for example.

IV

INDEX BY CAUSE OF FAILURE

While the chapters dealing with problems are set up by type of object or a functional category, e.g. boilers or springs, it was deemed useful to permit a study by failure mechanism. Accordingly, as set out below, the failure case histories are indexed within each functional category by the predominant mechanism responsible. This permits one to examine readily problems dealing with fatigue or stress corrosion or welding, etc., regardless of which chapter they happen to be in. These mechanism categories are set out below.

A **Mechanical**

1) Fatigue
2) Brittle Failure
3) Wear or fretting
4) Overloading
5) Overheating
6) Other.

B **Corrosion**

1) General and galvanic
2) Corrosion fatigue
3) Stress corrosion
4) Intergranular corrosion
5) Crevice corrosion
6) High temperature
7) Impingement and cavitation
8) Dezincification and others.

C **Manufacturing Defect**

1) Casting

2) Forging–rolling
3) Welding and joining
4) Machining and grinding
5) Heat treating
6) Plating and hard surfacing
7) Material choice
8) Design
9) Other.

PART 2

CASE HISTORIES

I

BOILER TUBES

A TUBE BURST is described in which clearly defined knife-edge thinning indicates freedom from prior faults. Another tube burst is identified as a result of excessive internal deposits and consequent overheating. Corrosion fatigue damage to boiler tubes and the value of radiography in its detection are discussed in the third example. A thick-walled burst and related cracking in several tubes is identified as a rare type of stress corrosion cracking occurring on the steam generating surface. Failure of superheater tubes resulted from high residual stresses and presence of caustic. A strange "wormy" appearance on the inside of a tube end is identified as the effect of corrosion on cold-worked metal.

The effect of high temperature oxidation is demonstrated when superheater tubes are overheated. The influence of manufacturing defects in giving subsequent failure of superheater tubes is clearly shown by microstructure and proven by electron microprobe analysis.

Exceptionally poor welded fabrication was responsible for crippling the economizer units of several boilers. The puzzling nature of some surface defects is clarified with radiography.

A-5 a) A fire row marine boiler tube failure by overheating.
b) Waterwall stationary boiler tube failure due to internal deposits.

B-2 Corrosion fatigue failures of fire row boiler tubes.

B-3 a) A boiler tube failure by stress corrosion cracking.
b) A superheater tube failure by stress corrosion cracking.

B-5 Corrosion of boiler tube ends at rolling folds.

B-6 Wastage of superheater tubes by external oxidation.

C-2 a) Failure of superheater tubes because of manufacturing defects.
b) Longitudinal manufacturing defects in boiler tubes.

C-3 Failure of economizer element tubes at weld defects.

C-9 Mechanical gouges caused when expanding tubes.

A-5 a) **A Fire Row Marine Boiler Tube Failure by Overheating**

Failure of a 2″ fire row tube in a controlled superheat marine boiler resulted in flooding of the furnace in several feet of water.* The incident occurred during ship trials after a total steaming time of only 7 hours. The tube burst open, as shown in Figure 1-1, just below the bend on the way to the steam drum. This tube was number 6 from the back wall, in "A" row.

Figure 1-1 Location of burst in tube bank.

Visual examination showed several features of interest:

a) The split was at the centre line of the tube—facing the fire and was consequently at a high heat input zone.

b) The thin edges of the split, practically a knife edge, and the "stretcher" marks on the inner surface at the break, as shown in Figure 1-2, are indicative of plastic flow of the metal.

c) No appreciable thickness of internal deposit was noted in the immediate area of the split.

The defective tube and the adjacent "A" row tube number 7 from the back wall, were made available for detailed laboratory study.

* R. D. Barer, "Marine boiler tube failures—three case histories", *Transactions of the Institute of Marine Engineers, Canadian Division Supplement,* No. **6**, Dec. 1961, p. 149.

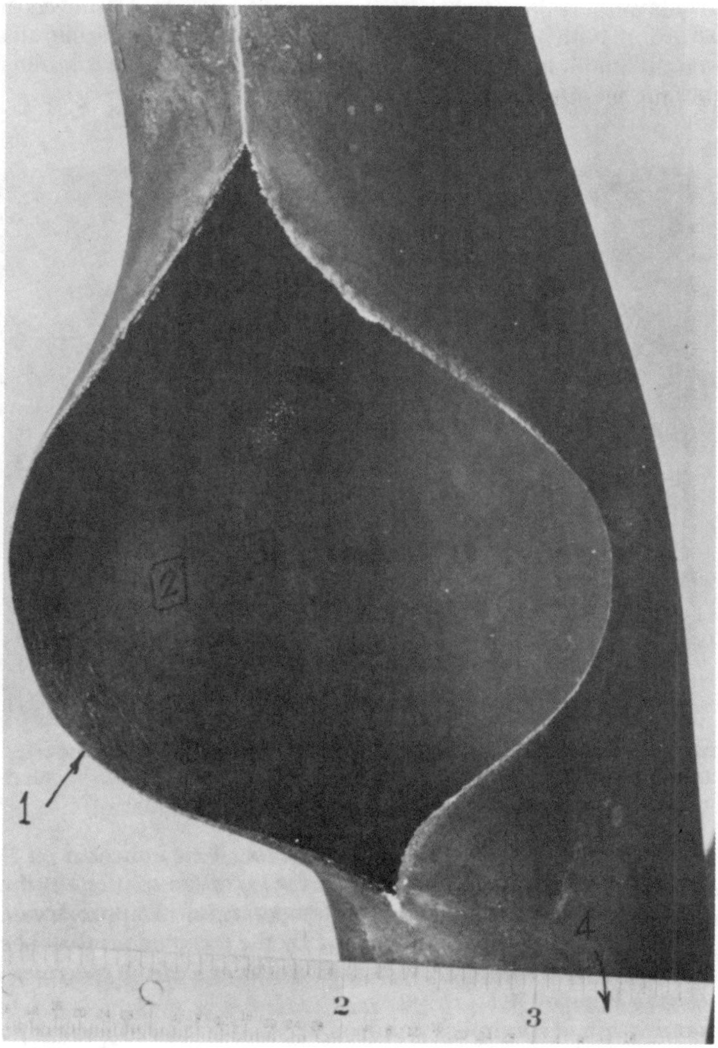

Figure 1-2 The burst tube. Note the thin edges and the "stretcher" marks on the inside surface.

Metallurgical Examination

Radiographic examination, as well as subsequent examination of sections under the microscope, failed to indicate any evidence of defective material. The metal of both tubes was sound.

Examination of specimens well removed from the rupture (approximately

4 ft.) and comparison with a specimen from tube number 7 showed that microstructure of both was the same (Figure 1-3). Spectrographic analysis also showed that both tubes were plain carbon steel, identical with each other, and containing no appreciable alloying elements.

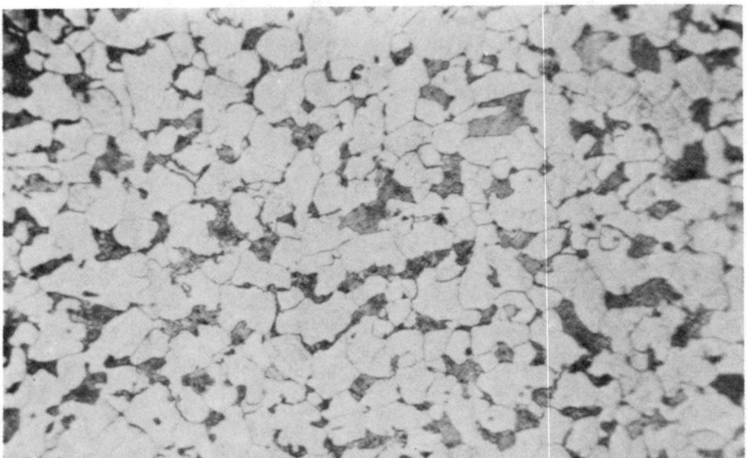

Figure 1-3 Normal microstructure of 0.25% carbon steel tube showing mixture of ferrite and pearlite. (×100)

Small pieces of metal were cut from the locations indicated on Figure 1-2. Examinations of specimen number 1 at the lip of the split established the fact that metal in this area had been to temperatures of approximately 1,700–1,800°F before being rapidly quenched by the tremendous flow of steam and water released by the burst. A typical microstructure from specimen number 1 is shown in Figure 1-4.

Examination of specimens numbers 2, 3 and 4 showed increasing amounts of bainite and pearlite, at the expense of the martensite. This indicates a decreasingly less drastic quench rate, and to a certain extent that temperatures were a little lower. It was established that the heat-affected zone, as evidenced in microstructural changes, extended well beyond 2 inches past the rupture, both above and below the opening. The heat effect did not extend fully around to the back of the split.

The internal scale was also analysed and measured. It was magnetite and was of negligible thickness on both tubes except within a foot or less of the split, where it became 0.0015″ thick. There would be little reason for this

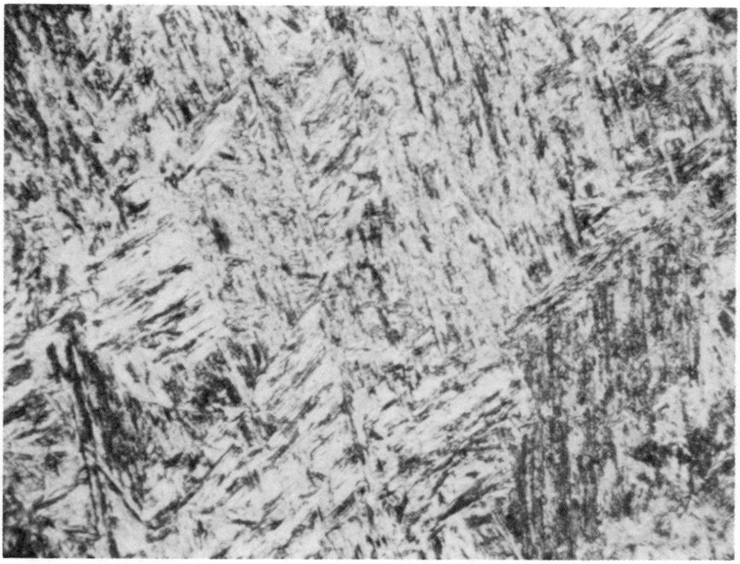

Figure 1-4 Typical microstructure of lips of burst (specimen No. 1, Figure 1-2). This is martensite and would only result from a drastic quench of the 0.25% carbon steel tube. (×400)

zone to develop an oxide film greater than elsewhere except as a result of the excessive heat input.

Reviewing all the factors and evidence available, the sequence of events was as follows:

a) Excessive steam generation because of the high heat input at this locality, combined with some temporarily poor circulation (cause unknown), impeded heat transfer.
b) Metal of tube wall became progressively hotter, eventually reaching 1,700–1,800°F. It is anticipated that this overheating took place in a short period of time—probably minutes.
c) At this temperature the metal has approximately one-twentieth of its usual strength and also possesses great ductility. The result was a gentle bulging until the wall became, by stretching, too thin for the internal pressure.
d) At the instant of rupture a great flow of water and steam (much of the water would flash to steam) effectively quenched the overheated metal (proven by the martensitic microstructure).

The exact cause of the initiation of step (a) above, could not be determined.

Conclusions

This boiler tube failure was essentially a simple case of local overheating. This overheating probably occurred in a matter of minutes, thereby avoiding the more complicated deterioration (as described later) from prolonged overheating in conjunction with other influences.

This failure presents a picture of a healthy boiler tube metal becoming red hot, and progressively hotter, then ballooning under the influence of internal pressure. The edges of the burst, smoothly tapered to a knife edge, confirmed the soundness of the metal.

Because it was so early in the boiler's life, the possibility was strong that something had been inadvertently left in the boiler, adversely affecting normal water circulation.

It was concluded that a failure of this type was not likely to occur again and the boiler was placed back in service. After thirteen years of service no further problems occurred in the generator tubes of this boiler.

A-5 b) **Water Wall Stationary Boiler Tube Failure Due to Internal Deposit**

An investigation was conducted into the cause of a failed water wall tube in a stationary boiler. It was also noted that several tubes in this boiler were slightly bulged.

Approximately two-foot lengths were cut out of the ruptured tube and from one of the bulged tubes. Also radiography in place was undertaken to assess the general condition of the other tubes.

The laboratory examination indicated that internal scale was undoubtedly a major factor. This scale, shown in Figure 1-5, was 0.045″ thick and would constitute a major impediment to heat transfer. It was found to consist primarily of calcium phosphate and a small amount of silica.

Metallurgical examination of the tube walls indicated that no deleterious effects could be detected at approximately 1 inch away from the bulge. No evidence of any other weakness or defect in the metal was indicated.

Radiography substantiated the findings on the two samples, i.e., the boiler tubes were in good condition and the scaling was general throughout the boiler. It was noted that the scale had tended to break away where the metal began to be overheated at the bulges (Figure 1-5).

Discussion

Examination of the ruptured boiler tube confirmed that the failure was due to overheating caused by scale buildup. The bulges on the other tubes, also in the lower sidewall region, showed that there was a general tendency for overheating in this area. The tube metal, weakened by excessive heat, stretched due to the internal pressure.

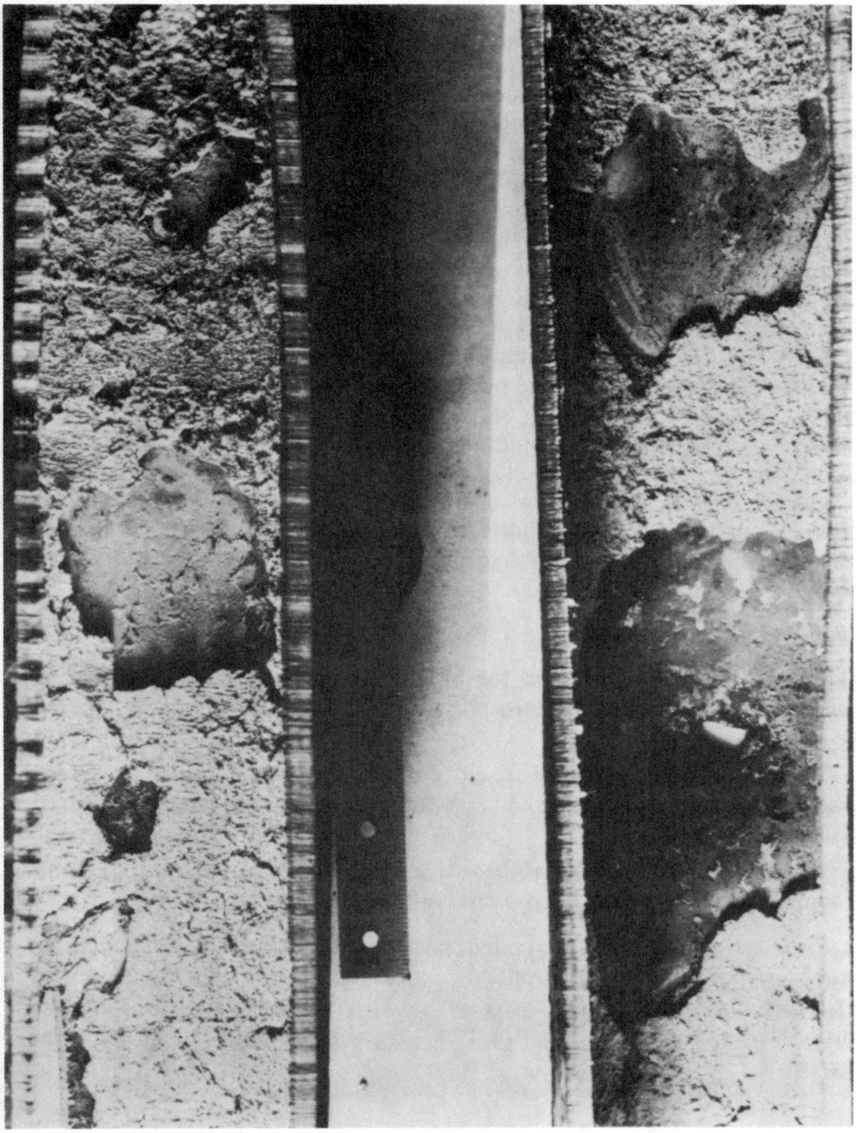

Figure 1-5 Inside of failed water wall stationary boiler tube showing thick (0.045 inches) internal deposits.

The fact that the scale had broken away at each bulge helped explain why there were numerous bulges and these were apparently arrested; the improved heat transfer resulting from the removal of the deposit prevented catastrophic burn-out.

Although excessive deposits were present in the tubes throughout the system, failure and near failures took place only in the sidewall region. It is possible that adjustment of oil sprayers could minimize the heat directed to these regions of the boiler.

The internal deposit was caused by using city water for make-up. It contained 31 ppm solids including calcium and magnesium oxides. Assuming an output of 20,000 pounds per hour steam and no return condensate, this would amount to an intake of approximately 15 pounds of solids per day. To this is added roughly an equivalent amount of boiler chemicals including disodium phosphate. Admittedly a large proportion of this is removed on blow-down. It is, however, readily apparent that if for some reason a tendency developed for these solids to deposit or bake on, there would be a considerable amount available for such depositiion.

In any case, the deposits did build up and would continue to do so in future unless remedial action was taken. Laboratory experiments aimed at removing the deposits showed with several chemicals at comparable concentrations, including hydrochloric, phosphoric and sulphamic acids as well as two proprietary compounds, that the hydrochloric acid was the most effective.

Conclusions and Recommendations

1) The boiler tube failure and the bulging were the result of overheating which is related to internal deposits. Heat flow to the sidewall areas appeared to be somewhat high.

2) The boiler tubes were free from injurious weaknesses or defects except in those areas where overheating had resulted in bulges. Simple repair appeared feasible.

3) The deposit build up was related to the intake of tap water and the chemicals supplied to react with the minerals in the tap water.

It was recommended that the boiler be chemically cleaned and that changes be made in quality of water supplied, i.e., use of greater condensate return and supplementing this with softened and deionized water which was available. Additionally, changing heat flow pattern by adjustment of oil sprayers was suggested.

B-2 **Corrosion Fatigue Failures of Fire Row Boiler Tubes**

A number of transverse cracks which developed over many years in certain fire row tubes in several ships with 3-drum marine boilers, have been studied.* A first step in these investigations was to determine the location and distribution of the cracks in leaky tubes submitted to the laboratory. Radiography proved to a particularly useful means of establishing severity and distribution of fissures, as shown in Figure 1-6.

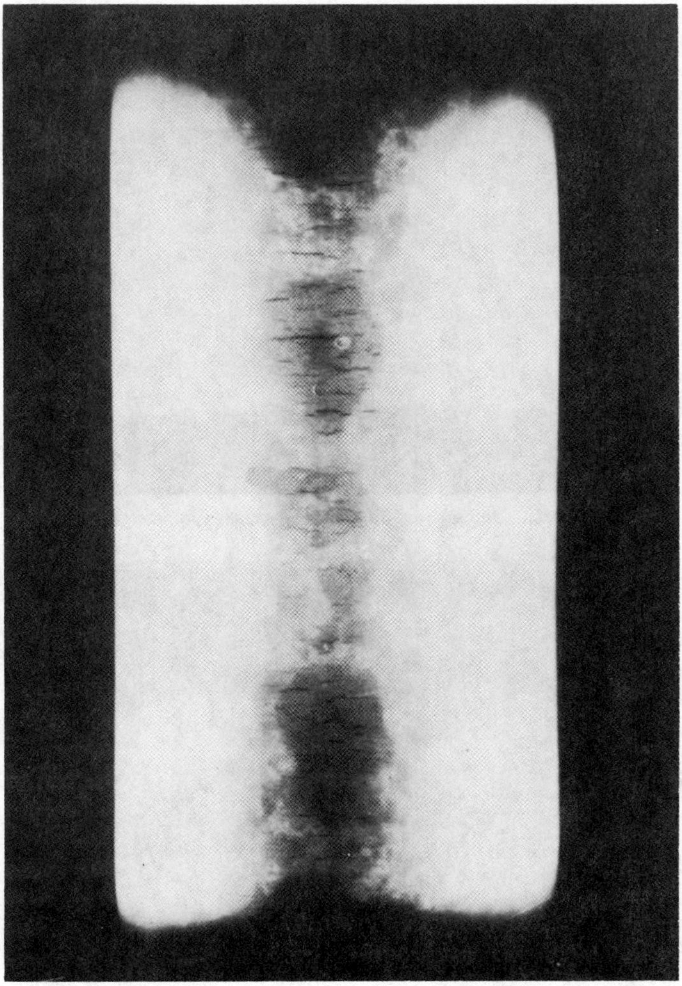

Figure 1-6 Radiograph of boiler tube showing many fine transverse cracks.

* R. D. Barer, "Marine boiler tube failures—three case histories", *Transactions of the Institute of Marine Engineers, Canadian Division Supplement,* No. **6**, Dec. 1961, p. 154.

Sections were cut from affected areas on these tubes to permit detailed examination of the nature of the cracking. Sections were also cut from non-affected tubes for comparison. Basic characteristics indicating a similar failure mechanism were noted, as follows:

i the incidence of cracks, usually rather numerous;

ii great distribution in depth (in one instance 5 cracks perforated in a 2 ft. length, with many others only part way through). (Figure 1-7);

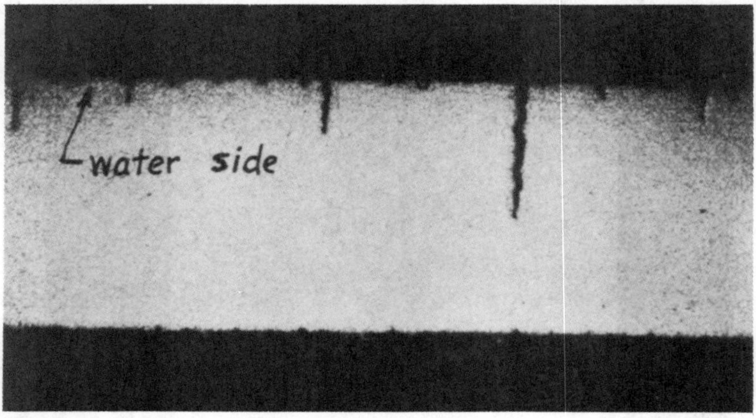

Figure 1-7 Section through tube wall showing transverse cracks. (×20)

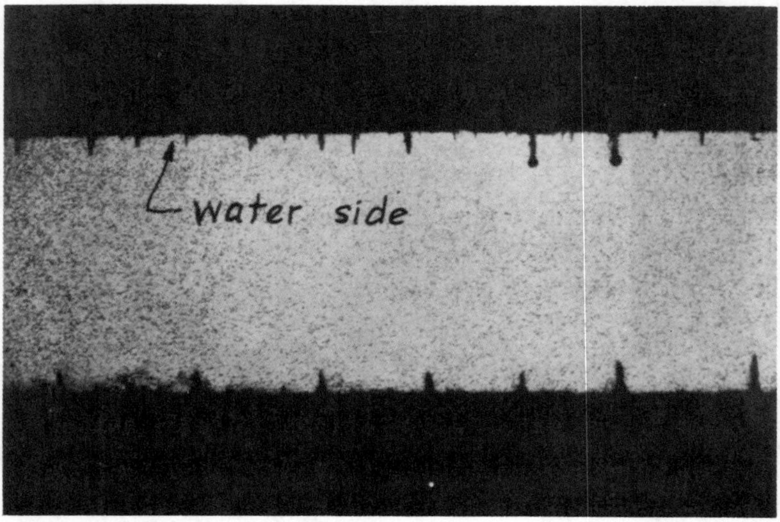

Figure 1-8 Section through tube wall showing transverse cracks on both waterside and exterior of tube. (×20)

iii) penetrating action on outer surface of affected tubes was minor except in one instance where flame impingement was especially severe. (Figure 1-8).

A close-up of one of the cracks is shown in Figure 1-9. From this, additional details may be noted:

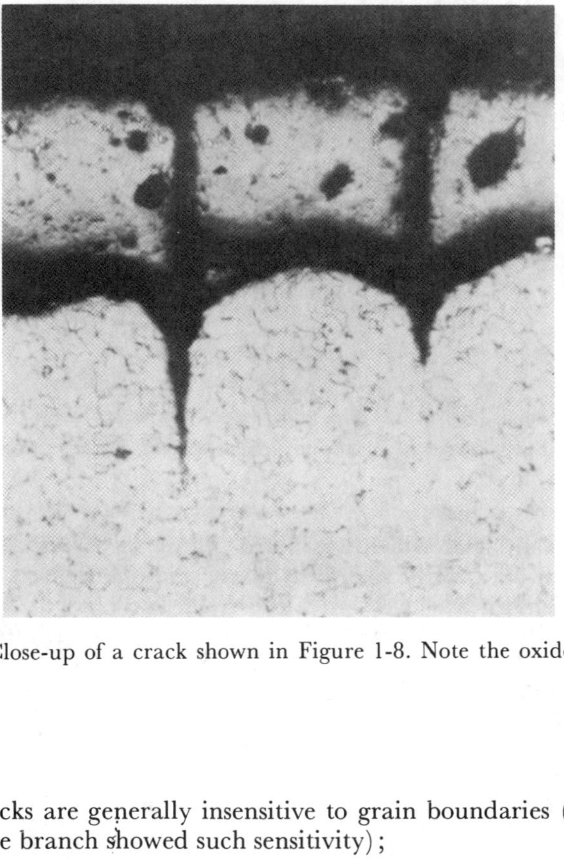

Figure 1-9 Close-up of a crack shown in Figure 1-8. Note the oxide layer on the tube surface. (×100)

iv) the cracks are generally insensitive to grain boundaries (although an occasional fine branch showed such sensitivity);

v) there is generally little, if any, oxide in the larger cracks;

vi) there is generally an oxide layer on the tube surface (water side) in which are scattered particles of copper. The layer was thin—approximately 0.0012″—and in some areas almost one-third of it was copper particles.

As shown in Figure 1-9, there appears to be some correlation between cracks in the oxide layer and the beginnings of a crack in the metal. This is confirmed in Figure 1-10, in which are shown "embryo" cracks—the earliest stages in corrosion of the tube, the formation of cracks in the oxides, and the apparent pointing-up of the general oxide layer at the location of these cracks—in fact, the initiating points of cracks in the tube wall. There is no copper in this microsection.

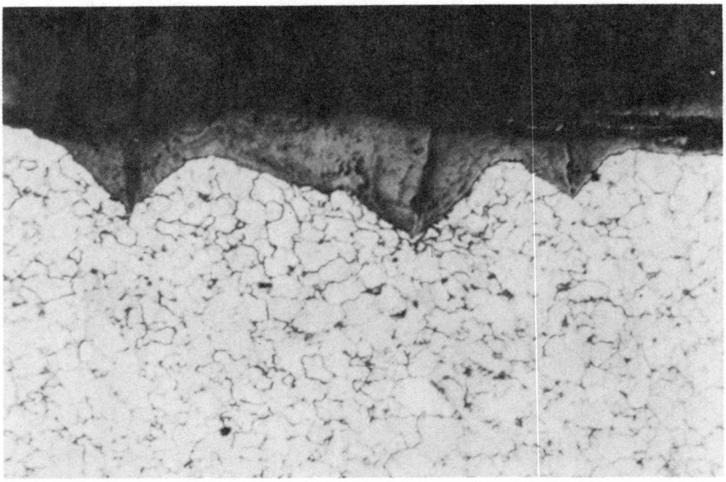

Figure 1-10 Early stages of corrosion fatigue cracks. Note cracks in oxide layer (×1000)

An additional factor which was noted during metallurgical examination was the absence of pearlite and in its place, small particles of "spheroidized carbide", as shown associated with the crack in Figure 1-11. The normal structure of these tubes based on examination of neighbouring sound tubes was that of ferrite plus pearlite. Metallurgically a spheroidized structure indicates lengthy exposure to temperatures of 900° F or higher. For example, approximately 15 hours at 1200° F will spheroidize the microstructure in these tubes; at 1000° F it would require a period almost one thousand times as long.

Overheating of defective tubes was readily established. The initial belief that internal deposits were the major factor in this cracking was modified to a certain extent. Similar deposits were found on adjacent tubes which were free of defects. That the deposits contributed to the overheating and cracking is a reasonable assumption; that their presence did not necessarily result in cracking was demonstrated.

It would appear that some additional factor was operating to give these tubes generally higher temperatures than their neighbours: temperatures sufficiently high and held long enough to produce spheroidization. Only one possibility suggests itself: that circulation of water (and steam) was in some way hampered for considerable periods. It is possible to visualize a situation

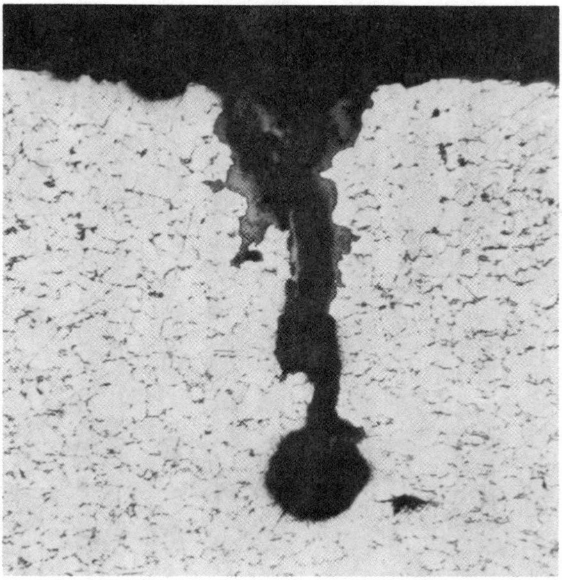

Figure 1-11 Details of one of the cracks shown in Figure 1-8. The rounded bottom suggests absence of stress after a time. (×100)

where, once slightly more than normal steam generation occurred (by slight overheating) an increasingly deteriorating situation would follow.

As a result of favourable definition of fissures by radiography in the laboratory, it was decided to use radiography as an inspection tool in ships of the class. These examinations were particularly informative, in that they occasionally found other defective tubes, not otherwise suspected, as shown in Figure 1-12. In addition, the extent of internal deposits was indicated.

It may be of some interest to refer briefly to the work on corrosion fatigue of McAdam and Geil* who obtained cracking identical with that shown here in Figure 1-9. Their work with rotating cantilevered specimens of mild steel in well water produced a great many detailed results. Among these, their photomicrographs show cracks, after a 10-day exposure at 9000 psi to 1400 cycles/minute, similar to those in the tubes here discussed. The fissures reported by McAdam and Geil all started at irregularly rounded corrosion pits. At a lower reversal rate, i.e., 100 cycles/minute, it required 13,000 psi to produce cracks—but these were thin, needle-like and not the same as those on the ship. Whether their results can be applied to the boiler

* McAdam, D. J., Jr., and Geil, G. W., "Influence of cyclic stress on corrosion pitting of steels in fresh water and influence of stress-corrosion on fatigue limits", *J. Research National Bureau of Standards*, **24** (1940), p. 685.

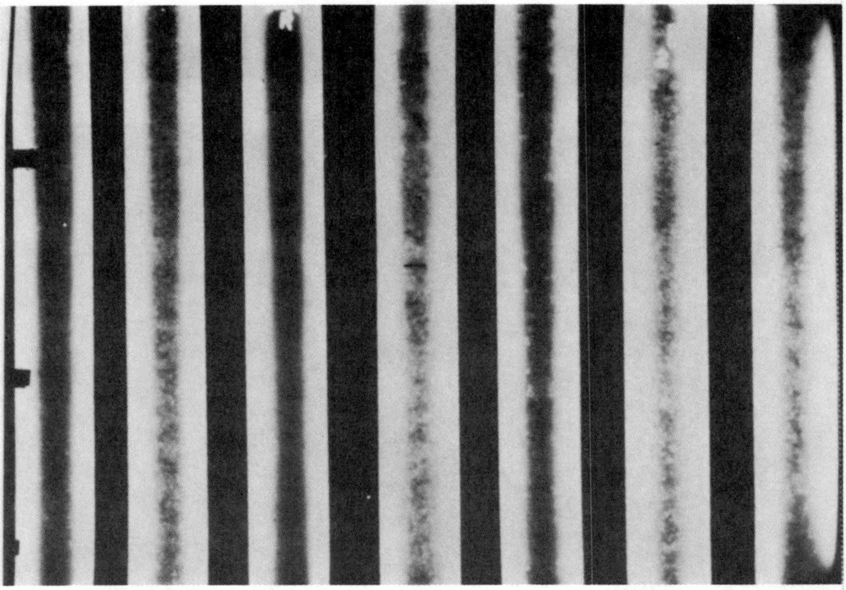

Figure 1-12 Radiograph of fire row boiler tubes showing cracking in the centre tube and internal deposits in most of the others.

tubes is questionable in view of different media, different temperatures, and that the specimens were rotating cantilever rather than long tubes subjected primarily to longitudinal pulsing. They tend to suggest, however, that corrosion fatigue was the phenomenon which caused failure of the tubes in the boiler.

Preliminary calculations showed that if one tube is overheated in relation to its neighbours, an upsetting effect results which can lead to stresses of a high level. For example, stress developed in an 8 ft. long tube under restraint, for a 200° F temperature rise approximates 36,000 psi. It seems reasonable that relatively few cycles would be necessary at this stress level.

Strain gauges were used to establish the magnitude of stresses present. Type A5 SR4 strain gauges were placed on several tubes (on a number of occasions) in the region of cracking (prior to removal) and the tube was severed above the strain gauge. In one instance, the tube sprang away from the cut by $3\frac{1}{2}$ inches indicating the severity of the residual bending stress. The strain gauge on this tube indicated a release of 32,000 psi on the tensile side.

An additional factor which may have been influential in the accumulation of the deposits is slight oil contamination. This role of oil is described by Rivers:*

"Anything that impedes rinsing is bad. Water-permeable deposits of any kind resist rinsing, and it is easy for boiler water to concentrate within the pores of such a deposit. For example, a boiler may remain quite clean under ordinary operating conditions. But suppose a small amount of oil contamination turns up sometime and causes a thin layer of oil-bound sludge to stick to certain steaming surfaces. Evaporation of boiler water within the pores of this adherent sludge might lead to complex silicate scale or even caustic attack of the metal. Similar effects might be produced by dissolved iron in the feedwater or the presence of any other binding substance that can form porous deposits. Failure through overheating or corrosion might thus occur at relatively conservative rates of heat input and in spite of excellent control over boiler water chemical concentration in general."

Conclusions

The evidence observed metallurgically indicates that the tubes failed by corrosion fatigue.

Overheating was a common factor and could have been responsible for the following:

a) It changed a non-aggressive environment to an aggressive one. On surfaces with an excessive rate of steam generation the net effect was that of concentrating the boiler water and it was then corrosive to steel.

b) With more overheating to the point where steam blanketing resulted, there occurred an additional slowing of mass flow, which in turn was harmful.

c) Expansion and contraction may have produced the cyclic stresses which caused the cracking. Contractions against restraint can produce stresses in the order of several 10,000 psi.

It was indicated that the remaining tubes could be expected to give good service, and that chemical cleaning and periodic checking of nozzle adjustment would be beneficial.

B-3 a) **A Boiler Tube Failure by Stress Corrosion Cracking**

This incident, in a light cruiser, involved failure by bursting of one generator tube as well as partial rupture of several others in an Admiralty 3-drum

* Rivers, H. M., "Concentrating films: Their role in boiler scale and corrosion problems", *Proc. 12th Annual Water Conf. of the Eng. Soc. of Western Pennsylvania* (1951), pp. 131–145.

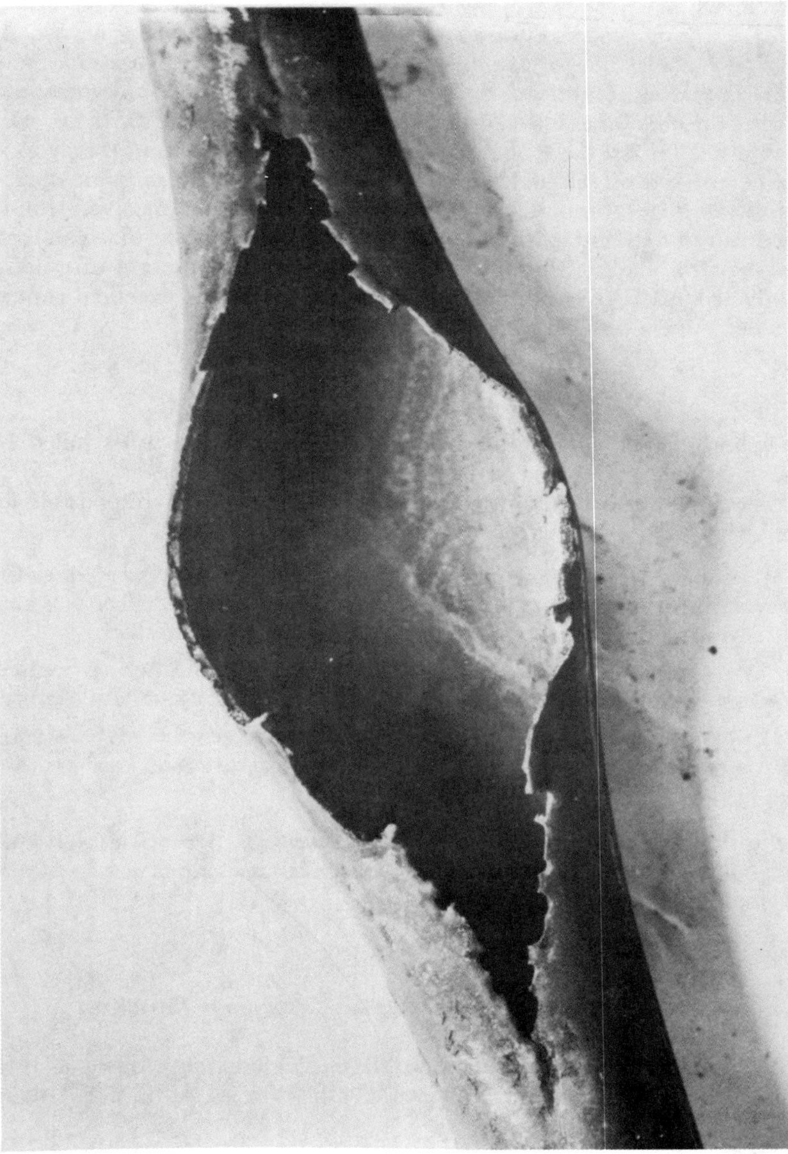

Figure 1-13 Burst boiler tube. Note the thick-walled rupture (compare with Figure 1-2).

boiler.* The ship was steaming steadily at half-load at the time, and records do not indicate anything unusual in operating conditions.

Visual Examination

The burst tube, A25† is shown in Figure 1-13. There appeared to be no metal wastage and this was confirmed by measurement. There was only a minor amount of stretching and the metal had parted along apparent "plates". Cracked metal would give a plate-like appearance on fracture. In fact, near the edge of the burst, many fine cracks $\frac{1}{8}$ inch to $\frac{1}{4}$ inch long, were to be seen.

Among other tubes in which cracking was noted were B39, B40, and B41, shown in Figure 1-14. Although these tubes had opened slightly, rather than bursting open, they had apparently been subjected to a similar mechanism. The typical internal cracking is shown in Figure 1-15.

The absence of fully ductile breaks was associated with the cracks, and the varied ruptures suggested that a change in operating conditions, presumably overheating, had occurred at a particular instant.

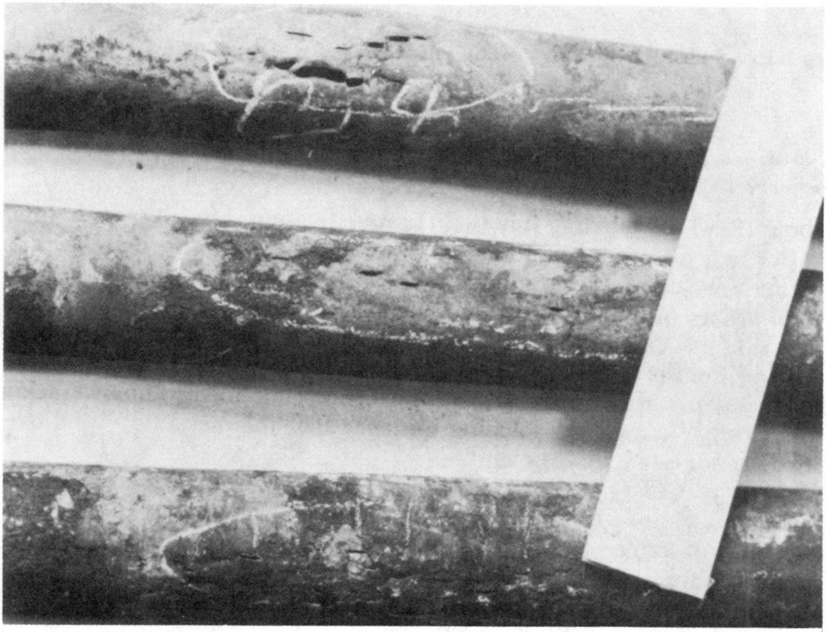

Figure 1-14 Exterior of three leaky boiler tubes.

* Barer, R. D. "Case history of failure of marine boiler tubes by stress corrosion cracking, *Corrosion*, Vol. **11**, No. 4, April 1955, p. 148T.

† Fire row tubes number A1 to A50. B row is second row from fire.

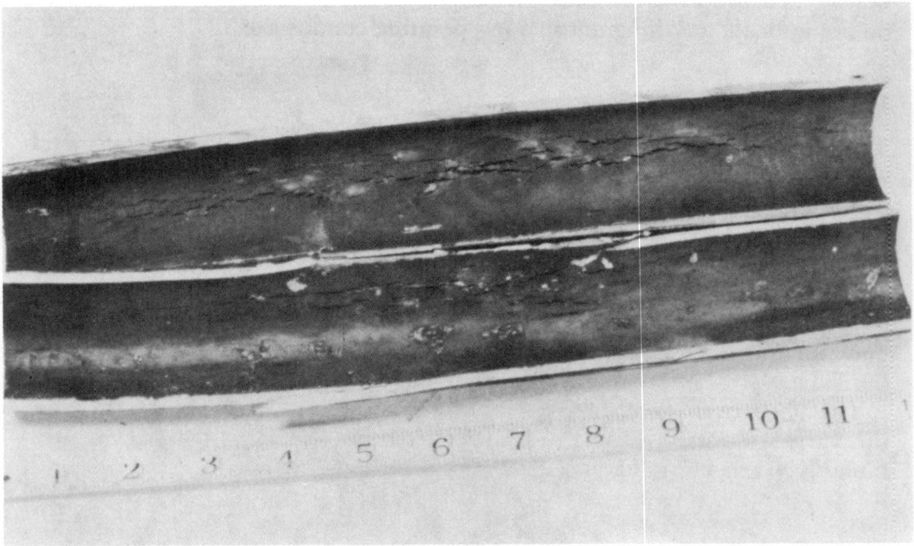

Figure 1-15 Split tube showing cracking inside.

Micro-examination

Tube A25, whose bursting initiated the investigation, was given a thorough and detailed micro-examination. Sections were cut from the tube near the burst. A particularly interesting section, Figure 1-16 shows the presence of several cracks in addition to the main one which is a continuation of the burst. Evidence of normal ductility of the metal may be noted in the outside 25 per cent of the tube wall where the final break was on a 45° plane at the necked-down region. Figure 1-17 shows some of the intergranular cracking found in A25, as well as including evidence that the metal had been overheated, and then rapidly cooled.

Tubes B38 to B41 inclusive were also given a detailed examination. An excellent example of the typical intergranular cracking found in practically all sections examined is shown in Figure 1-18. This cracking is also indistinguishable from cracking exhibited by the steel of an autoclave exposed to 50 per cent caustic soda at 250° F and 400 psi, as reported by Copson.*

The binocular microscope was then used to examine slices from other randomly selected tubes. The cracking was found only in some tubes of "A" and "B" rows, of either side. Sound tubes were found next to cracked ones.

* Copson, H. R., "The influence of corrosion on the cracking of pressure vessels", *Welding Journal*, Feb. 1953, pp. 75s–91s; *Corrosion*, **10**, 1954, pp. 124–139.

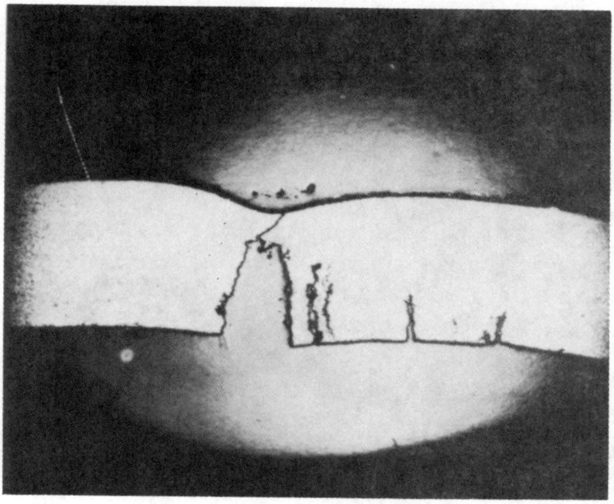

Figure 1-16 Section through tube wall near burst showing ductile necking down and final 45° rupture. Also shown are a number of intergranular cracks. (×7)

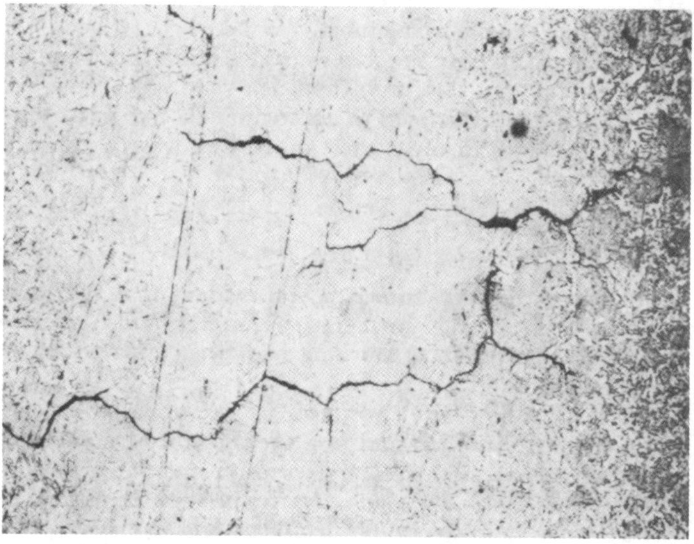

Figure 1-17 Fine intergranular cracking in non-fire, inside surface of tube A25 near the burst. Note normalized-type structure indicating that the tube had been overheated. (×200)

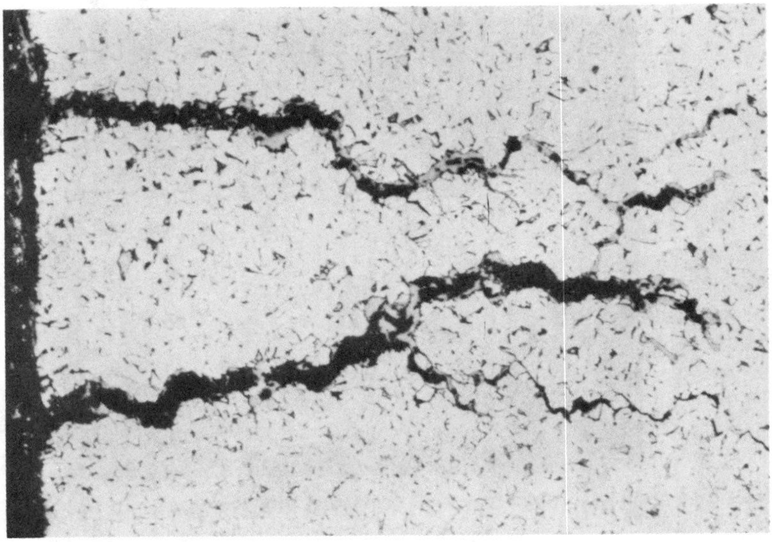

Figure 1-18 Section through tube wall of tube B39 showing intergranular cracking typical of caustic cracking. (×100)

In the defective tubes, the cracking was in a zone approximately centred at two to four feet from the top of the tubes.

Tubes from other boilers on the ship were also examined. Four tubes from each boiler, i.e., A25 and B25, from both left- and right-hand sides, were sectioned at 8-inch intervals down to approximately five feet from the top. No cracking was detected in any of these tubes. This was, of course, a limited sampling.

Discussion of Results

From the microstructures examined, it was evident that A25 and several other tubes had been severely weakened by intergranular cracking before final failure. This intergranular cracking had been initiated on the water side of the generating tubes.

Many metals and alloys are susceptible to intergranular cracking when they are in a stressed condition and in a specific corrosive environment. It should be emphasized that both stress and corrosive environment are necessary.

In the boiler tubes under discussion the stresses were the normal operating stresses resulting from steam generation at 400 psi and 650° F, and the residual stresses. The high residual stresses from fabrication by extrusion and the local stresses from bending the tube would be relatively small after several years' operation, but would still be significant in relation to operating stresses.

The corrosive environment in this case, could only have been caustic from the boiler water. The caustic in this particular instance was by mistake at least twice the usual concentration. However, this increase is not considered too significant. It is true that the caustic concentration in boiler water is usually very low and is expressed in parts per million, commonly 200 ppm. This concentration is approximately 0.1 per cent of that required for aggressive attack.

That a hot spot can occur at a particular area on a tube was shown by Davidson and associates* who used thermocouples to study various conditions. Their findings of a zone of severe overheating is shown in Figure 1-19. They considered this resulted from "steam blanketing" following flame impingement. Figure 1-20 indicates how rapidly the caustic content in the concentrating film boiler water can rise with a small increase of temperature above that of the over-all boiler water. A 100° F rise corresponds to 50 per cent caustic in weight per cent. The net effect may be "equivalent to continuous exposure

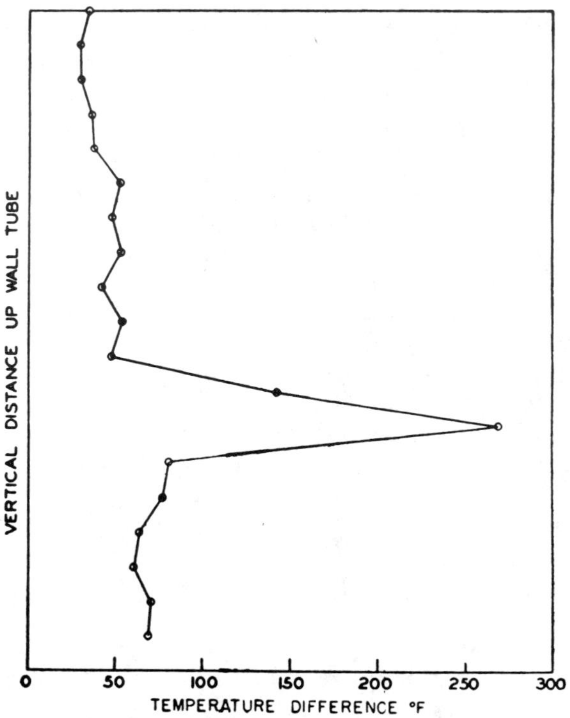

Figure 1-19 Evidence of a hot spot in a 50 ft. vertical water wall tube.[35]

* Uhlig, H. H., Ed. *The Corrosion Handbook,* John Wiley and Sons, 1948, p. 529.

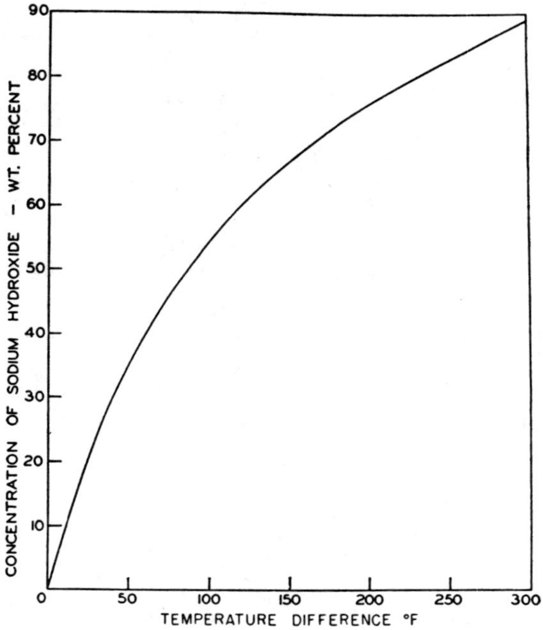

Figure 1-20 Caustic content attainable in concentrating film of boiler water. [Based on data from *International Critical Tables,* **3**, 370 (1928)[35].]

of the hot steel to a corrosive caustic liquor, even though the main body of boiler water contains but a few parts per million hydroxide."

The intergranular attack which results from these conditions is commonly known as "caustic cracking". Although it is difficult to reproduce operating conditions in the laboratory, instances of cracking taking place in hours rather than days are recorded.

The study of boiler-seam cracking by the Joint Research Committee on Boiler Feedwater Studies and the U.S. Bureau of Mines, as reported by Berk and Waldeck,* indicates a danger zone above 180°F and 15 to 43 per cent caustic. The probability of crack development in this zone was 40 per cent. Weir† reported failure of specimens exposed to solutions of less than 10 per cent caustic, although he states that more concentrated solutions are more effective.

Conclusions

Failures by stress-corrosion cracking of marine boiler tubes are commonly

* Berk, A. A. and Waldeck, N. F., "Caustic danger zone", *Chem. Eng.* **57**, June 1950, pp. 235–238.

† Weir, C. D., "Caustic cracking: Stress-corrosion tests in sodium hydroxide solutions at elevated temperatures". *Proc. I. Mech. E.*, 1950, pp. 18–23.

the result of a concentrated caustic solution in contact with stressed boiler metal. Usually the concentration of the caustic results from leakage between seams and around rivets, or capillary action and subsequent evaporation. A rare example of the same type of failure is one where no seams or crevices exist for concentrating the boiler solution by leakage; that is, failure resulting from concentration of the boiler solution right on the steam generating surface of the boiler tube.

It has been shown how stress and corrosive environment together result in severe cracking. That this was an example of stress-corrosion cracking, i.e., electrochemical action aided by stress, is indicated. The particular conditions leading to this failure were channelling of hot gases and subsequent concentration of caustic in the boiler water.

Recommendations to prevent recurrence included attention to external

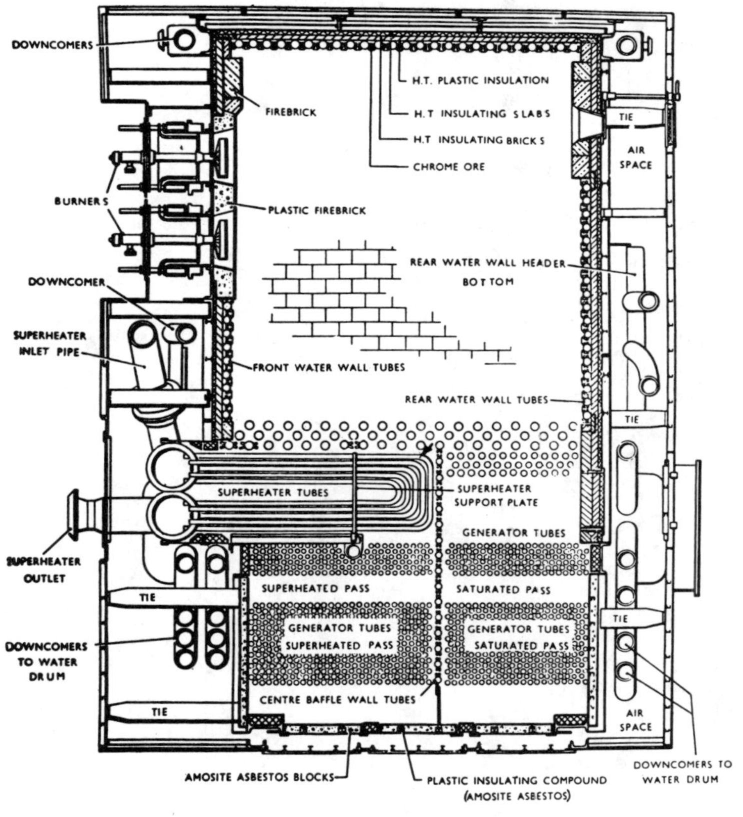

Section through boiler—plan view

Figure 1-21 Section through Y100-type boiler—plan view showing horizontal superheater tubes behind three rows of vertical generator tubes.

water washing of generator tubes and consideration of alternative boiler water treatments, e.g., use of trisodium phosphate, to avoid caustic.

B-3 b) A Superheater Tube Failure by Stress Corrosion Cracking

The superheater bank in this marine boiler consists of horizontal tubes behind the front 3 rows of vertical generator tubes, as shown in Figure 1-21.

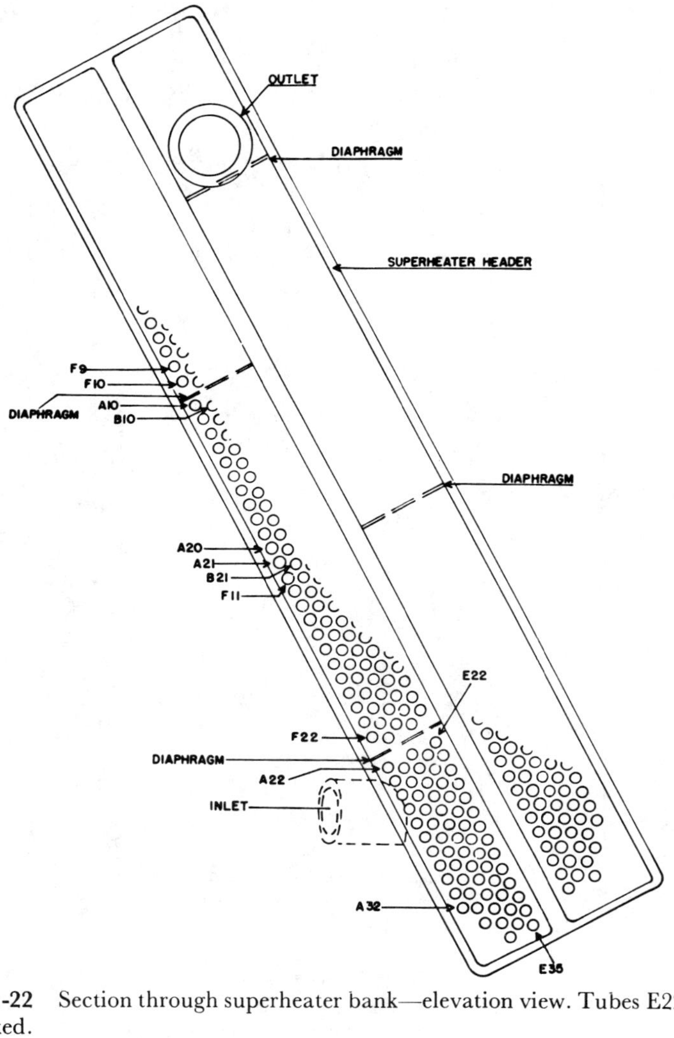

Figure 1-22 Section through superheater bank—elevation view. Tubes E22 and E35 are marked.

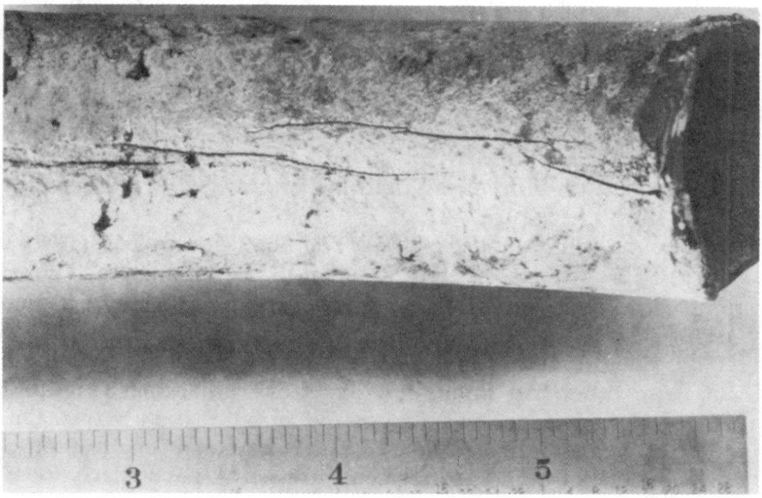

Figure 1-23 Failed superheater tube E31. The rupture took place on the inside of the bend.

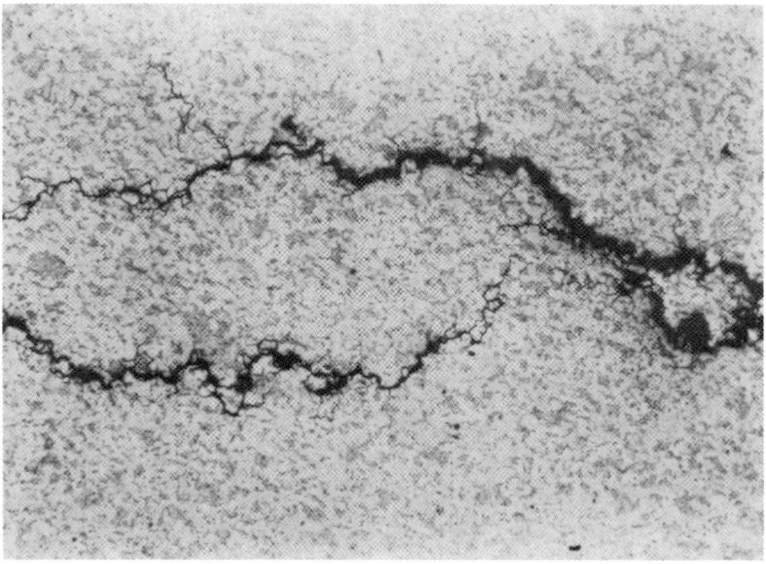

Figure 1-24 Intergranular cracks, typical of caustic cracking in superheater tubes. ($\times 200$)

A series of failures plagued this ship: three inner row superheater tubes, E26, E27 and E29 (as shown in Figure 1-22) failed early in its history. Subsequently in the course of an Atlantic trip, additional failures, namely E30, E31 and E35, resulted in boiler shutdown. Investigation and repair took this ship out of service for almost 2 months.

Visual examination of failed tubes showed that they had ruptured longitudinally on the inside of the bends, and the cracks were mainly confined to the bends, as shown in Figure 1-23.

Microscopic examination of transverse sections taken from the two failed tubes confirmed that the cracks had initiated in the tube bores. The failure mechanism was clearly intergranular stress corrosion cracking as shown in Figure 1-24.

It was also noted under examination at low magnification that the inner surface of these tubes had numerous "score-like" markings, Figure 1-25, which seemed at first glance to be suggestive of fabrication marks. On sectioning these proved to be fissures formed by a corrosion mechanism probably under the influence of high residual hoop stresses. A particularly deep one

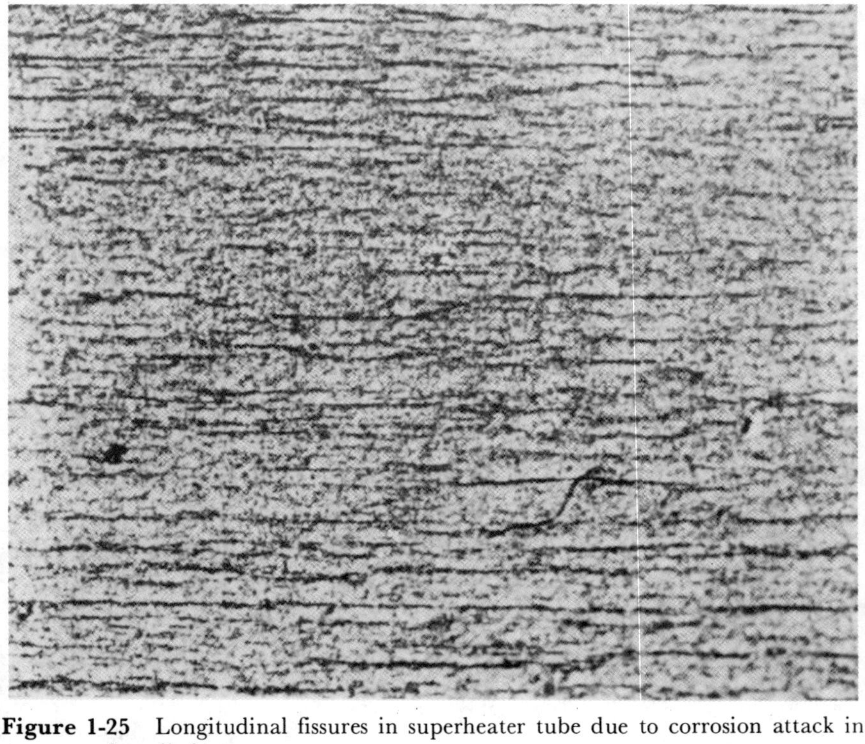

Figure 1-25 Longitudinal fissures in superheater tube due to corrosion attack in presence of tensile hoop stresses.

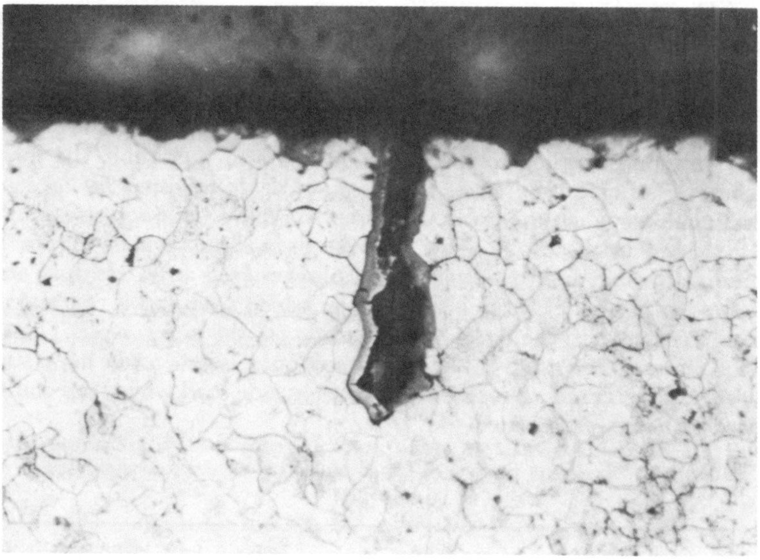

Figure 1-26 Deep longitudinal corrosion fissure in an unfailed superheater tube. (×200)

Figure 1-27 Strain Gauge (SR4) attached to inside of bend of superheater tube bore for residual stress determination.

is shown in Figure 1-26. Although some manufacturing scores were noted they were shallow and did not seem significant.

Strain gauge measurements confirmed that tensile hoop stresses approaching the metal's yield point were produced in the bores of the tubes on the inside of the bend by the cold bending operation. Figure 1-27 shows a small SR4 strain gauge applied to the inside of a superheater tube, the location of highest residual tensile stress. (Stresses were determined by measuring the strain change when the tubes were split to relieve the hoop stresses.)

As indicated in the table below and the drawing shown in Figure 1-28, the manufacturing and bending operations resulted in a residual tensile hoop stress in the bore on the inside (of the bend) wall and a residual compressive hoop stress in the bore on the outside (of the bend) wall. The stress corrosion failures took place in the inside wall only, where *tensile* hoop stresses were present. No cracking took place in the outside wall where the compressive hoop stresses were located.

Table 1-1 Residual hoop stresses in unfailed inside row superheater tubes (psi).

Tube	Exterior, outside wall	Bore, outside wall	Bore, inside wall	Exterior, inside wall
E22	6000 (T)	1600 (C)	5000 (T)	3500 (T)
E23	16,000 (T)	10,000 (C)	16,000 (T)	5000 (C)
E24	15,000 (T)	—	19,000 (T)	15,000 (C)
E27	38,000 (T)	33,000 (C)	8000 (T)	14,000 (C)
E28	11,000 (T)	13,000 (C)	3500 (T)	—

Straight length

Tube	Exterior	Bore
E	19,200 (C)	17,750 (T)
E	—	15,000 (T)

T = tensile stress
C = compressive stress.

A concentration of caustic, high enough to produce cracking in the stressed tubes, was also apparently present. There was some question regarding the source of the caustic. A strong caustic solution used in de-preservation would probably have been removed in the flushing procedures normally used. Analysis of the bore deposits indicated that phosphates were present along with some caustic. This suggests that carry-over, or possibly water-wedging,*

* Filling a boiler with treated water for idle periods.

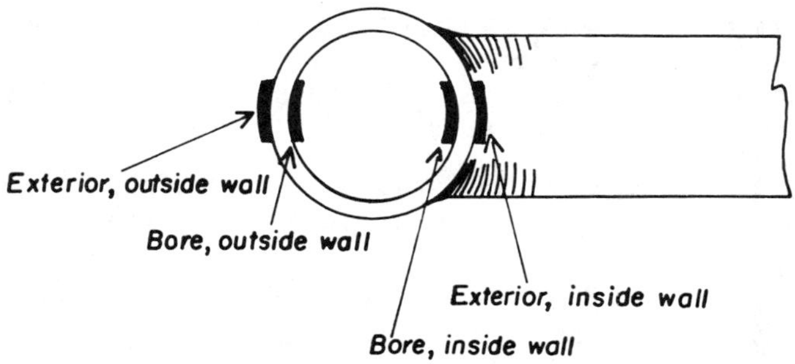

Figure 1-28 Drawing of E row superheater tube (sectioned at bend) showing locations of the strain gauges referred to in Table 1-1.

was responsible for the caustic. Carry-over is known to occur during high rates of steaming.

Discussion

Two main factors are necessary for stress–corrosion cracking to take place in the superheater tubes: namely, a suitable corrosion environment, and high residual tensile stresses.

The strain gauge evaluations have confirmed that residual tensile stresses near the yield point were present. It is also apparent from the failures that a concentration of caustic, high enough to produce cracking in the tubes, was present.

The presence of high residual stresses can be readily explained: the 650° C stress relief specified for superheater tubes was not, in fact, carried out. Such a stress relief would eliminate all but a small residual; for example, one hour at 650° C for SAE 4130 steel changes a 62,500 psi residual to 1900 psi. Actual service stresses are only in the order of 2500 psi. These stress–corrosion cracking failures would not have taken place had the maximum tensile stress only been at approximately the 4000–5000 psi level.

Once the corrosion fissures formed, because of the high residual hoop stresses, they acted as stress raising notches in the initiation of the caustic cracking. It is also noteworthy that these stress-raising fissures would not have formed to any extent in the absence of the high residual stresses.

Why the failures at the bends were confined to the E row tubes is not fully understood. While it is true that the E row tubes were the most severely cold worked, some cold working had taken place in all the tubes and the residual tensile stresses would be no higher than the yield point of the cold worked metal. Only a slightly increased yield point would be expected as a result of additional cold work. It might be argued that the outer row tubes are hotter, and, as a result, a greater amount of stress relief is produced in service; however, failures might then have been expected at the bend in the A row tubes furthest from the fire.

Two factors are believed to be of prime significance in the failures of the superheater tubes by stress corrosion cracking:

a) The superheater tubes were not given a stress relief after being cold bent with the result that high residual tensile hoop stresses were present in the bores of the tubes.

b) A higher than normal caustic content was probably present in the superheater tubes as a result of carryover or water wedging.

B-5 **Corrosion of Boiler Tube Ends at Rolling Folds**

Unusual markings had been noted in the tube ends, from the water drum, for almost $1\frac{1}{2}$ inches in a marine boiler. Concern was expressed that these might be a form of cracking and hence affect reliability of the boiler.

Visual examination in place suggested that these unusual "wormy" grooves, Figure 1-29, were probably from corrosion and did not appear too serious. A few tubes were made available for laboratory examination. Sectioning of an end of a tube showed that the depth of the grooves was only approximately $\frac{1}{8}$th of the wall thickness, Figure 1-30.

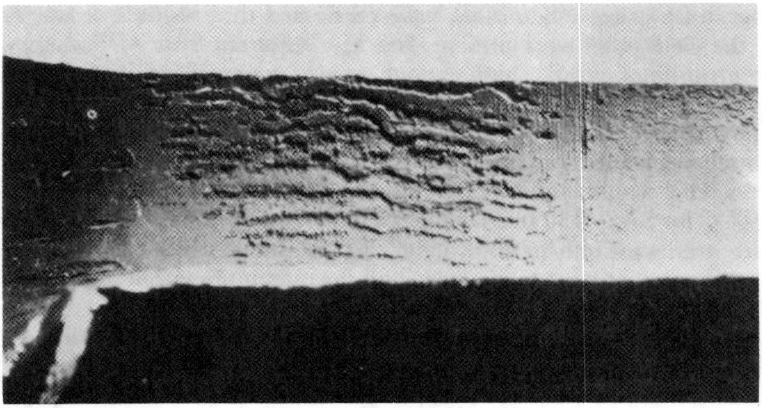

Figure 1-29 "Wormy" grooves inside boiler tube caused by crevice corrosion.

At higher magnification, Figure 1-31, microscopic examination showed that the grooves were associated with severely cold worked metal which was folded over during the tube rolling operation. It is particularly interesting to note the corrosive action initiating in the crevice so formed. It appears that with continued corrosion the end result is the "wormy" grooves, and what was shown in Figure 1-30.

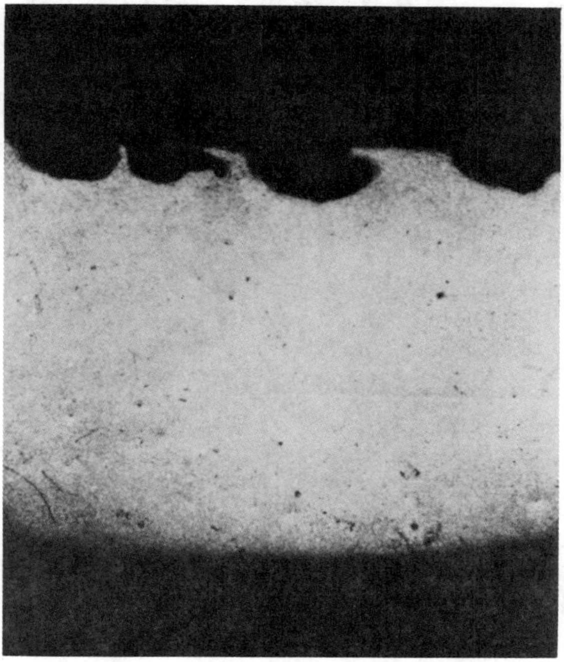

Figure 1-30 Section through tube (in Figure 1-29) showing depth of the grooves. (×20)

It was, therefore, confirmed that this corrosion attack did not represent a serious condition. It would be avoided by control of tube rolling operations to ensure that excessive surface cold working is prevented.

B-6 **Wastage of Superheater Tubes by External Oxidation**

A number of failures of superheater tubes have occurred in the tubes closest to the fire (Figure 1-21) and at specific locations (Figure 1-22). These are tubes A32, and A20 and A21.

These failures invariably result in loss of steam and the tube or tubes responsible are usually plugged off by operational necessity at the first opportunity. Subsequently, these tubes are burned out by the hot furnace gases and it is only rarely that one of these tubes becomes available for assessment. During a boiler inspection, leakage of water was noted and the tube was removed for examination. This tube is shown in Figure 1-32. The rupture showed some scale on the outside (approximately 0.006″ thick) and excessive scale on the inside of approximately 0.033″ (Figure 1-33). This excessive

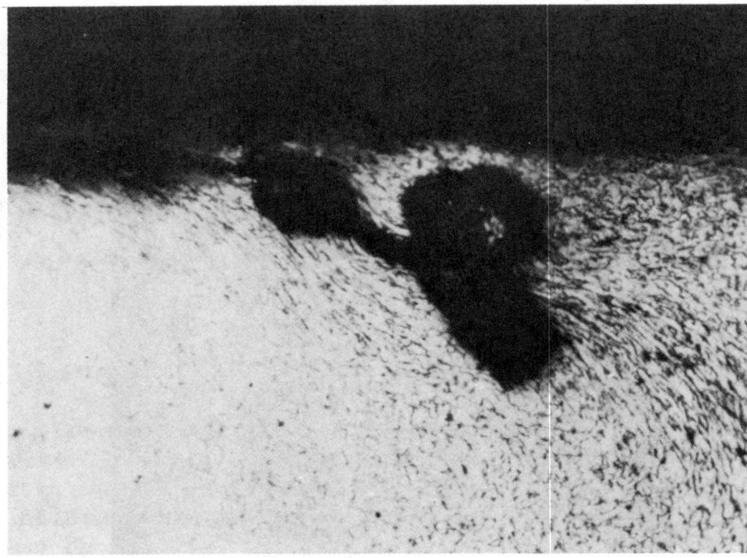

Figure 1-31 Fold in cold-worked metal of the end of the boiler tube shown in Figure 1-29. Note that corrosion attack has initiated in the fold. (×200)

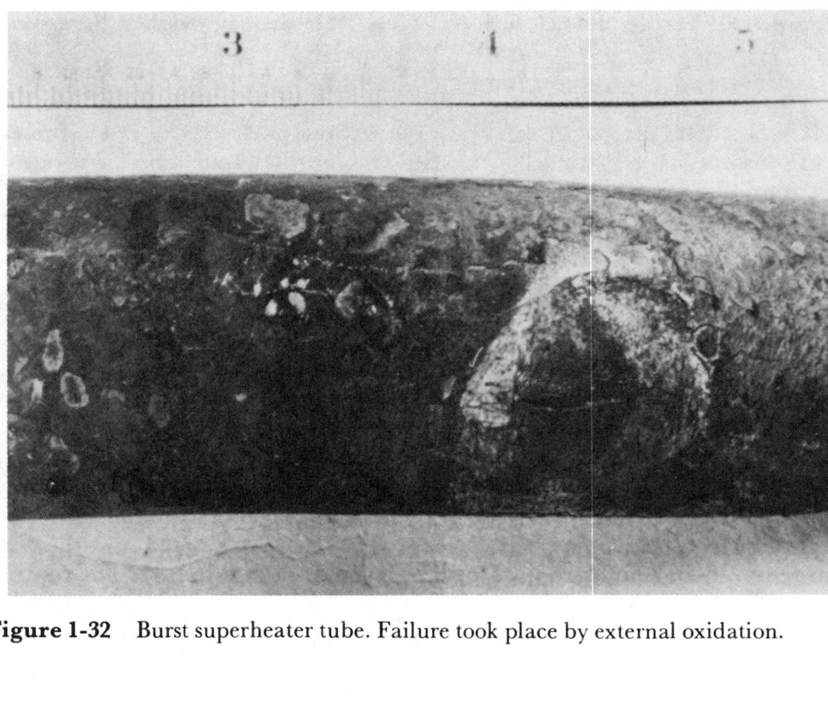

Figure 1-32 Burst superheater tube. Failure took place by external oxidation.

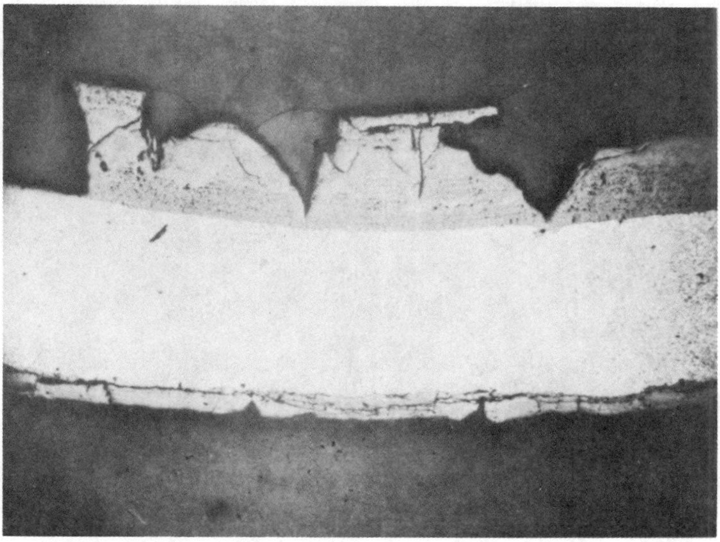

Figure 1-33 Section through wall of tube in Figure 1-32. Note the thick scale on the inner surface. (×20)

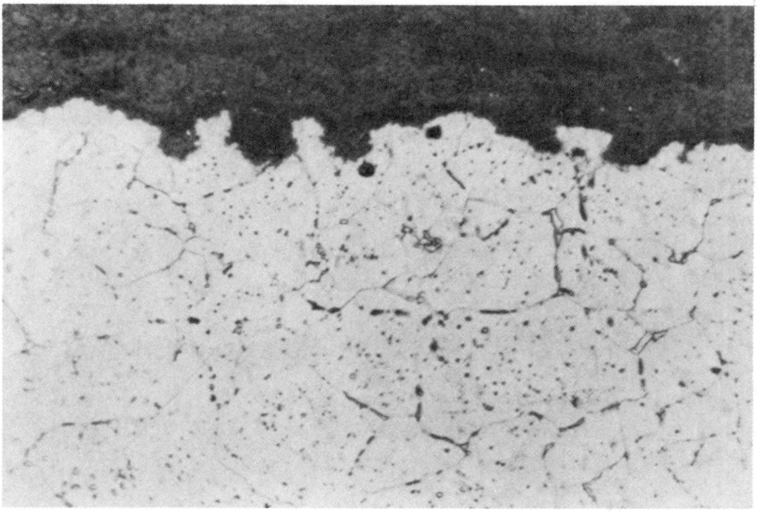

Figure 1-34 Close-up of scale-metal interface of Figure 1-33. There is no evidence of any deleterious action except oxidation. Note also the spheroidization of the steel. (×645)

scale was only in the immediate area of the rupture. The metal thickness was also affected: adjacent to the rupture it was thinned to approximately 0.042″; at 3 inches away from the rupture but still on the fire side, it was 0.093″. This compares with 0.127″ midway around the tube, and 0.144″ at the back of the tube.

Examination under the microscope did not indicate any effect other than overheating: the microstructure showed ferrite grains with spheroidization of the carbides (Figure 1-34). A sample of metal from the back of the tube (i.e. non-fire side) and about 3 inches away from the rupture had a quite different structure: only partial spheroidization of the pearlite, which was still recognizable in separate grains. All this confirms that the fire side at the bend was overheated.

Once overheating started, and with it an extra thickness of oxide, it would subsequently be easier for the metal to overheat and scale, because of the insulating effects.

Experience with other superheater tube failures suggests that the fire side bend is the one that overheats most readily. This overheating may occur when a ship is stopped suddenly after running at full power. At such a time little steam is being drawn off and tremendous heat is still available.

The first failures of superheater tubes took place in boilers that had been severely choked by vanadium-rich slag (Figure 1-35). Gas laning was then suggested as the apparent reason for some of the failures. However, wastage has followed the same general pattern in clean superheater banks. (The vanadium-rich slag can be removed by a peroxide chemical treatment.*)

As a matter of interest, in a later study of superheater tube wastage, a number of tubes were made available for examination. A preliminary radiographic examination revealed a severely wasted tube, thinned to the point of bulging, Figure 1-36. This study also confirmed that wastage was

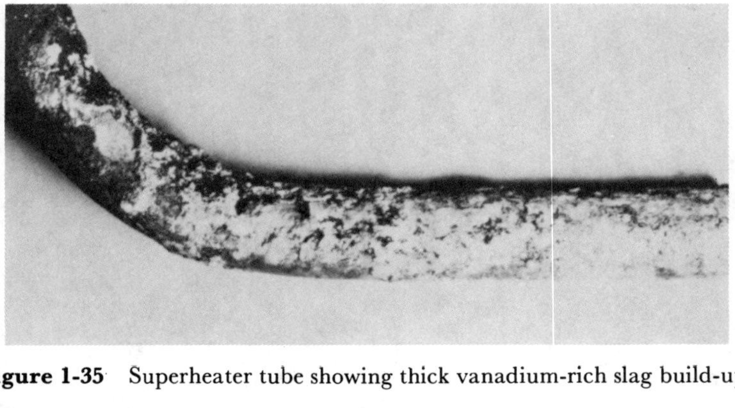

Figure 1-35 Superheater tube showing thick vanadium-rich slag build-up.

* Contact authors for additional information.

confined to a small zone and led to development of a nuclear gauge† and an ultrasonic method of inspection in place.

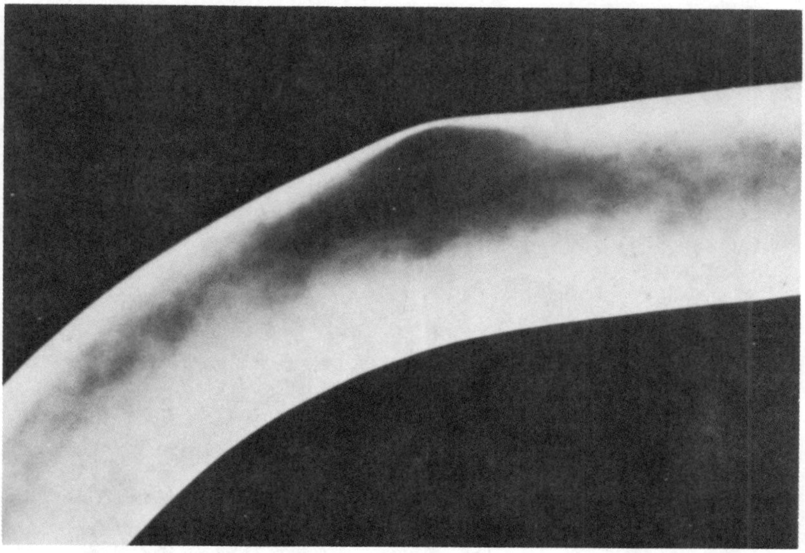

Figure 1-36 Radiograph of fire row superheater tube bend showing thinning (0.040 inches) and bulging at beginning of bend.

The cause of these superheater tube failures has been high temperature oxidation due either to a design factor leading to partial steam starvation, or to a heat flow pattern of the combustion gases.

C-2 a) Failures of Superheater Tubes Because of Manufacturing Defects

The failures of four superheater tubes in a marine boiler (Figures 1-21 and 1-22) were reported to our laboratory. These failures were unusual in that they occurred in the inner rows away from the more intense furnace radiation (compare with B-3(b)). Also, the failures occurred in that section of the superheater which had recently been fitted with new tubes, and these tubes had therefore been subjected to relatively few steaming hours.

It was reported that this boiler was being steamed auxiliary when excessive

† Peters, B. F., "A thulium 170 nuclear gauge for measuring superheater tube wall thicknesses in a marine boiler", *Naval Engineering Journal,* Dec. 1964, p. 887.

use of feed water, and steam issuing from the funnel, indicated a tube leak. Operation prior to the most recent failure was considered normal.

Three tubes were made available for examination, including two which had failed some months earlier and been plugged off. These plugged tubes are often badly oxidized and relatively valueless; in this instance, however, some information was derived from them. The most recent failure (E24), not exposed to steaming following rupture, is shown in Figure 1-37.

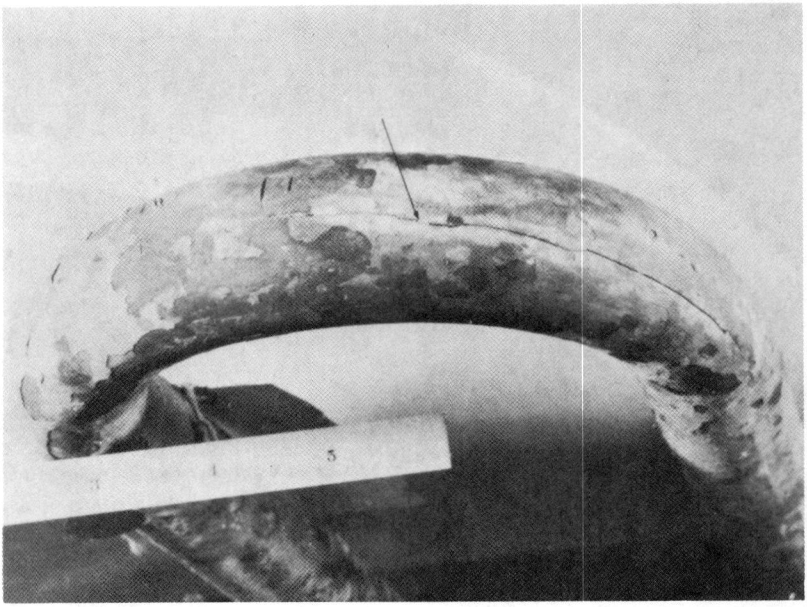

Figure 1-37 Superheater tube rupture caused by longitudinal defect in tube.

All three tubes were analysed and met specifications of approximately 1.5% Cr, 0.6% Mo, and 0.15%Ni.

Tube E24 was sectioned at the location marked in Figure 1-37. Metallurgical examination suggested that failure had initiated in a manufacturing defect or flaw, as shown in Figure 1-38. At higher magnification, Figure 1-39, it was noted that the metal surrounding the defects was low in carbon content; that is, no carbides were present. The fact that the steel was high in carbide content away from the defect supports the view that the defects were inherent; that is, metallurgical inhomogeneities in the tubing would be expected in a product containing defects which originated in the ingot stage.

Both the subsurface defect (at the right in Figure 1-39) and the extension of the rupture had associated with them a mottled grey metallic substance which was not detected in the unetched condition. The composition of this material was apparently different from that of the ferrite matrix; however,

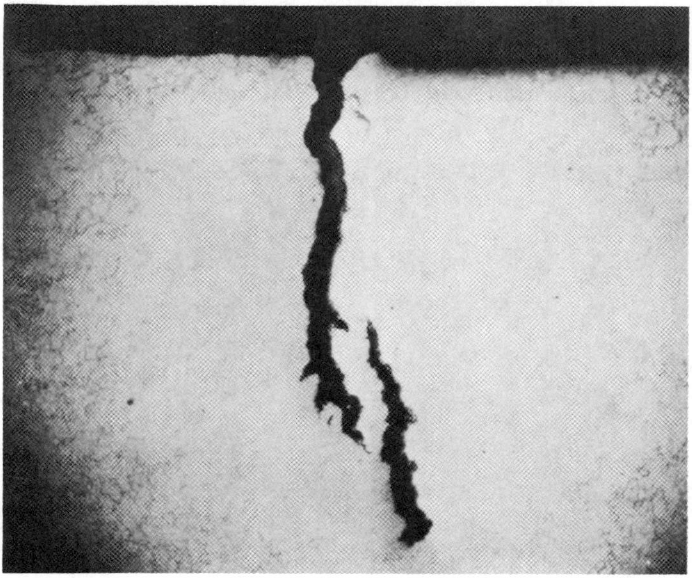

Figure 1-38 Section at edge of burst of tube shown in Figure 1-37. Note both the extension of the rupture and the subsurface defect. (×80)

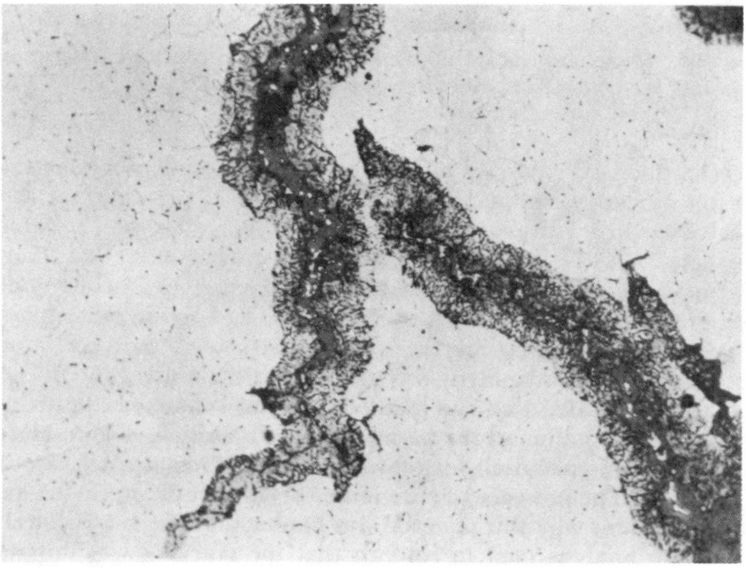

Figure 1-39 Defect similar to that in Figure 1-38. Note the large ferrite grains and the mottled grey structure associated with the defects. (×500)

spectrographic analysis of the edge of the burst showed essentially no difference in composition to that of the parent metal. The darker grey non-metallic material in the rupture extended onto the inside surface of the tube, and was apparently magnetic iron oxide. This material was not present in the internal defect, however.

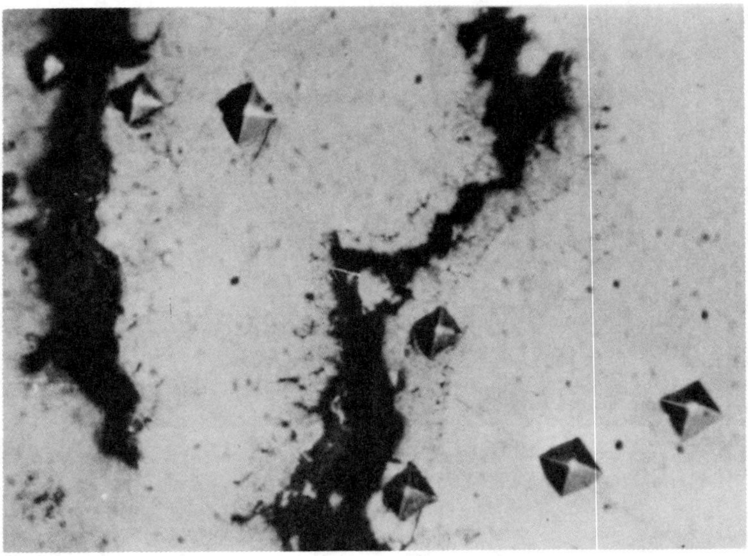

Figure 1-40 Microhardness indentations in mottled grey material showing that the grey material is considerably harder than the ferrite matrix. (×500)

Microhardness indentations made both in the mottled grey material and in the soft ferrite grains are shown in Figure 1-40. It is apparent that the material associated with the defects is considerably harder than is the nearby parent metal. It was estimated that the mottled grey material had a hardness equivalent to approximately Rockwell C20. Other areas containing the mottled grey material were also detected in tube E24 and in other tubes.

Because of interest concerning the origin of this strange "mottled" structure, an electron probe microanalysis was obtained at the University of Toronto. This analysis indicated that the mottled material contained approximately twice as much chromium as the normal tube composition. The results of the electron probe microanalysis scan for chromium, across a crack, are shown in Figure 1-41. The presence of chromium carbides in the steel (the mottled structure) explains why this material was harder than the matrix steel. The results of this analysis tend to confirm that the cracking was initiated by inhomogeneities which were due to original fabrication.

The metallurgical examination has shown that inherent defects associated

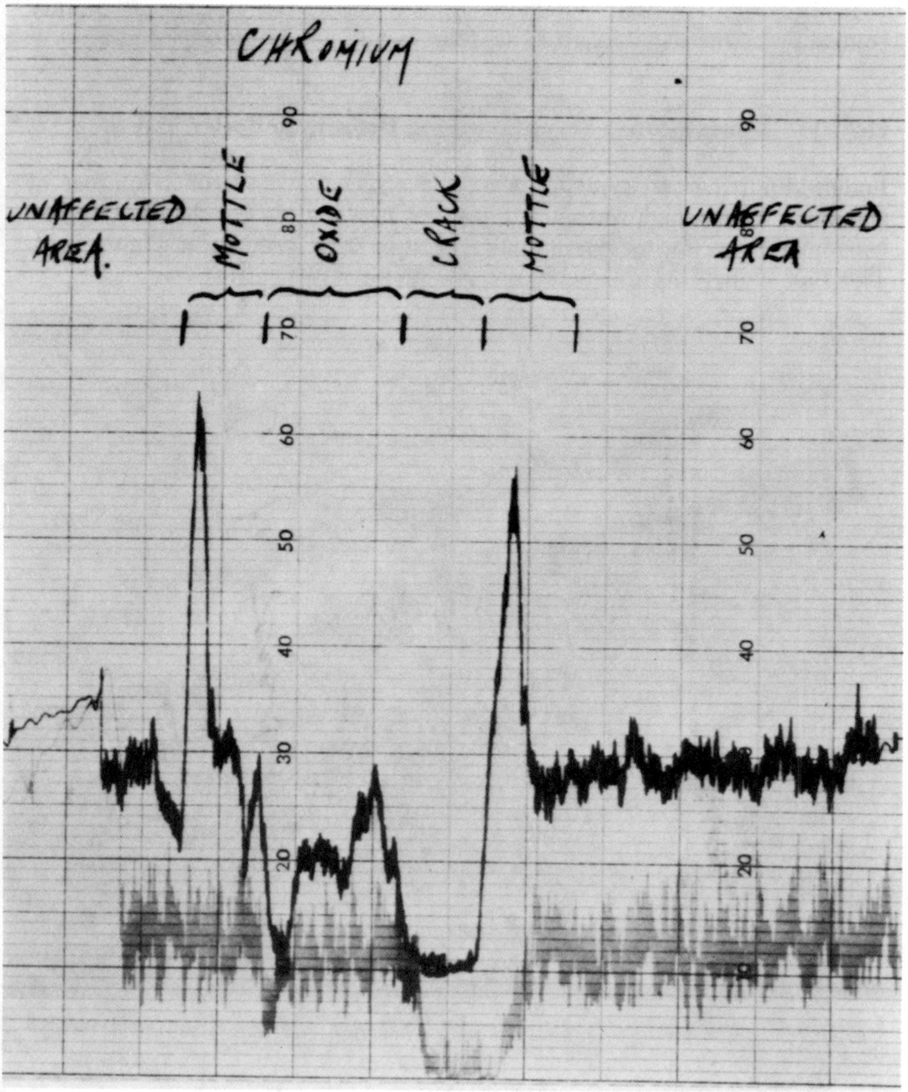

Figure 1-41 Electron Probe Microanalysis Scan for Chromium across a crack in superheater tube (Figure 1-39). Analysis shows that "mottled" grey area was rich in Chromium. [University of Toronto Report, October 1964].

with the original fabrication were present in all the failed tubes. The metal adjoining the defects was much harder than that in the remainder of the tubes. It is probable that, on cold bending, the defects split open leaving cracks in the inside surface at the bends.

It was apparent that the manufacturer needed to tighten his quality control procedures so that defective tubes would not get into service.

C-2 b) **Longitudinal Manufacturing Defects in Boiler Tubes**

Longitudinal defects were noted on the external surface of a number of generator tubes which were being installed in a boiler. The defective region became obvious during expanding of the tube ends as shown in Figure 1-42. This type of marking was suggestive of a manufacturing defect.

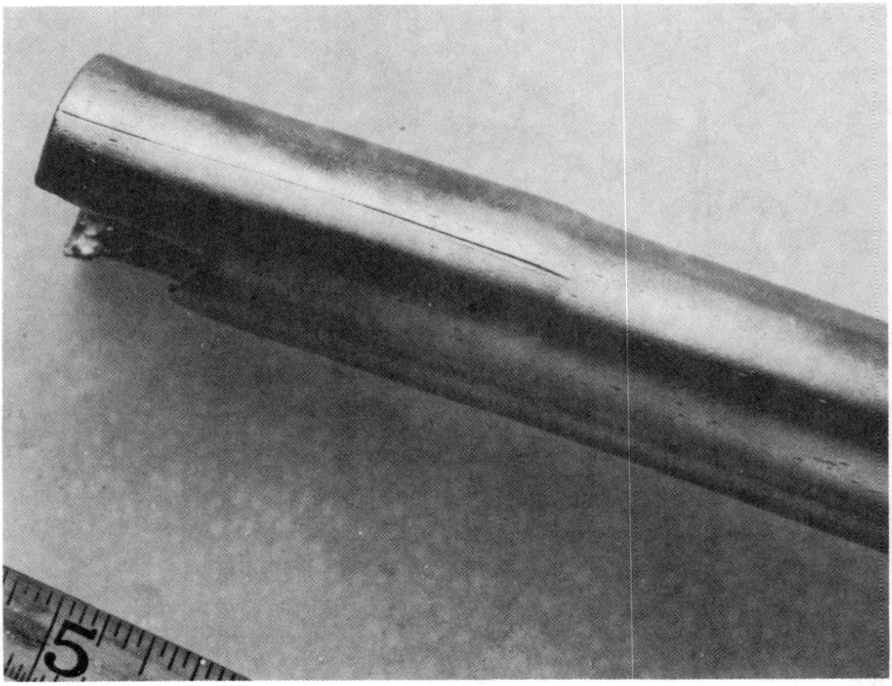

Figure 1-42 Longitudinal fold in surface of boiler tube. Expanding operation opened the fold.

Metallurgical examination of the worst area showed that the defect was severe, extending almost one-quarter of the way through the tube wall as shown in Figure 1-43. Another section cut at 18 inches further along the tube, at an area having no visible defect, was most revealing, Figure 1-44. These photomicrographs establish the flaw as an inclusion or fold, likely incorporated at an early stage in tube manufacture.

A magnetic particle inspection was made on numerous other tubes in this

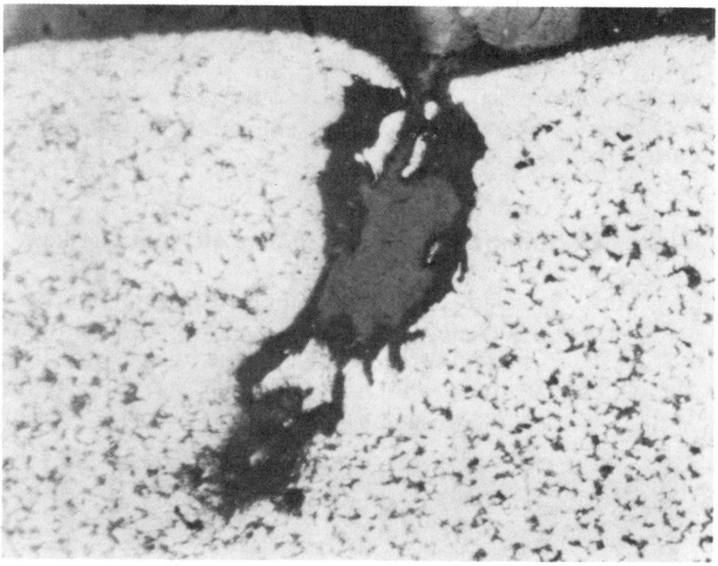

Figure 1-43 Section through obviously defective area in Figure 1-42. (×80)

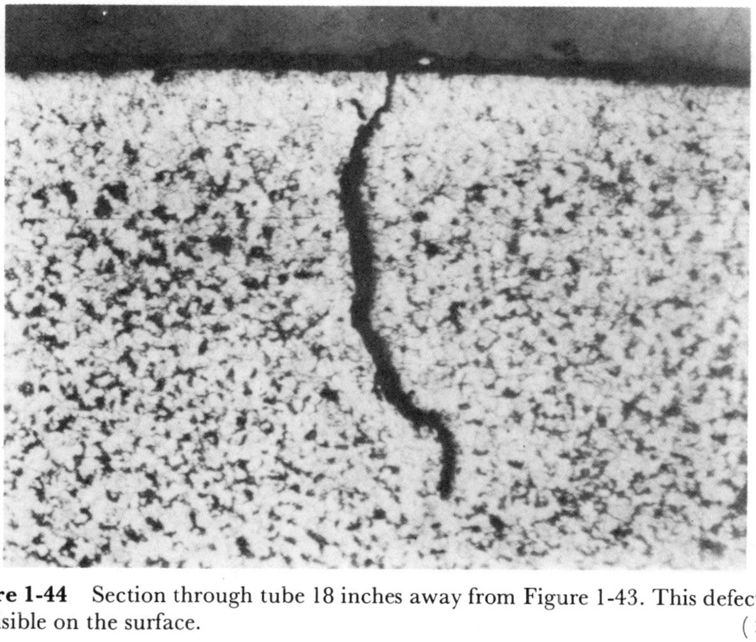

Figure 1-44 Section through tube 18 inches away from Figure 1-43. This defect was not visible on the surface. (×80)

and subsequent boiler work and several tubes were set aside for further examination. Defects were generally shallow and could be removed by light dressing. While these shallow markings are not structurally detrimental, they nevertheless can introduce a lack of pressure tightness.

Again, tubes with these defects, shallow or severe, should not leave the manufacturers' plants.

C-3 Failure of Economizer Element Tubes at Weld Defects

Following reports of leaking economizer tubes in several marine boilers, an investigation was requested. These economizers are studded tubular heat exchangers, as shown in Figure 1-45, widely used in power boilers, which preheat the boiler water just before it reaches the water drum of the boiler proper. In one instance it was understood that engine room personnel had

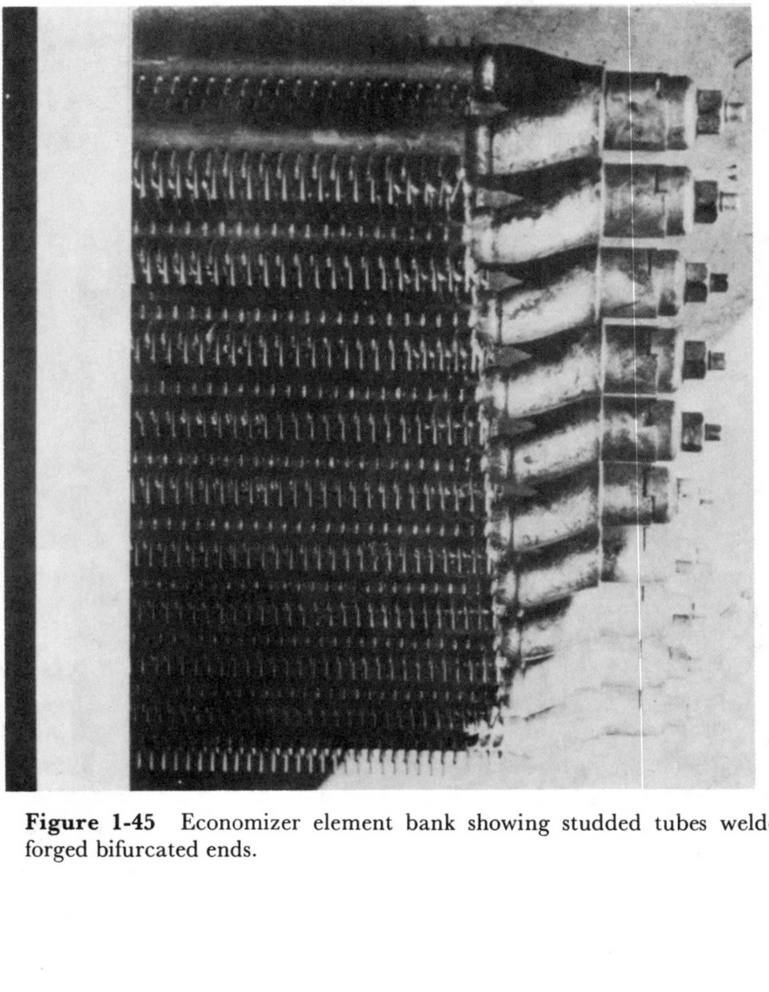

Figure 1-45 Economizer element bank showing studded tubes welded to thick forged bifurcated ends.

difficulty in maintaining even one of two boilers in operation with the number of leaks which developed in both boilers.

The company representatives, considering it isolated porosity, attempted repair on a rather unsatisfactory patchwork basis. However, it was only after radiographic examination of the welded joints that the magnitude of the problem became apparent. A typical example of the weld defect revealed is shown in Figure 1-46. This lack of fusion, confirmed by numerous sections such as in Figure 1-47, was widespread—a systematic defect—and seriously weakened the economizer elements. In some of the welds as little as 25% of the wall thickness of the tube existed at the joint. In addition, the lack of fusion served as a sharp notch which weakened the remaining metal still further. It was rather surprising with the stress-raising effect of the notch and in the presence of thermal stresses, that the economizer elements had not failed sooner. Welding a thick forging to relatively thin tubing requires appropriate thinning of the thicker piece.

The widespread nature of this inferior welding was confirmed by radiography of portions of a new economizer element, available as a spare.

It would appear that exercise of quality control by the manufacturer was

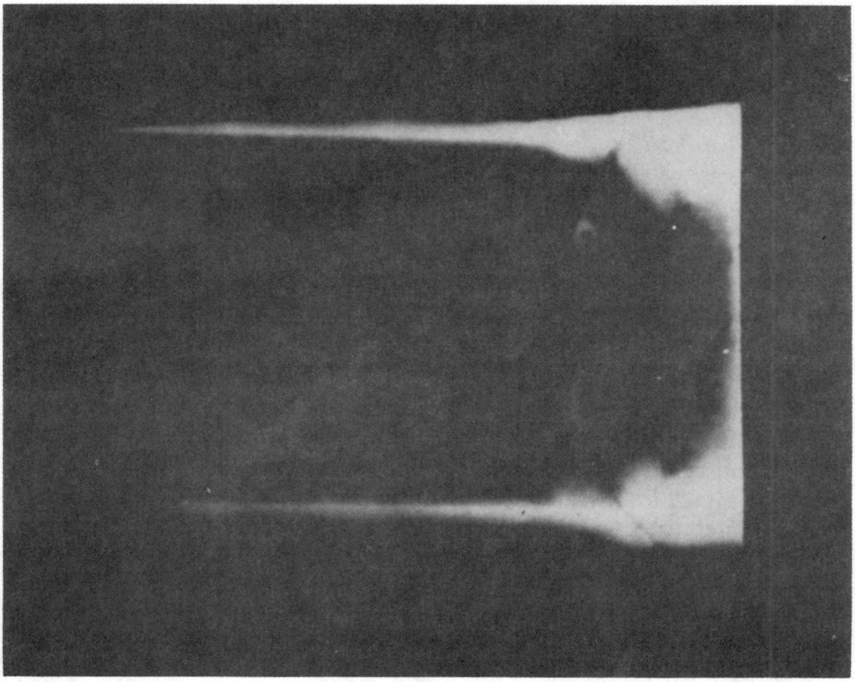

Figure 1-46 Radiograph of a weld (Figure 1-45) showing lack of fusion of weld.

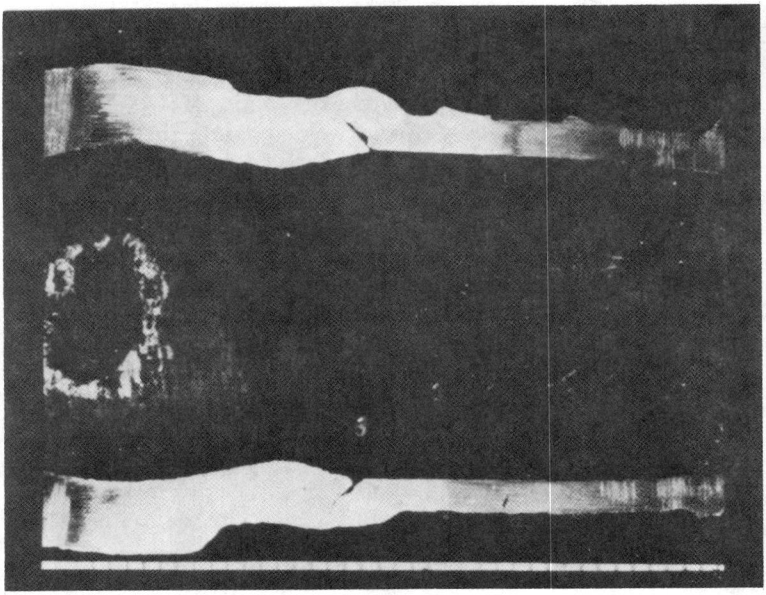

Figure 1-47 Section of economizer element showing lack of fusion in weld.

inadequate. It was learned from a representative of the company that radiography, although available, had not been used and that the supervisors had relied on sectioning of occasional samples.

It was recommended that all these economizer element welds be repaired because they were not only unreliable in their existing condition but also unsafe.

C-9 Mechanical Gouges Caused when Expanding Tubes

The presence of some pits of indeterminate depth in a fire row boiler tube was a matter of concern to the boiler engineer. Even though this apparent weakness in the tube was only 8 inches down from the steam drum, little information could be obtained regarding nature and depth of the apparent corrosion attack from visual examination.

This boiler tube was radiographed in place. A copy of the radiograph, Figure 1-48, shows that the defective area in the tube consisted of a number of circumferential gouges which appeared to have been caused by mechanical means. Densitometer measurements on the radiograph indicated that the deepest groove extended no more than 0.030″ into the 0.125″ tube wall.

It appears that these gouges were caused by a slip of a tool during original installation. Subsequent corrosion was insignificant.

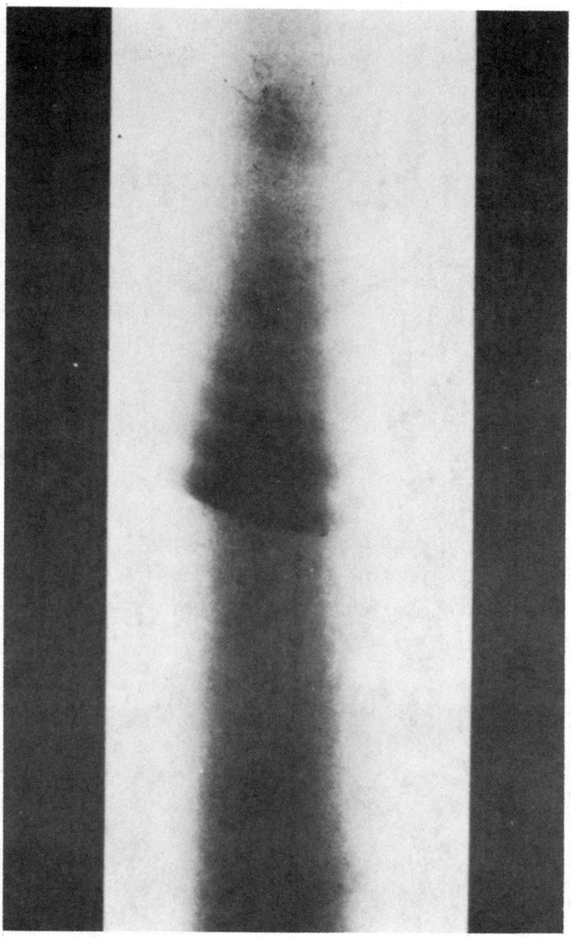

Figure 1-48 Radiograph of tube showing gouge marks caused when expanding tube.

II

BOILERS—GENERAL

IN THIS CHAPTER several boiler-related problems are examined. A badly mangled oil thrower ring on a turbine shaft was able to furnish clues on its fracture surface to indicate failure by fatigue. The probably unreliable service which can result from metal spray on a steam piston rod is demonstrated.

Unusual pitting of a turbine and its evaluation is described in the third example. The mechanism responsible for severe pitting in a labyrinth gland area is described.

Corrosion fatigue in diverse items such as an actuating control unit and in a condensate line is discussed. Stress corrosion cracking in gaskets made of stainless steel is demonstrated in samples which disintegrated.

Troubles with type 310 stainless castings in the difficult exposure conditions associated with superheater support plates are described. Influence of shrinkage cracks in the welds of stainless steel flame rings and importance of mechanical notches on service life are explained.

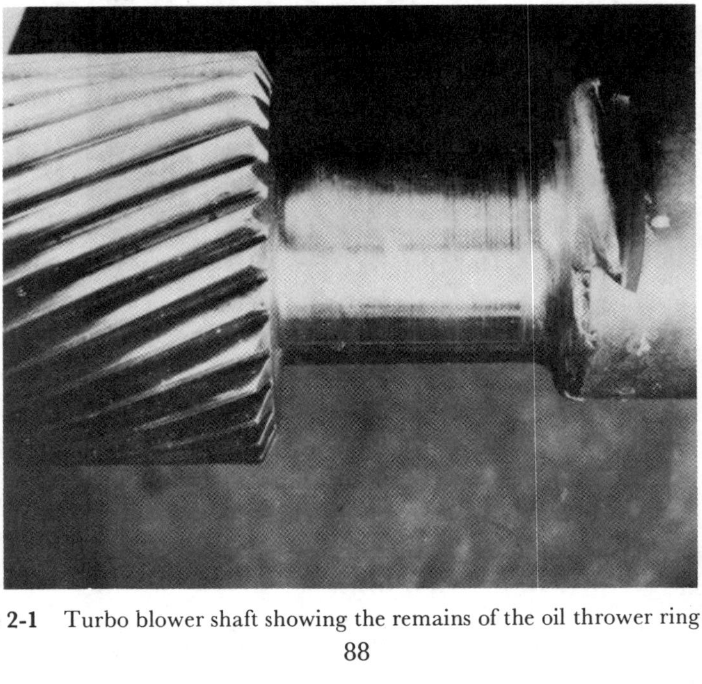

Figure 2-1 Turbo blower shaft showing the remains of the oil thrower ring at right.

A-1 Turbine shaft oil thrower ring failure by fatigue.

B-1 a) Corrosion failure of steam piston rod with spray-on metal.
b) Pitting corrosion of turbo-alternator turbine.
c) Galvanic corrosion in labyrinth glands of turbines.

B-2 a) Thermohydraulic generator—failure by corrosion fatigue.
b) Condensate line corrosion fatigue failure.

B-3 Failure of stainless steel gaskets by stress corrosion cracking.

B-6 Failure of 310 stainless steel superheater tube support plates by oxidation.

C-3 Flame ring—failures at weldments.

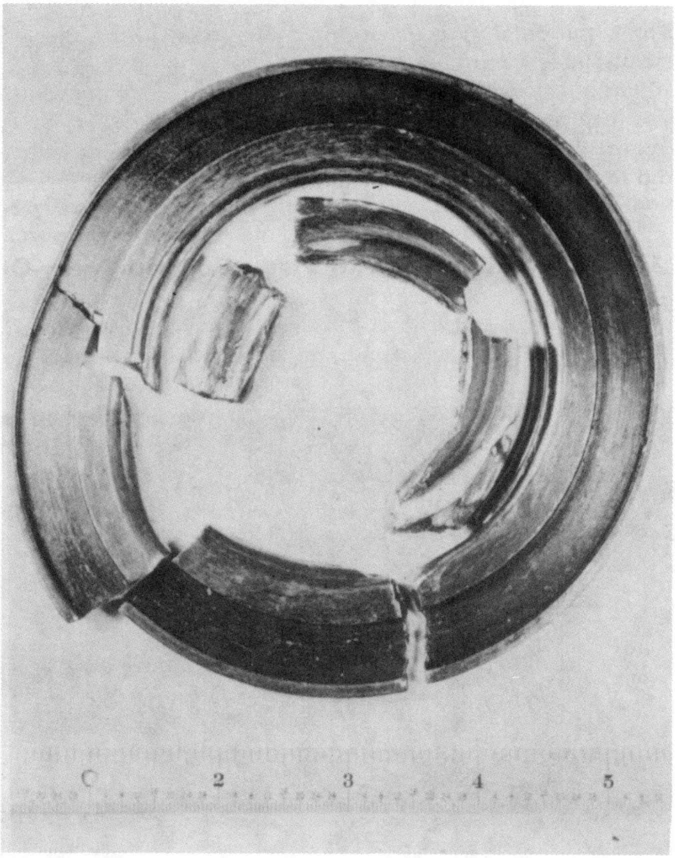

Figure 2-2 Turbo blower oil thrower ring fragments. One section of the inner part of the ring is still on the shaft in Figure 2-1.

A-1 Turbine Shaft Oil Thrower Ring Failure by Fatigue

The failure of the top oil thrower ring on a turbo blower was brought to our attention and the failed parts were subsequently submitted for examination. The turbo blower shaft is shown in Figure 2-1 and the recovered parts of the ring are shown in Figure 2-2. The turbo blower appeared to be free from any other damage.

Examination of the numerous fracture surfaces showed most of them to be badly worn and smeared. Much of the evidence was therefore destroyed. However one piece of the inner ring showed, almost hidden on the fracture surface, markings which could be considered to represent a rapidly advancing fatigue crack, Figure 2-3. The origin of this crack could be traced to the sharp corner (Figure 2-4) machined at the junction of the inner and outer ring portions.

In addition, the outer ring portion had separated into 3 pieces, and on one of these fractures, clear fatigue markings were evident, Figure 2-5.

While it was not possible to arrive at an exact sequence of events, there is nevertheless little doubt that a generously filleted corner at the junction shown in Figure 2-4 would have averted the failure. It is noteworthy that the designer did not stipulate a radius at this junction.

B-1 a) Corrosion Failure of Steam Piston Rod with Spray-On Metal

A defective steam piston rod from a Fire and Bilge Pump was submitted for examination. At some previous repair the shaft had been turned down and

Figure 2-3 Fracture surface between outer and inner parts of the ring. The markings at the bottom edge indicate that the fracture progressed from left to right.

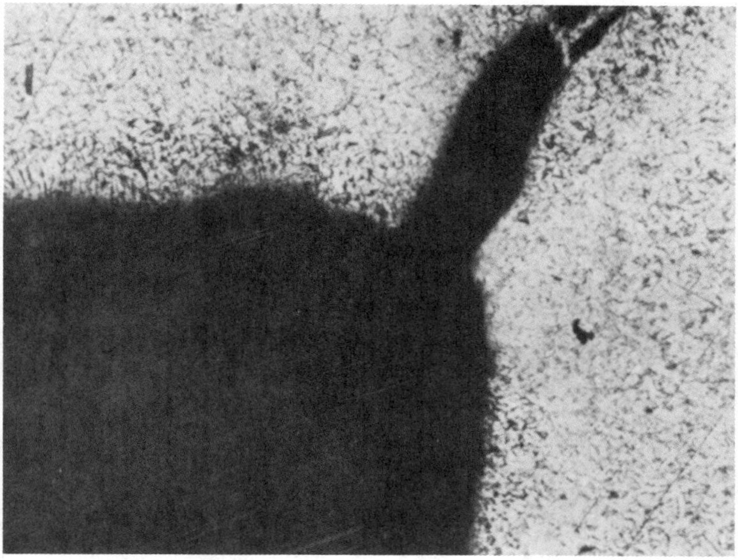

Figure 2-4 Cracked sharp corner junction of outer part of ring and oil thrower ledge. Note the absence of a radius even at 80 × magnification.

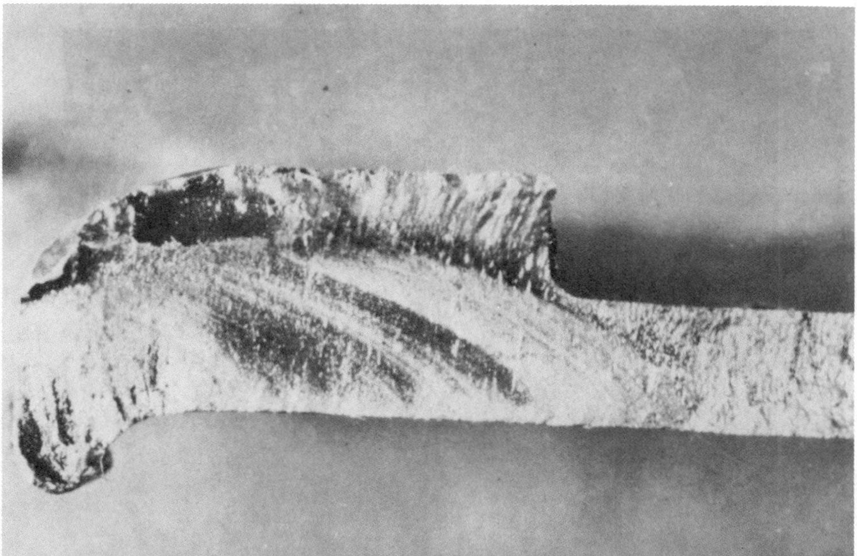

Figure 2-5 Fractured surface of radial crack in outer part of oil thrower ring showing classical fatigue markings.

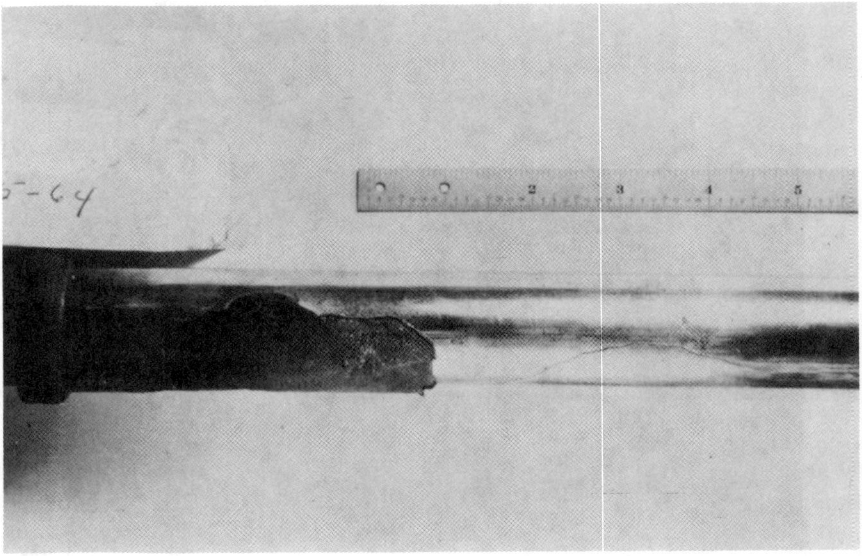

Figure 2-6 Steam piston rod corrosion at crevice between shaft and the spray-on metal.

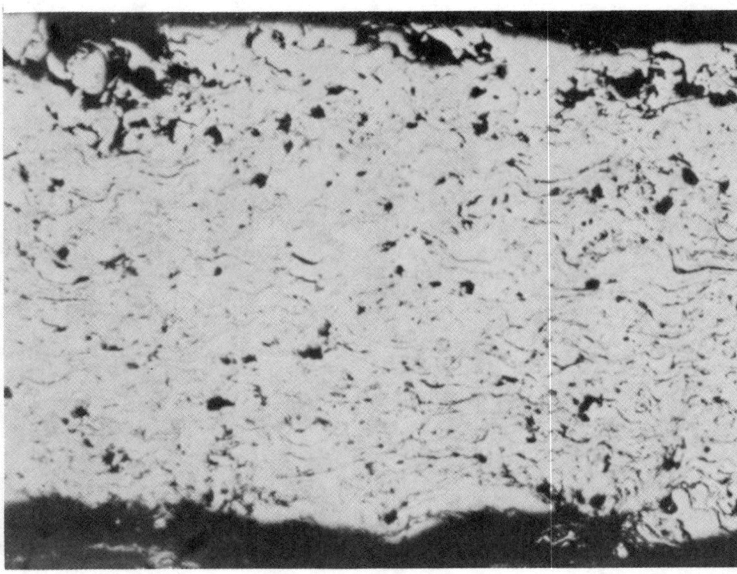

Figure 2-7 Section through type 410 stainless steel spray-on metal. Note oxide inclusions and porosity.

built up with metal spray. The spray coating was cracked over the full length of the repair and in some areas had flaked from the shaft. The extent of deterioration is illustrated in Figure 2-6.

The piston rod and spray coating were spectrographically analyzed and were found to be plain carbon steel and type 410 stainless steel respectively.

Visual examination showed voluminous corrosion products which, combined with thermal stresses, were undoubtedly involved in the rupturing of the relatively weak spray coating.

Figure 2-7 is a cross-section of the spray coating showing the extent of porosity and oxide inclusions. It is likely that failure was caused by penetration of moisture to the interface between spray and base metals, where it resulted in the severe corrosion of the carbon steel rod.

It was recommended, for this type of repair, that the metal spray filler rod be of the same composition as the base metal. Under these circumstances any corrosion taking place should be visible at the surface. It is possible that improved spraying techniques might result in a less porous coating, in cases where a particular corrosion resistant surface is desired; however, plating from a reliable source would be more likely to give durable results.

B-1 b) **Pitting Corrosion of Turbo-Alternator Turbine**

Severe pitting of the turbine body from a turbo-alternator was a matter of concern to inspecting officials. The unusual type of pitting is shown in Figure 2-8.

Figure 2-8 Turbo-alternator turbine showing corrosion attack confined mostly to the first stages.

Close examination revealed a fine white, powdery deposit on rotor and blades at the high pressure end. Analysis proved this deposit to be essentially ordinary salt (NaCl). The fact that only a limited part of this turbine had this minute deposit suggested that this was a single occurrence; if recurrent, it would likely have built up more uniformly and extended further down the turbine.

A close-up of one of the worst areas is shown in Figure 2-9. The localized corrosion evident appears to have occurred at some time in the past—it is now essentially dormant. The appearance of the pitting and distribution suggest that it occurred as a result of the presence of moisture (condensate) containing chloride during a lengthy idle period. Some of the pitting extends to approximately $\frac{1}{2}$ inch width and 0.04 inch depth.

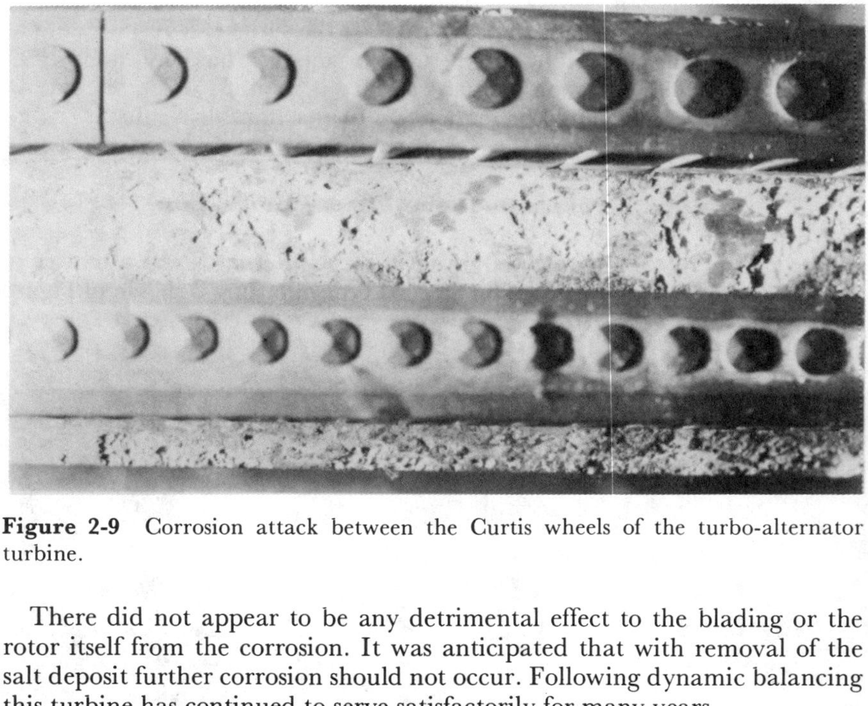

Figure 2-9 Corrosion attack between the Curtis wheels of the turbo-alternator turbine.

There did not appear to be any detrimental effect to the blading or the rotor itself from the corrosion. It was anticipated that with removal of the salt deposit further corrosion should not occur. Following dynamic balancing this turbine has continued to serve satisfactorily for many years.

B-1 c) **Galvanic Corrosion in Labyrinth Glands of Turbines**

The shaft labyrinths on auxiliary turbine rotors have suffered severe corrosion as shown in Figure 2-10. It may be noted that the gland tips coincide with the corroded grooves.

Figure 2-10 Arrangement of the labyrinth gland around the shaft labyrinth. Note that the gland tips coincide with the corroded grooves.

Spectrographic analysis of the two parts involved gave the following results:

labyrinth gland: bronze containing 88% copper, 10% tin, 2% zinc.

shaft labyrinth: low alloy steel containing approximately 3–4% nickel, 0.6% molybdenum.

Examination of the pits at higher magnification tended to eliminate other possible mechanisms except corrosion by a series of local cells. This grouping of the pits in a distinct pattern is shown in Figure 2-11. The detailed mechanism suggested by the shape and disposition of the pits is as follows:

a) During shutdown of the turbine the remaining steam condenses, forming small drops of slightly acid solution (pH approx. 6.5).

b) The drops tend to gather at the edges of the labyrinth seal, forming small corrosion cells between the labyrinth gland and the shaft labyrinth.

c) Coupling of the gland and the shaft through various metallic paths produces eventually a galvanic cell between the bronze (cathode) and the steel shaft (anode). The result is an intensive pitting action.

The answer adopted to eliminate attack in this area was nickel plating of the shaft labyrinth. Considerable experimentation to ensure a well-bonded

Figure 2-11 Series of corrosion pits along the shaft labyrinth. Note Figure 2-10.

nickel plate was conducted prior to acceptance of this method. It might be mentioned that there are chemical methods of treating the boiler water to minimize this corrosion.

B-2 a) **Thermohydraulic Generator Failure by Corrosion Fatigue**

Failure of an actuating control unit for marine boilers by cracking was evaluated. A section of one of these units, with arrows indicating location of the cracks is shown in Figure 2-12.

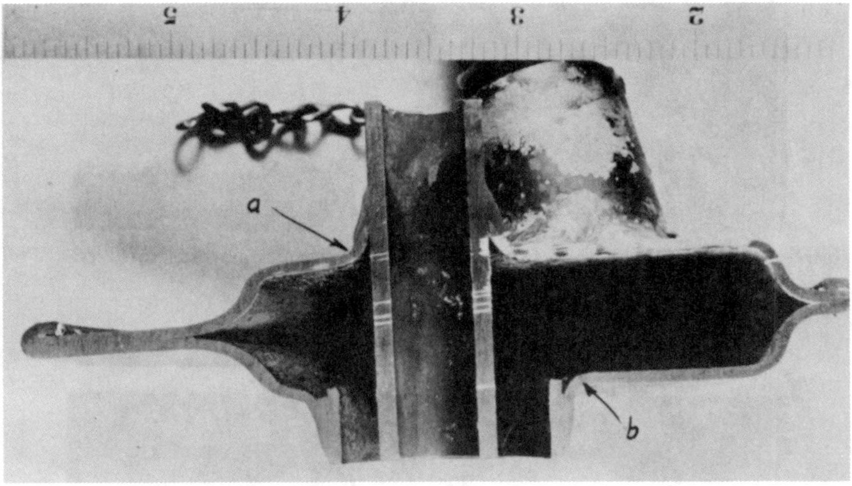

Figure 2-12 Sectioned thermohydraulic generator head. Arrow "a" points to rupture and "b" points to partial penetration of another corner.

A close-up examination of the crack at (a), shown in Figure 2-13, indicates the predominant corrosion factor in this corrosion fatigue failure. A similar examination of the corner at (b), shown in Figure 2-14, also indicates that corrosion was the major factor involved. This attack could have been expected to cause rupture there also in time.

It is interesting that both the rupture and the partial penetration took place at a relatively tight bend and at a zone where the relatively rigid junction changes to a more flexible part of the generator head.

Various ways of repairing these cracked generator heads were considered. Because corrosion was a predominant factor the possibility of fabricating new heads using a corrosion-resistant metal, such as phosphor-bronze, seemed to offer a reasonable improvement. Alternatively, it was believed that a head differently designed, and welded to minimize rigidity at areas where some flexing is needed, might have offered a sufficiently long service life.

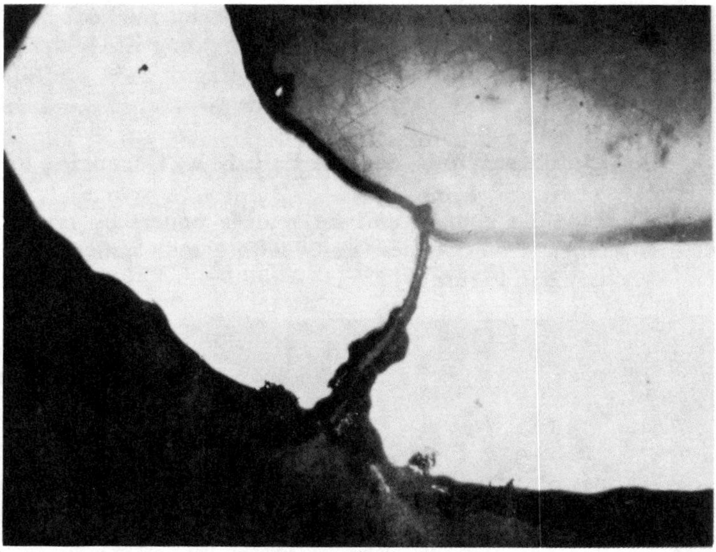

Figure 2-13 Corrosion fatigue crack ("a" in Figure 2-12) which initiated on the water side (bottom). (×20)

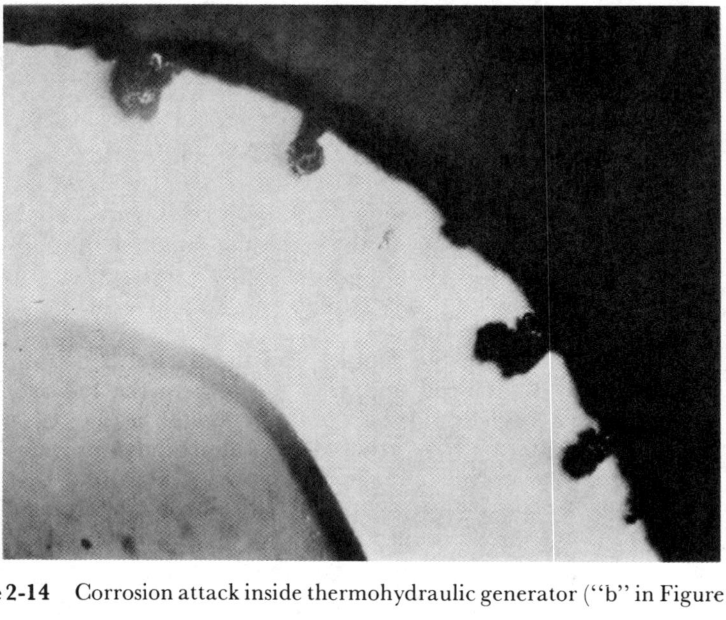

Figure 2-14 Corrosion attack inside thermohydraulic generator ("b" in Figure 2-12).

B-2 b) **Condensate Line Corrosion Fatigue Failure**

Leaks in a saturated steam line to the pressure gauge of a marine boiler were brought to our attention for investigation. This was the fourth failure of this small diameter tubing.

Radiographic examination, a useful preliminary prior to sectioning, showed that a number of circumferential cracks were present in the area of the leak.

Following this the tube was split without disturbing the circumferential cracks shown in Figure 2-15.

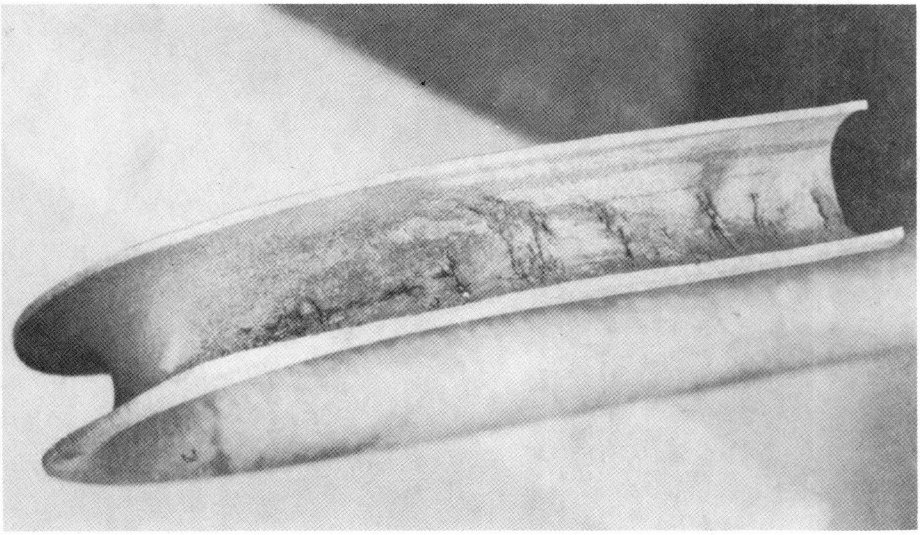

Figure 2-15 Transverse corrosion fatigue cracks inside steel condensate line.

Microscopic examination revealed cracks similar to those shown in Figure 2-14 but generally somewhat deeper.

Corrosion fatigue cracking of this type is often associated with a high residual tensile stress. This condition would result if the tube had to be bent while being fitted. Superimposed on the residual stresses are the fluctuating stresses due to service conditions. Corrosive attack acting under the above conditions produces cracks of the type noted.

This problem might be mitigated by stress relief after bending to shape. Alternatively, this and other corrosion problems in condensate systems could be eliminated by suitable treatment of the condensate, as with filming amines.*

* E. E. Galloway "Filming Amines Control Corrosion in Utility Plant Condensate System" *Corrosion,* Vol. **15** No. 8, August 1959, p. 99.

B-3 Failure of Stainless Steel Gaskets by Stress Corrosion Cracking

The disintegration of spiral wound gaskets used in the hand holes of marine boiler headers was brought to the attention of our laboratory. These gaskets, Figure 2-16, included samples from the economizers, the rear wall headers, and the superheater headers.

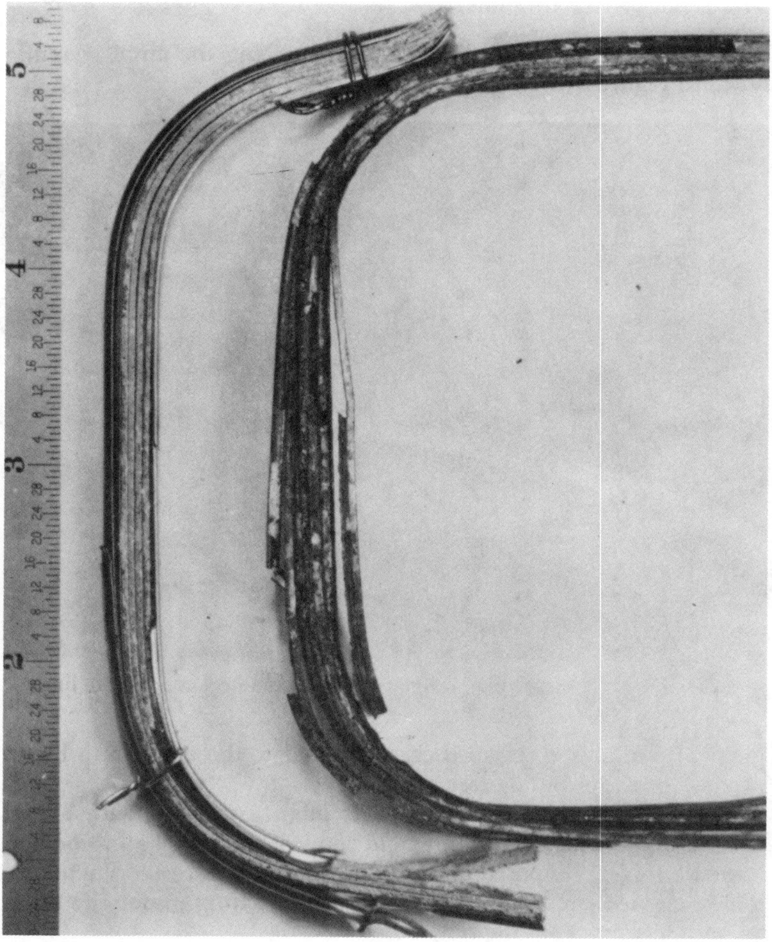

Figure 2-16 Stainless steel spiral wound gaskets. The right-hand one has disintegrated by stress corrosion cracking.

Spectrographic analysis confirmed that the metal in these gaskets was 18-8 stainless steel.

A piece from one of the disintegrated gaskets was examined under the

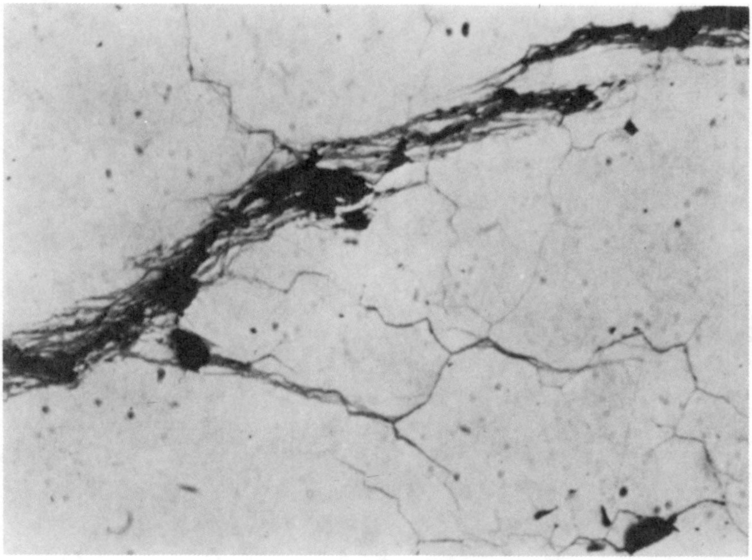

Figure 2-17 Transgranular stress corrosion cracking of stainless steel spiral wound gasket, Figure 2-16.

Figure 2-18 Radiograph of cast 310 stainless steel support plate showing shrinkage voids and cracks between tube openings.

Figure 2-19 Type 316 stainless steel flame ring.

microscope. It was readily apparent, as shown in Figure 2-17, that stress-corrosion cracking was responsible.

Information about the exposure conditions included the fact that some of these gaskets had been exposed to a chloride-containing chemical used in cleaning one of the boilers. A leach for chloride performed on the samples showed that generally only traces of chloride were detected but a high chloride content was found in asbestos from the disintegrated gaskets.

The hazards in terms of destroying a stainless-containing gasket material need to be borne in mind when considering chemical cleaning with chloride-containing solutions. However, because chemical cleaning is so infrequent, simple replacement of these gaskets may be the economic solution.

B-6 Failure of 310 Stainless Steel Superheater Tube Support Plates by Oxidation.

The superheater tubes shown in Figure 1-21 are held in place and spaced

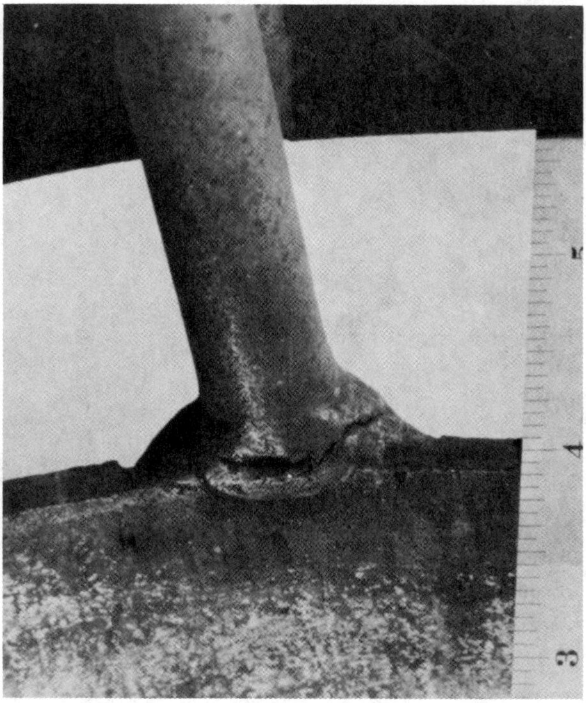

Figure 2-20 Crack in flame ring which had propagated from a defect in the transverse weld.

by means of a support plate. These support plates are cast from type 310 stainless steel and have not resisted oxidation too well. Severe wastage at the leading edge resulted in some superheater tubes sagging.

Investigation confirmed that the cast support plates were of the correct composition.

Radiography of these plates proved to be unusually informative. As shown in Figure 2-18, two gross defects were common to almost every cast opening: a large triangular shrinkage void and a crack at the narrow neck.

It would appear that heat transfer from the leading edge would be markedly impeded by these defects, leading to premature wastage. It was consequently recommended that plates be fabricated of wrought type 310 stainless and that some thickening of the leading edge be considered.

C-3 Flame Ring Failures at Weldments

Flame rings shown in Figure 2-19 were failing at frequent intervals by cracking through the weld joining the support lug to the ring. Typical

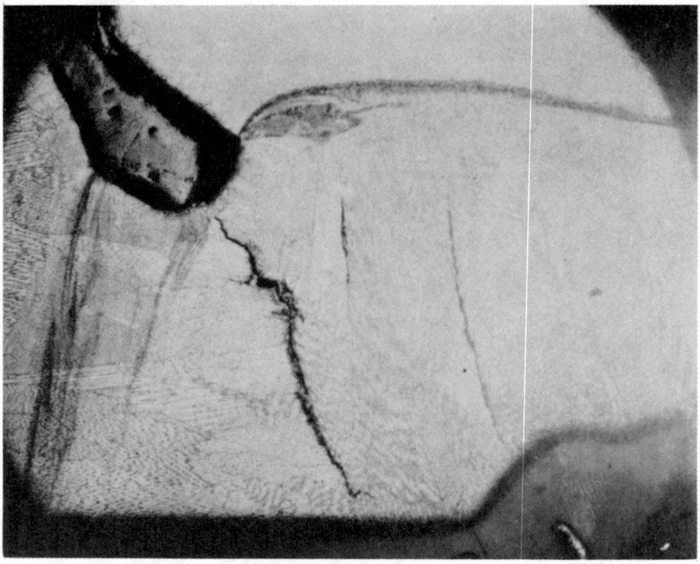

Figure 2-21 Section through flame ring weld. Note the cracks in the weld bead. (×30)

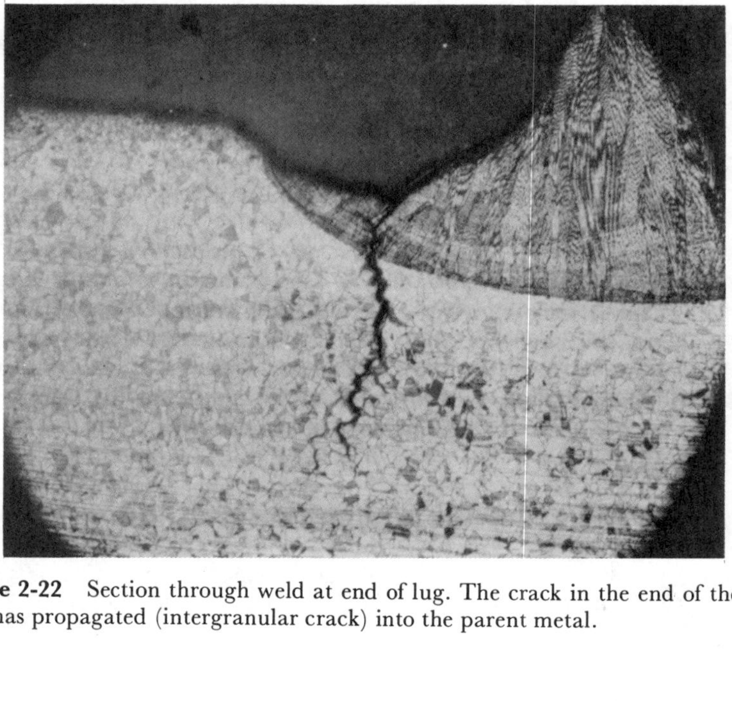

Figure 2-22 Section through weld at end of lug. The crack in the end of the weld bead has propagated (intergranular crack) into the parent metal.

cracking of this junction is shown in Figure 2-20. In addition, many lugs were cracked at the location shown by arrow in Figure 2-19.

Radiography showed cracks at the end of the lug (where it merged into the body of the ring) and also incomplete penetration of the two main butt welds joining the two parts of the ring.

Metallurgical examination of sections cut through the weld of Figure 2-20 showed cracks, Figure 2-21, attributed to original fabrication. Also a section through the weld at the end of the lug showed how a small shrinkage crack had subsequently propagated into the parent metal of the ring (Figure 2-22). Similar weld cracks were found in a new flame ring (not exposed in service).

In their service in marine boilers, the rings were subject to fluctuating stresses produced by temperature differentials across the ring. The defects in the welds grew into cracks of serious proportions under the influence of these stresses.

Spectrography showed that both the lugs and the rings were type 316 stainless steel, welded with type 310 stainless rod.

In addition to the defective welds, it was believed that the thick rigid joint of the lug to the more flexible ring tended to intensify or concentrate the stresses at the edge of the weld. Tapering this joint would undoubtedly alleviate the tendency to crack.

III

HEAT EXCHANGERS

An interesting assortment of failures mostly by corrosion in a variety of heat exchangers comprises a large part of this chapter. Corrosion fatigue of aluminum brass condenser tubes is outlined, as well as replacement in an alloy immune to the corrosion component. Failure of a copper–nickel–iron resistant to moderate flow of sea water was attributed to inadvertent crevices incorporated in design. Impingement or corrosion–erosion is a common failing of brass components exposed to moderate flow of sea water.

Instances of dezincification of tubes and tube sheets made of uninhibited brasses are described.

Systematic defects when thin monel is welded, and their detection by radiography are described. Failures in small complex heat exchangers were attributed, on the basis of metallurgical examination, to defects in the soldered joints and not to the corrosion generally suspected.

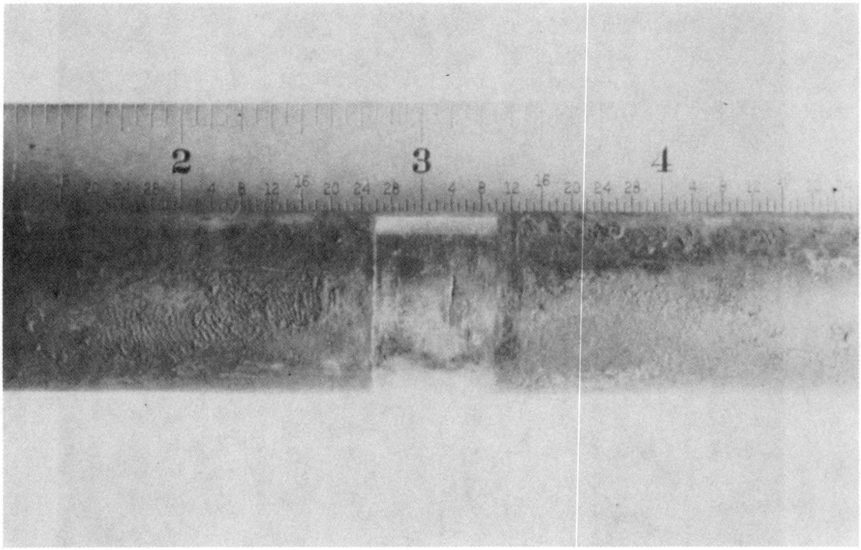

Figure 3-1 Brass condenser tube showing cracking at the location of the support plate.

B-2 Aluminum brass condenser tube corrosion fatigue problem.

B-5 Unit cooler corrosion failure problem.

B-7 Deterioration of brass condenser components by impingement.

B-8 a) Brass condenser tube failure by dezincification.
b) Dezincification of tube sheets of air ejectors.
c) Dezincification and impingement of condenser tube sheets.

C-3 a) Monel evaporator basket failures at welds.
b) Heat exchanger failures by fatigue of soldered joints.

B-2 Aluminum Brass Condenser Tube Corrosion Fatigue Problem

A leaky condenser tube was brought in for examination, together with another adjoining tube. The external crack is shown at the location of the support plate in Figure 3-1.

The tube was split open longitudinally and revealed a number of cracks

Figure 3-2 Inside of condenser tube (Figure 3-1) showing corrosion fatigue cracks.

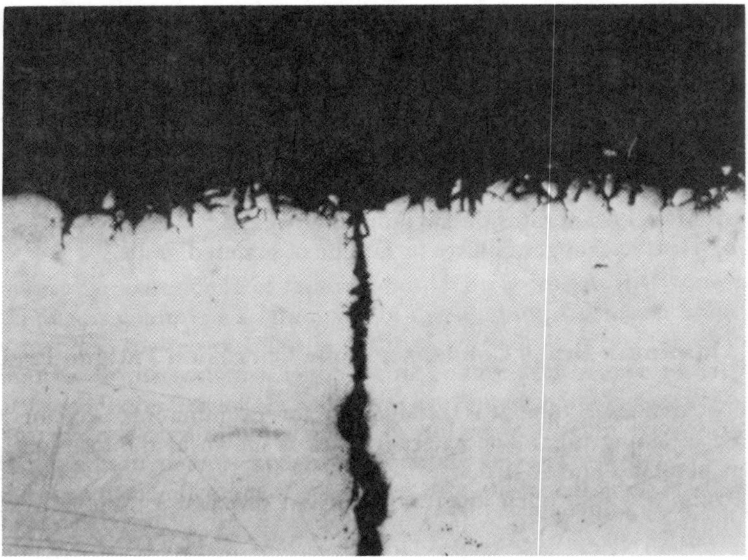

Figure 3-3 Section through brass tube (Figure 3-2) showing corrosion fatigue crack.

Figure 3-4 Transgranular corrosion fatigue crack in aluminum brass tube. ($\times 400$)

as shown in Figure 3-2. The width of these cracks suggested corrosion fatigue and this was confirmed by metallurgical examination, as shown in Figure 3-3. Extension of the cracking was predominantly mechanical (fatigue) as shown by the transgranular cracking in Figure 3-4.

Analysis of both tubes submitted showed that they were inhibited (0.1% arsenic) aluminum brasses.

It is interesting to note that a tube of a composition usually expected to be free from corrosion troubles, has in fact, corroded and thereby initiated at the corrosion pits, the failure here shown.

The propagation from the pits was apparently due to the vibration of this tube. This heat exchanger had only one support plate, thereby allowing some tubes freedom to vibrate.

Most of these condensers have had tubes replaced in cupro-nickel and this type of failure has not recurred.

B-5 Unit Cooler Corrosion Failure Problem

A rash of leaks was experienced early in the life of the unit coolers used to cool rooms of electronic and electrical apparatus on board ships.

A unit cooler consists of fourteen finned tubes emanating from a header, bent twice at 180°, and returning to another header. Inside the straight portion of these tubes were "turbulators", thin spiral bands as shown in Figure 3-5. Sea water is circulated through these heat exchangers.

These tubes and the "turbulators" were made of a reasonably corrosion resistant alloy, namely 95–5 copper–nickel–iron (1.4% iron). It was therefore surprising that they should fail in their first year.

When the defective coolers were examined it was noted that the perforation in the leaky tubes was invariably associated with either pits or a groove as shown in Figure 3-6. In either case, the weakening of the tube wall was in a

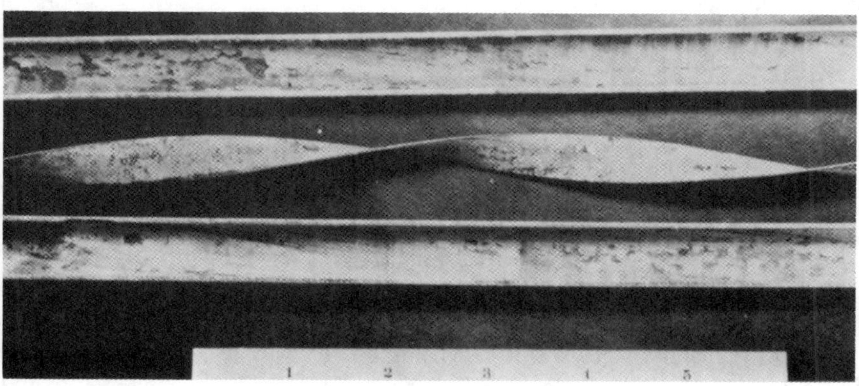

Figure 3-5 Split unit cooler tube showing spiral "turbulator".

Figure 3-6 Corrosion in unit cooler tubes. In the lower sample the corrosion was a pitting action whereas in the upper sample mechanical action had augmented the corrosion, resulting in a uniformly shaped groove.

spiral pattern. The spiral band or "turbulator" appeared, therefore, to be responsible for the leaks.

The most probable mechanism of the attack is what is known as "contact" or "crevice" corrosion. This is a rapid pitting type of attack which occurs on a metal surface in sea water when there is some obstruction to uniform exposure of the metal surface to the sea water. If a small bit of sand is placed on the surface of a metal normally resistant to sea water, it will be found that a severely corroded zone has developed beneath the deposit.

This type of corrosion appears to be the mechanism for the pitted groove, but the relatively smooth groove had, in addition, some rubbing or vibration which removed successive corrosion product films and resulted in the wearing of the ridges between successive pits. Vibration of the spiral turbulator would also, by removing the corrosion products, tend to speed up the corrosive action. An obvious means of correcting the trouble was the removal of the "turbulator".

Since the prime function of the "turbulators" was to increase heat transfer by increasing turbulence, there was some concern whether efficiency would be seriously affected by the removal. Accordingly, Reynolds Numbers were calculated from these flow conditions (2–5 ft/sec) and proved that the flow was turbulent (RN 7000–17000) without the "turbulators". Subsequent experiments proved that the turbulators had in fact, no significant effect on heat transfer.

B-7 Deterioration of Brass Condenser Components by Impingement

Several examples of deterioration of brass components and tubing associated with condensers have been noted over a period of years.

Composition included inhibited Admiralty metal (70% Cu–30% Zn) and red brass (85% Cu–15% Zn).

An example of the type of corrosion experienced by red brass ferrules is shown in Figure 3-7. Incidentally, ferrules of aluminum brass in the same service withstood the action much better. An example of condenser tubing which perforated from severe corrosion is shown in Figure 3-8.

A section cut from this condenser tube showed the interesting profile of Figure 3-9 with its smooth rounded pits undercut somewhat on the downstream side. This is a known characteristic of corrosion–erosion or impinge-

Figure 3-7 Impingement attack on inside of red brass ferrules.

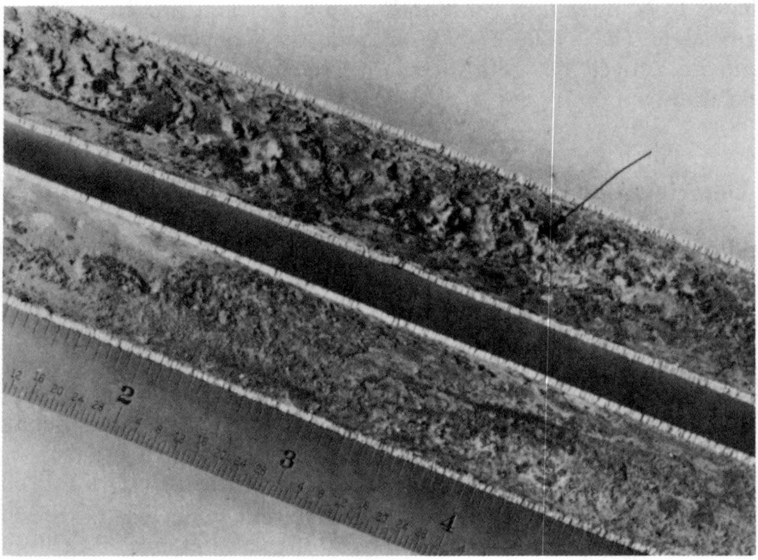

Figure 3-8 Split condenser tube showing area of severe impingement attack on the bottom half of the tube. Arrow points to perforation.

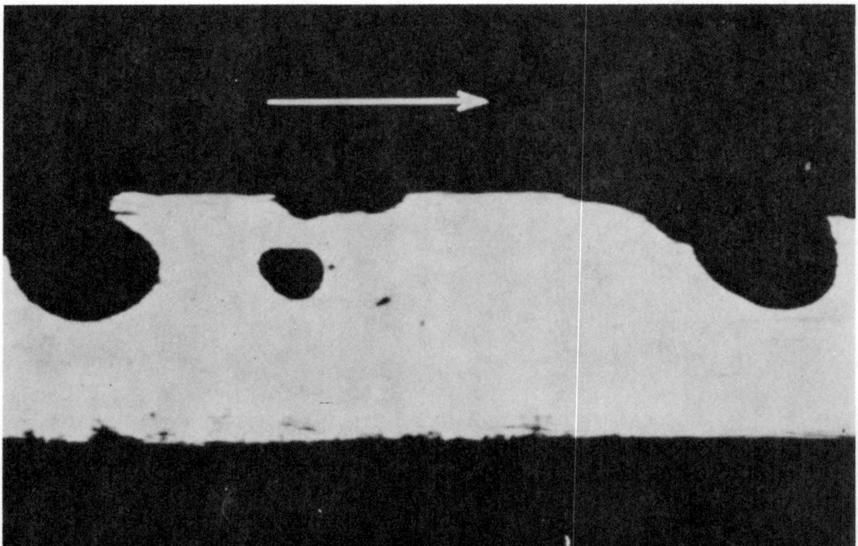

Figure 3-9 Longitudinal section through condenser tube showing impingement type pitting. The arrow indicates the direction of the sea water flow. (×35)

ment. Another important characteristic of impingement attack is the horseshoe shaped pitting shown in Figure 6-18.

Figure 3-10 Pitted section of condenser tube resulting from "plug-type" dezincification of uninhibited aluminum brass.

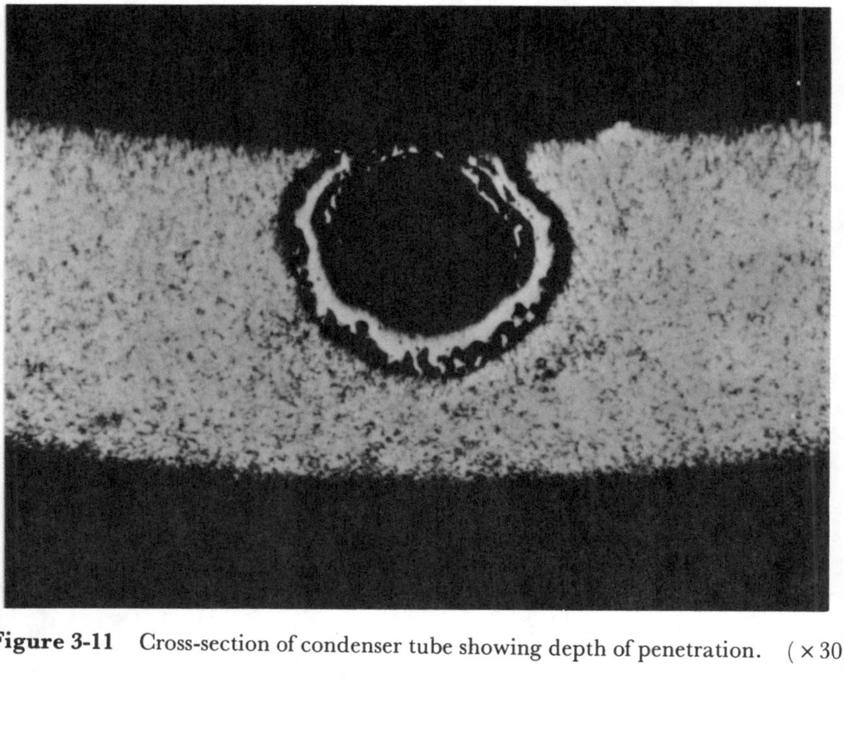

Figure 3-11 Cross-section of condenser tube showing depth of penetration. (×30)

B-8 a) Brass Condenser Tube Failure by Dezincification

Seven tubes from a ship's condenser, including one that leaked, were submitted for examination. Radiographic examination showed that all except the leaky one were in satisfactory condition. Visual examination on sectioning confirmed that a number of pits were present over a length of approximately four inches, three feet from one end, Figure 3-10.

Some of the pits had perforated the metal. Figure 3-11 shows a cross-section of one of the pits, and the depth to which the tube wall has been penetrated. The pit is filled with spongy copper, indicating that corrosion was caused by dezincification of the brass.

Spectrochemical analysis revealed that a corrosion-inhibiting element was absent, and that the tube was therefore susceptible to dezincification. The other six tubes each contained about 0.4% arsenic, ample for corrosion inhibition. All tubes were made of aluminum brass of the following approxi-

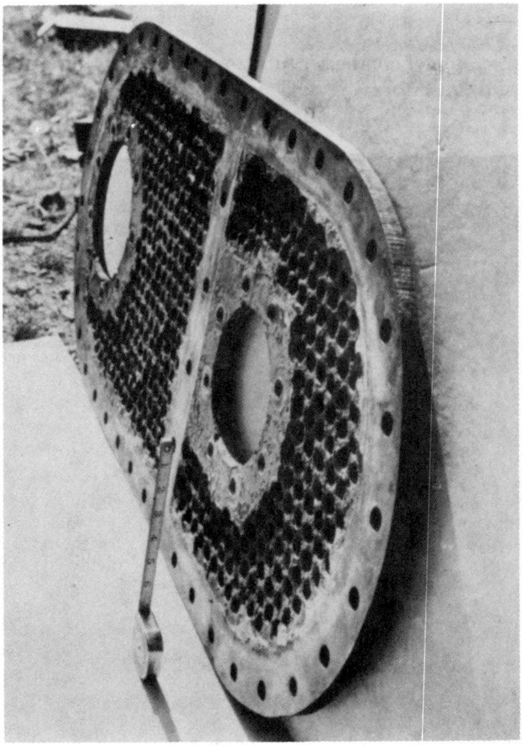

Figure 3-12 Bottom tube sheet from air ejector. Note the depth of attack in the area of the tubes.

mate composition: copper—77%, aluminum—2%, zinc—remainder, and conform to ASTM specification B111-52.

B-8 b) **Dezincification of Tube Sheets of Air Ejectors**

During an overhaul, corrosion was noted on the bottom tube sheets of two air ejectors (a type of de-aerator) from an older type destroyer. One of these tube sheets is shown in Figure 3-12.

A piece was cut out of each of the corroded tube sheets for metallurgical examination. This examination immediately indicated that the layer covering the corroded brass was metallic copper. As shown in Figure 3-13, the brass appears to be disintegrating and what re-deposits between the grains of brass is another porous material (copper). Metallurgical examination of the sound metal in all four tube sheets showed them to be alike, i.e., a two-phase structure with small amounts of β-phase between the grains of α-phase.

Chemical analysis of the four tube sheets showed that, with only minor variations, all were 60% copper, 39.7% zinc, 0.3% lead, and none were inhibited. Brasses which contain small amounts of arsenic, antimony or phosphorus are considered to be inhibited, i.e., their susceptibility to this type of attack is greatly reduced. The fact that the analyses and the 2-phase structure seen under the microscope established the four tube sheets as being identical, indicated that service conditions must be held accountable for the difference in behaviour of top and bottom tube sheets.

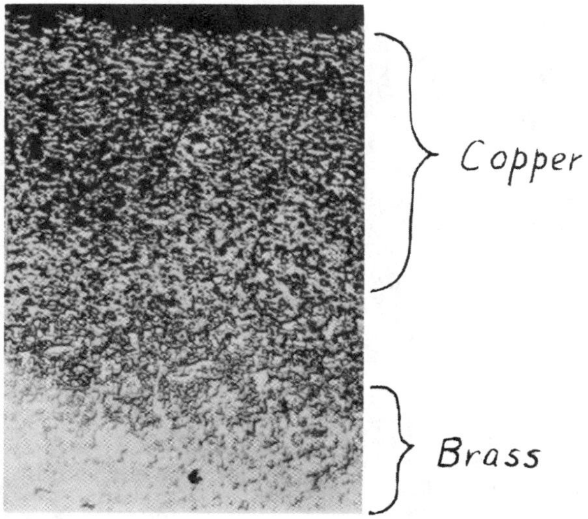

Figure 3-13 Section cut through dezincified zone of tube sheet (Figure 3-12) showing depth of dezincification. (×30)

It has been established that where flow of water is slow or stagnant, dezincification is favoured. In this unit, condensate flow was such that almost stagnant conditions prevailed on the bottom tube sheet and not on the top tube sheet. In addition the bottom tube sheet was somewhat hotter.

This, and similar problems, have in large part been eliminated by use of materials resistant to dezincification (such as tin bronzes, cupro-nickels) or, where replacement was not convenient, by the application of cathodic protection, e.g., steel anodes.

B-8 c) **Dezincification and Impingement of Condenser Tube Sheets**

Unexpected pitting under a division plate of a ship's condenser was investigated. The pitted tube sheet and the divider plate are shown in Figure 3-14.

Figure 3-14 Naval brass condenser tube plate suffering from dezincification and impingement attack.

A microphoto of a section through one of the pits is shown in Figure 3-15. One interesting feature of the dezincification is evident from this section: the tube plate to the left of the pit showed negligible dezincification, whereas dezincification had progressed to a moderate depth on the right. It is probable that this dezincification paved the way for the impingement type pitting.

It is noteworthy that the side with the dezincification was not cathodically protected, whereas a steel anode was fitted on the other side.

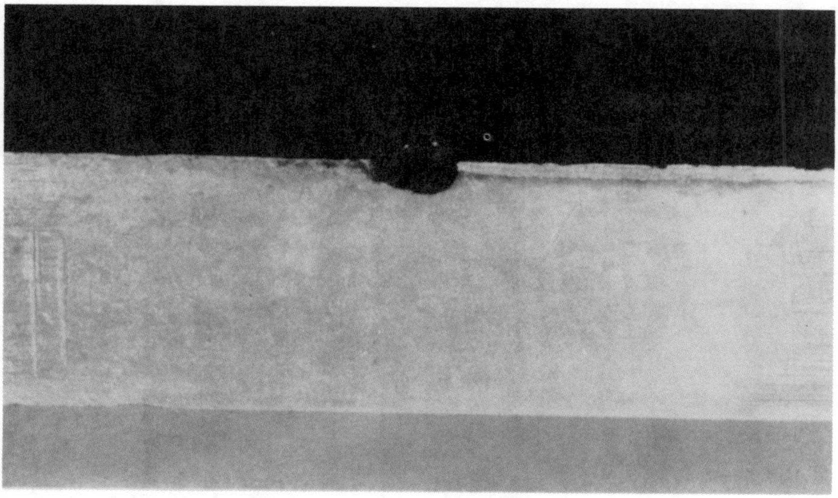

Figure 3-15 Section from Naval brass tube plate showing dezincification and impingement attack. Note that dezincification took place only on right of pit. (×2)

It has been noted on many other occasions that where sufficient zinc or mild steel anodes are fitted in waterboxes, or where the water boxes themselves are cast iron, the resulting sacrificial corrosion prevents dezincification and impingement attack of Naval brass tube plates and ferrules. This intentional or inadvertent cathodic protection may explain why in some quaters Naval brass is still considered to be a good material for such components.

C-3 a) **Monel Evaporator Basket Failures at Welds**

An evaporator basket which failed in service was submitted for examination. These baskets are essentially corrugated heat exchangers, part of which is shown in Figure 3-16, for distilling sea water. Monel is used to resist concentrated brines.

Failure as indicated in Figure 3-16, was along a vertical weld.

Radiographic examination of the failed weld, Figure 3-17, indicated that the entire length of the weld showed incomplete penetration. It is apparent that the failure was caused by the defective production weld.

The circumferential weld and the other longitudinal welds were also radiographically examined. Many of the longitudinal welds showed lengths of incomplete penetration. A few short lengths of incomplete penetration were also indicated in the circumferential weld.

As part of the operation of these baskets they were subjected to variations in pressure, i.e., a cycling which can readily result in fatiguing. The incom-

Figure 3-16 Crack in monel evaporator basket along longitudinal weld.

plete penetration, shown in Figure 3-17, provided the initiating notch for cracking.

It is interesting that considerable effort and expense was put into choice of Monel to ensure trouble-free operation, and yet similar attention to detail was neglected for the joining.

In spite of this being very thin gauge, 0.04″, and somewhat difficult to weld because of configuration (access from one side only), nevertheless these welds were satisfactorily repaired by local welders.

C-3 b) **Heat Exchanger Failures by Fatigue of Soldered Joints**

Small heat exchangers, of a specific type, were used to cool (a) fresh water and (b) oil on a marine diesel engine. These two coolers operated in tandem: sea water cooled the fresh water, and then the fresh water cooled the oil. Both coolers were of similar construction, consisting of a core assembly and an enclosing case. The core assembly consisted of a number of oblong tubes,

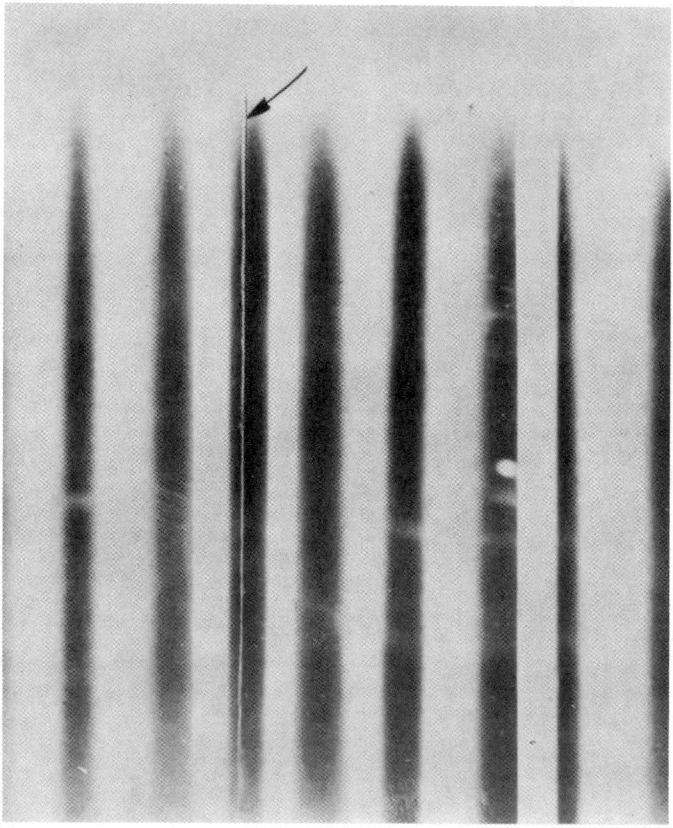

Figure 3-17 Radiograph of longitudinal weld of evaporator basket showing incomplete penetration.

the ends of which were soldered into a header plate. The header plate was in turn soldered into the casing as in Figure 3-18.

These heat exchangers failed frequently at the joint to the header plate.

Metallurgical examination of a section cut in the affected area, Figure 3-19, showed that cracking had initiated inside the fresh water cooler, that is, on the salt water side, and had propagated halfway through the joint.

This microphotograph also indicates the way the cooler is fabricated. The tinned 95–5 cupro-nickel tube (at right) is first soldered to the 95–5 cupro-nickel header plate with the dark solder, a lead base—2.5% silver alloy which has a 579° F melting point. The lighter metal is a cover pass of 60–40 lead–tin solder which has a melting range from 360° F to 460° F. (Two solders are used to ensure a leak free heat exchanger.)

The cracks initiated on the inside of the cooler in both the water and oil coolers. This means that in one case the initiating site was on the salt water side and in the other, on the fresh water side. Since the coolers which used salt water failed no more frequently than the oil coolers, it would appear that the failure mechanism was not corrosion.

Figure 3-18 End of fresh water cooler showing three rows of tubes soldered into header plate. Bottom left area has been repaired.

Figure 3-20 shows, at high magnification, the end of the crack indicated by the arrow in Figure 3-19. The crack has associated with it a light colored inclusion. The "loop" of inclusion shown in Figure 3-20 suggests that the crack propagated along this defect. It should be noted that the top part of the "loop" has no crack associated with it, but rather appears to be directly related to the cast structure of the solder. These inclusions were noted in the solder of both the oil cooler and the water cooler.

In other instances examined, cracks followed inclusions at the interface of the cupro nickel and the solder.

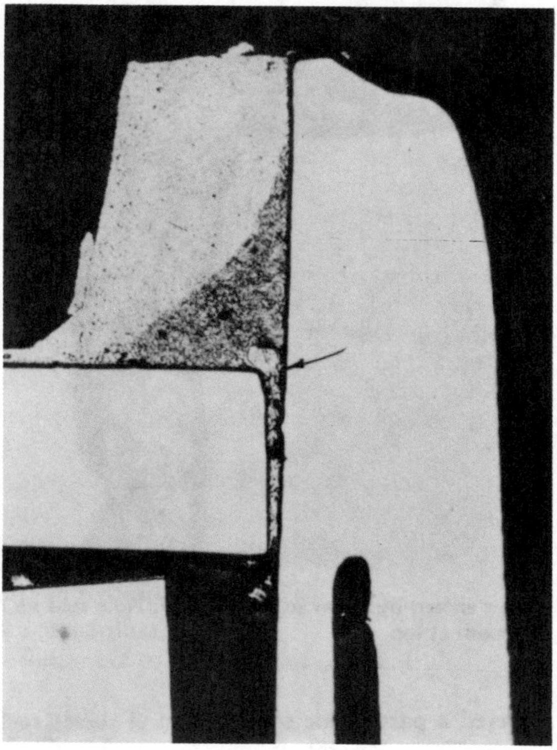

Figure 3-19 Section through soldered junction of cooler. Tube is at right soldered to header plate (left) with two solders. Arrow indicates extent of crack which initiated below. (×34)

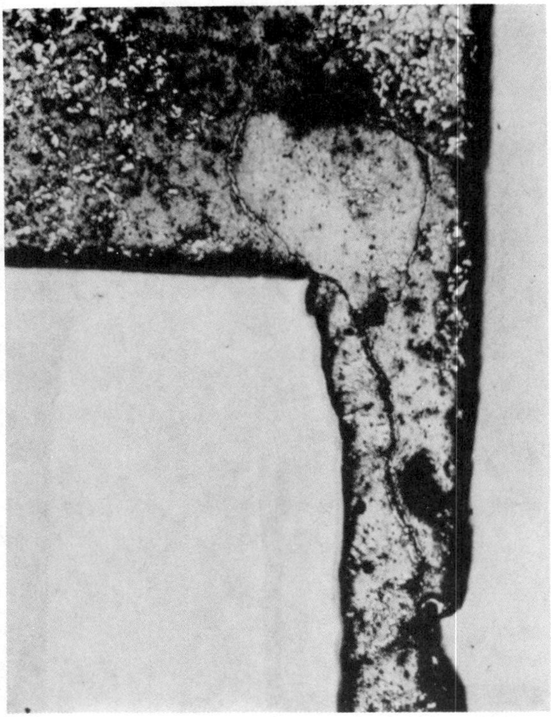

Figure 3-20 Area shown by arrow in Figure 3-19. Note that no crack is associated with "loop" of deposit at top. (×200)

Vibration played a part in the propagation of these cracks because some tubes soldered to an internal support plate had broken free.

Because of the difficulties involved in dismantling and resoldering the joints these units were considered non-repairable. Until this investigation, the manufacturer of these units had believed that corrosion was the cause of failure.

IV

PRESSURE VESSELS

Glass-lined hot water tanks failed in large numbers because the internal lining had to be disrupted for fittings. A better choice for this service is the silicon bronze tank, but it too can fail (rather more rarely) as shown in the next example.

Compressed air bottles used by divers need to be manufactured with more attention to forging flaws. Requirements of personnel safety also demand high standards of quality in butt welds of recompression chambers. Most unusual laxity in design and materials used were primary factors in fatalities associated with rupture of a drying drum. Poor design of end covers used in moderate pressure steam service was the cause of cracking and occasional rupture.

B-1 Failure of glass-lined domestic hot water tanks.

B-2 Silicon bronze hot water tank failure.

C-2 Forging defects in aqua lung cylinders.

C-3 Weld quality in recompression tank.

C-8 a) Drying drum failure.
b) Fatigue failures of valve chest end covers.

B-1 Failure of Glass-Lined Domestic Hot Water Tanks

Of a total of over 200 glass lined hot water tanks installed in a housing development in Victoria, Canada, essentially all failed within seven years. Of these tanks, six were made available for destructive examination. Five of these failed immediately above the hot water outlet as shown in Figure 4-1, and one failed just below the cold water inlet. Also all the tanks showed some minor pitting at the junction between the dished end and cylindrical shell. Two of the tanks showed many tiny rust pits throughout.

There appeared to be a problem where the continuity of the coating was disrupted for fittings, and this problem was called to the attention of the manufacturer.

For trouble-free service, our recommendation was for the installation of silicon bronze (often called "copper") tanks. In local experience these give reliable service in excess of fifteen years. It was appreciated that economics might favour the glass-lined tank (or even the cheaper galvanized tank which

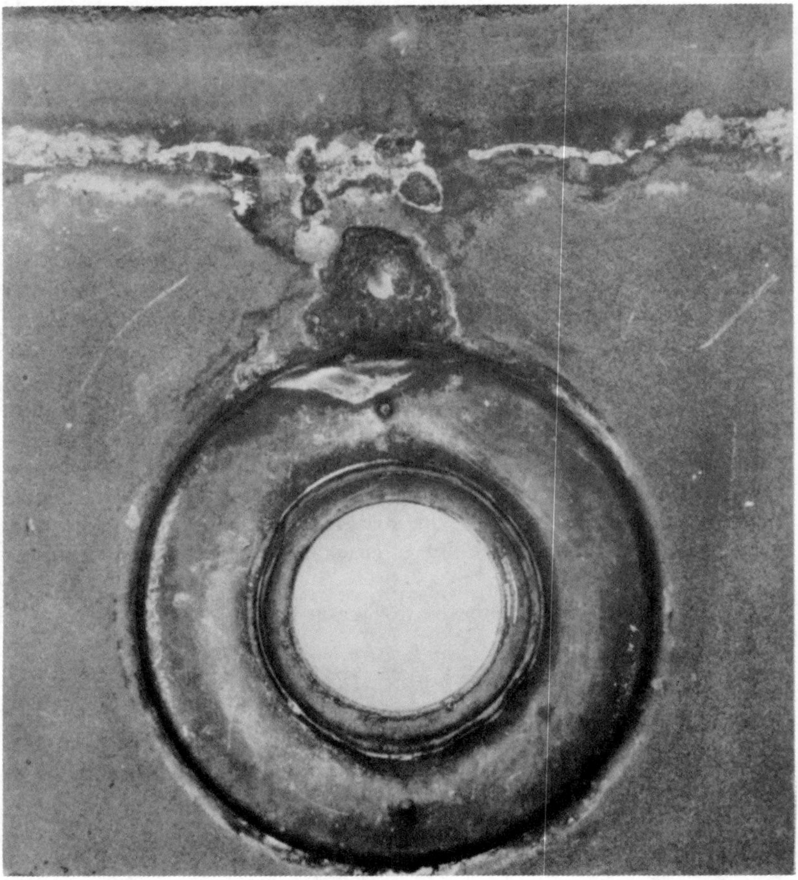

Figure 4-1 Corrosion attack in glass-lined hot water tank at the hot water outlet.

lasts approximately three years), over the bronze tank. A purely economic assessment, however, overlooks the value of reliability, not to mention possible damage to furnishings.

B-2 Silicon Bronze Hot Water Tank Failure

A large (24 inch diameter) silicon bronze hot water tank was brought into a plumbing shop for repair. The tank had what appeared to be a simple perforation about an inch from the longitudinal weld.

As a result of the authors' interest in failure mechanisms in these tanks,

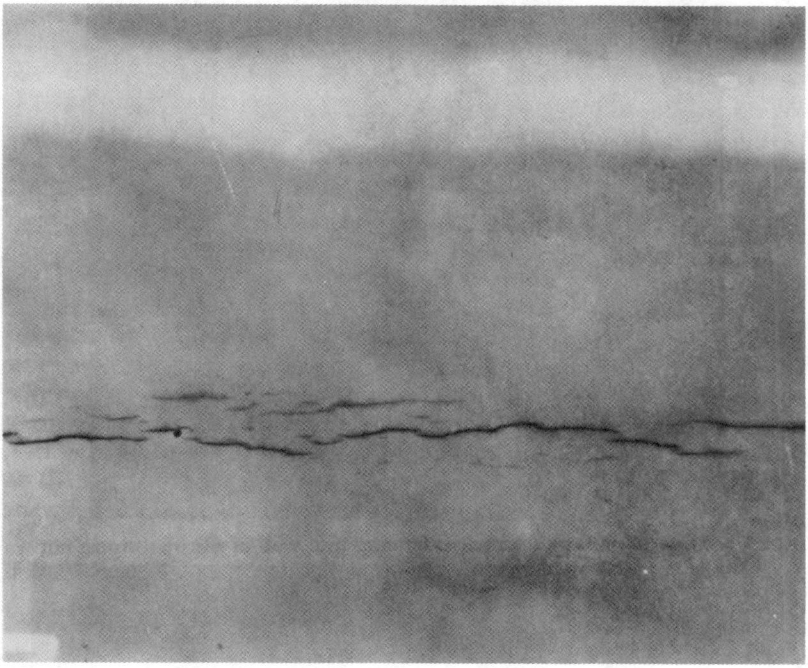

Figure 4-2 Radiograph of silicon–bronze hot water tank showing extensive cracking parallel to longitudinal weld. (×1)

it was arranged that this tank be made available for non-destructive examination.

Radiographic examination proved most revealing as shown in Figure 4-2. Extensive cracking had taken place parallel to both longitudinal welds. A number of these cracks were on the verge of penetrating the wall. (All the cracks were on only one of the two sheets of metal used to form the cylindrical shell.) The radiographs indicated that the cracks had originated in grooves, suggesting a corrosion fatigue mechanism. Transverse cracks were also shown to be present in the welds.

When it was shown that cracking was more serious than originally suspected, permission was obtained for destructive examination. Metallurgical examination of the cracks (Figure 4-3) confirmed that they had originated in areas of corrosion attack. Numerous corrosion grooves or fissures were present and, as shown in Figure 4-4, these had initiated under a coating. Spectrographic analysis indicated that the copper–3% silicon tank had been tin coated on the inside before fabrication.

From the above evidence it appears that the failure was due to corrosion fatigue. Evidence has indicated that tin is cathodic to the bronze at hot water

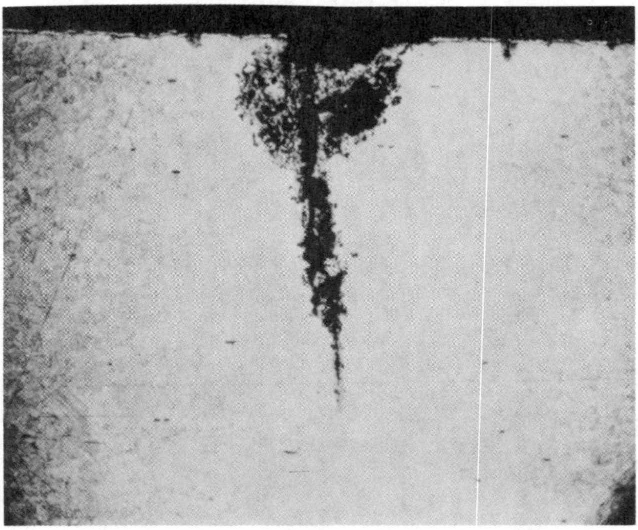

Figure 4-3 Corrosion fatigue crack extending into wall of silicon–bronze hot water tank. (×100)

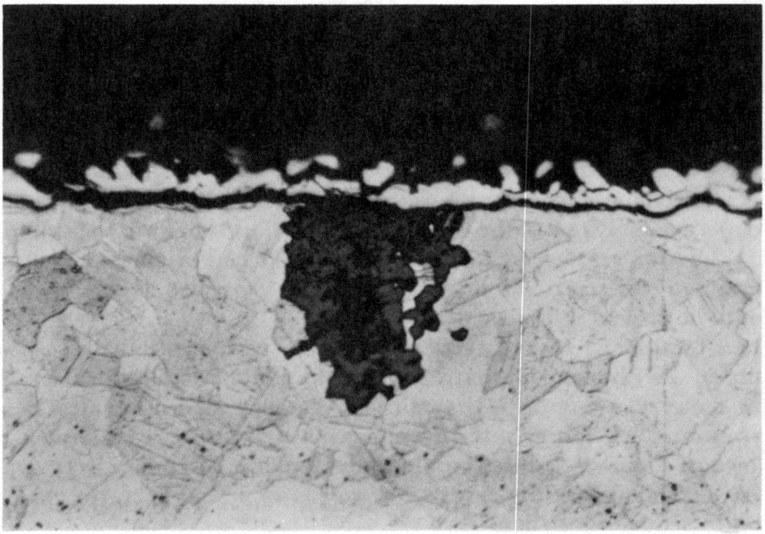

Figure 4-4 Corrosion attack under thin layer of tin plating. This type of corrosion initiated the attack shown in Figure 4-3. (×400)

temperatures. Galvanic corrosion of the alloy at breaks in the tin coating would result in stress raising notches. These notches could then propagate under the combined action of stress and corrosion. With water pressures up to 120 psi, the hoop stress in the $\frac{1}{8}$ inch thick wall would be around 11,500 psi. Periodic water hammer effects from the sudden shut-off of taps could give much higher stresses. These relatively high stresses would make the fatigue life of this material very short under corrosive conditions.

C-2 **Forging Defects in Aqua Lung Cylinders**

In the course of pressure testing at 3000 psi divers' 70 c.f. compressed air bottles, a workman noted a groove on the inside surface of the small open end. Because of concern about the nature of this groove, we were permitted to destructively examine this bottle.

On cutting the bottle open, numerous fine grooves were seen extending into the bottle proper from the narrow neck. These are shown in Figure 4-5 somewhat enhanced by magnetic particle indications.

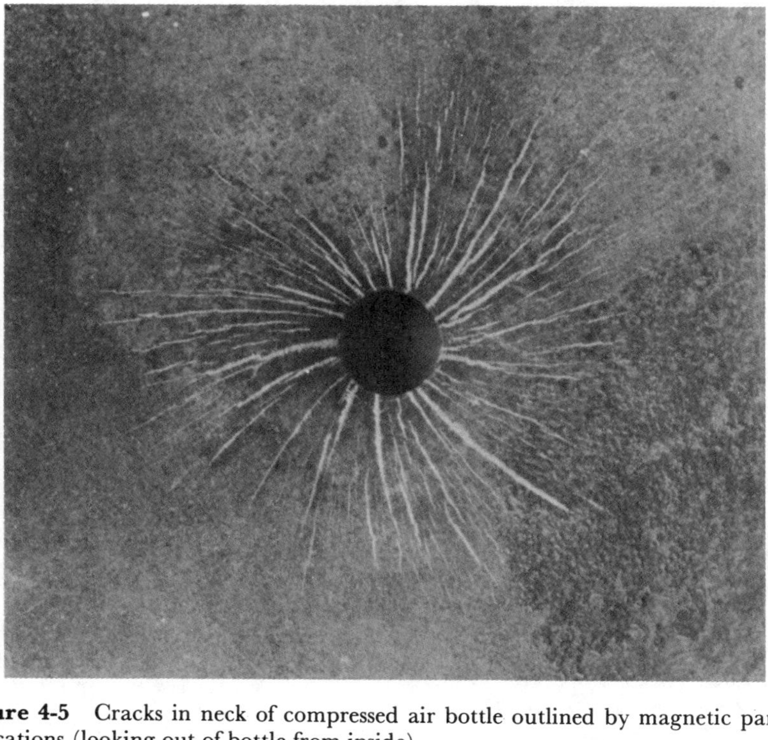

Figure 4-5 Cracks in neck of compressed air bottle outlined by magnetic particle indications (looking out of bottle from inside).

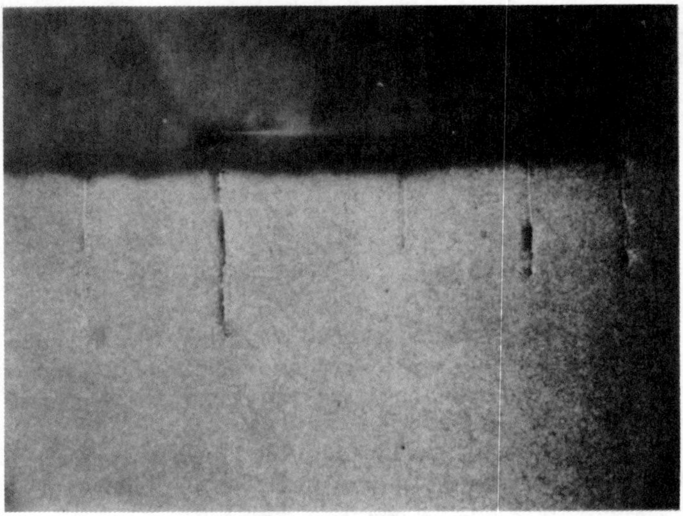

Figure 4-6 Sectioned aqua lung bottle showing cracks (forging folds) extending into the steel. (×30)

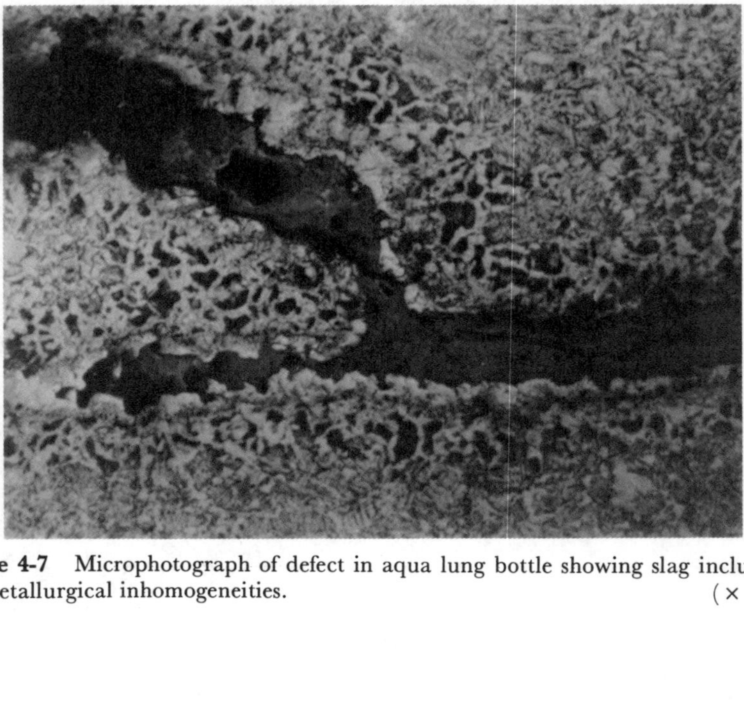

Figure 4-7 Microphotograph of defect in aqua lung bottle showing slag inclusion and metallurgical inhomogeneities. (×100)

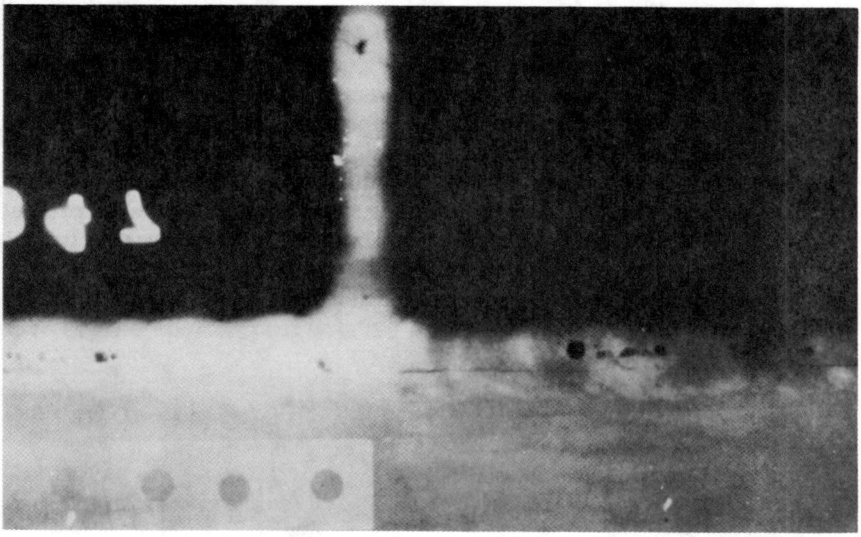

Figure 4-8 Radiograph of recompression tank showing incomplete penetration and cracking.

Metallurgical examination of the metal indicated that the cracks were invariably associated with forging folds which were formed when the neck was fabricated. Figure 4-6 shows a number of fine defects extending into the metal from the inside surface. Some of these forging discontinuities have not yet opened up. Figure 4-7 shows, at higher magnification, slag inside one of the defects. The metal associated with the crack is decarburized, indicating that it was surface metal before the neck was forged. It is noteworthy that the cracks associated with the folds had not yet propagated into the sound metal.

Although these forging folds had been exposed to considerable service and had not necessarily propagated, nevertheless, in terms of possible sudden loading, such as by dropping, some hazard existed from these flaws serving as initiating sites for brittle failure.

This bottle was not considered fit for service, and as a consequence approximately 400 other bottles were inspected from the small end by borescope. Of these, 15% had these fine folds and were set aside pending word from the manufacturer.

It should be borne in mind that the seriousness of these flaws in terms of continuing use is really governed by the impact toughness of the steel. If the steel should prove to have toughness such as indicated by say, 30 ft. lbs. Charpy V Notch at 0° F, then the hazard from possible brittle rupture is lessened.

C-3 Weld Quality in Recompression Tank

Visual examination, following purchase, of a portable aluminum recompression chamber revealed a surprising transverse crack in a large fillet weld. Although this was not in a structurally important area, there was concern that other welds of structural importance might be somewhat unsound.

Radiography of longitudinal and circumferential welds of the 30 inch diameter pressure vessel was most revealing: transverse cracks were present in both longitudinal and circumferential welds; incomplete penetration (Figure 4-8) was present in the circumferential weld.

In addition, more careful visual examination revealed several more cracks in fillet welds.

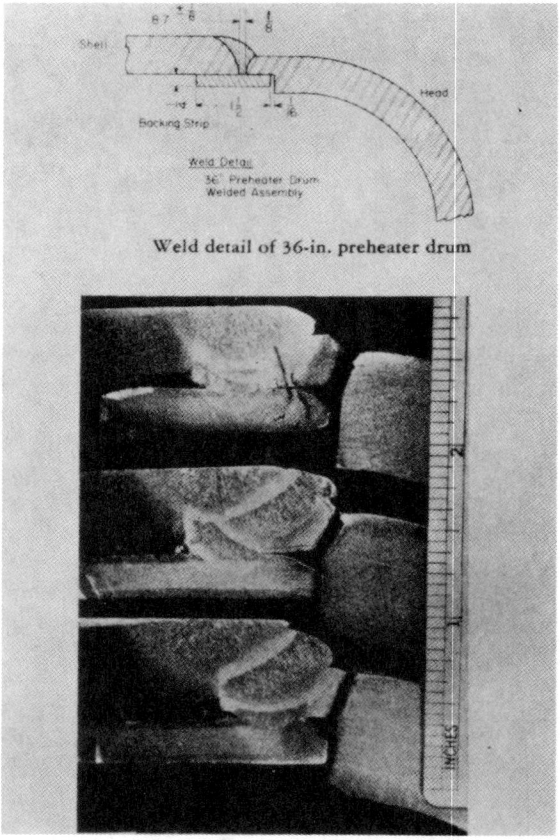

Weld detail of 36-in. preheater drum

Figure 4-9 Section through welded area of drying drum showing unacceptable (by ASME code) joint. Micrographs of the fractured section at three different places on the periphery. Top left, end of the drum shell; lower left, backing strip; right, head.

It may be pointed out that this portable recompression chamber is occasionally carried in aircraft while performing its function—that is to re-pressurize divers who are suffering from the bends—and therefore, the problem of safety and reliability is magnified: sudden change in pressure by rupture would be fatal to the diver and of course the aircraft would be vulnerable.

These defective welds were re-worked to our radiographic standards.

As a result of this experience two older steel re-compression chambers were examined. Similar defects, including gross incomplete penetration and lack of fusion, were revealed and subsequently repaired.

As a result of this and other experiences, it appears sound policy to remain somewhat sceptical that quality is always what one expects it to be, even from sources generally accepted as being reliable.

C-8 a) **Drying Drum Failure**

A drying drum used in paper manufacture ruptured, killing three people.* A diagnosis of the available evidence suggested the following factors were involved:

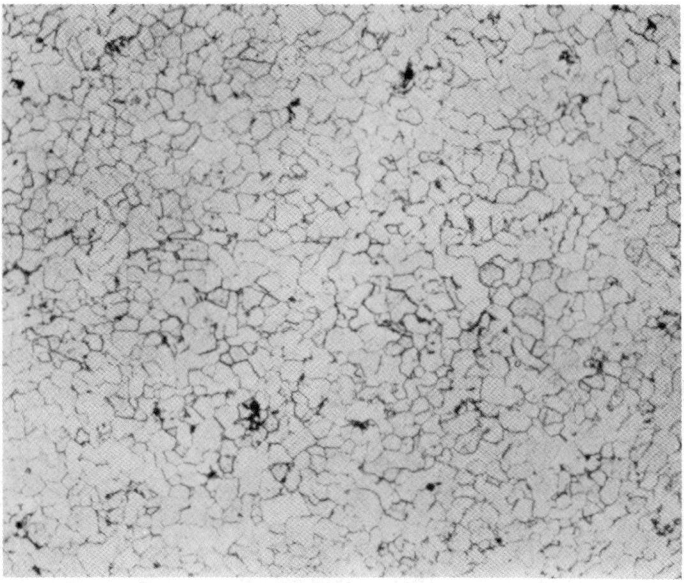

Figure 4-10 Micrograph of the shell metal in a region at a distance from the weld. Nital etch. (×100)

* G. Sines and E. C. McLean "The failure of a welded drying drum by caustic embrittlement", *Mechanical Engineering*, Vol. **78**, December 1956, p. 1105.

1) internal pressure approximately 250 psi,

2) the design, as shown in Figure 4-9, was a flagrant violation of the ASME pressure vessel code,

3) metals used for cylindrical shell were substandard, Figure 4-10; namely, essentially ingot iron instead of at least mild steel.

It was most surprising that responsible authorities could have inspected this pressure vessel and permitted its certification in engineering service. It is tragic that lives had to be lost before this engineering blunder was noted and other units like it were withdrawn from service.

C-8 b) Fatigue Failures of Valve Chest End Covers

The brittle failure of the end cover of a valve chest powered with steam at

Figure 4-11 Failed cast iron end cover from auxiliary feed pump.

200 psi, was reported (Figure 4-11). Shortly thereafter a similar end cover which was cracked and leaking was made available (Figure 4-12).

There was considerable hazard to personnel involved when the end cover of the valve chest of one of these pumps failed. The internal pressure in the valve chest was over 200 psi; that is, the total internal force on the cover was better than two tons. At failure, the ruptured portion of the end cover becomes a dangerous projectile. In addition there was the hazard associated with release of steam.

A close examination of the fractured surface of the end cover of Figure 4-11 showed that a crack had extended into the cover from the outside for some time before the final rupture. This mechanism—i.e., fatigue from the outside corner—was confirmed on sectioning the end cover of Figure 4-12 as shown in Figure 4-13.

Because of the seriousness of this failure, some 200 of these end covers were

Figure 4-12 End cover showing magnetic particle indications at bell-flange junction.

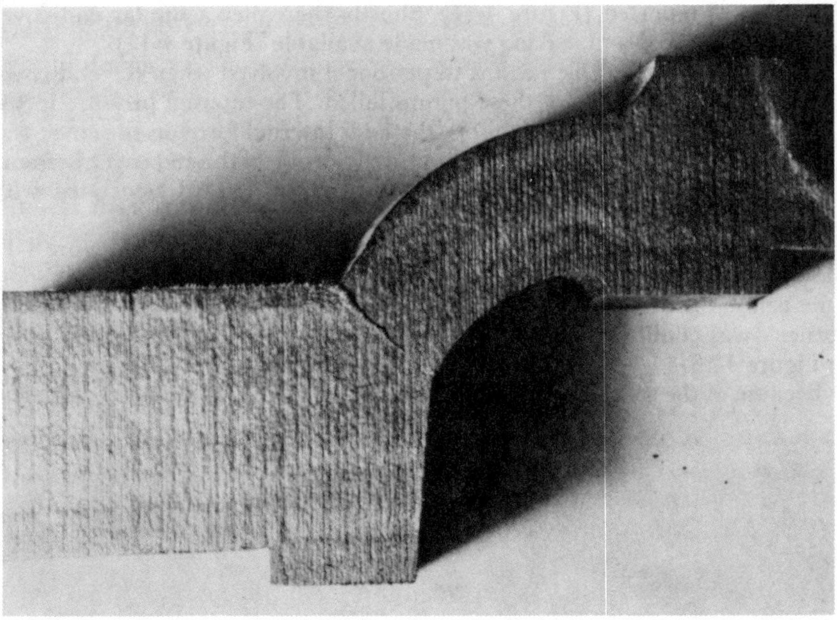

Figure 4-13 Sectioned end cover showing extent of fatigue crack which initiated at bell-flange junction on the outside.

made available for examination over a period of time. It soon became obvious that the sharp corner shown in Figure 4-13 had acted as a stress-raiser. Also magnetic particle inspection readily differentiated the covers with fine cracking from those without. About 30% of the 200 covers were defective.

Many of the end covers had been in service for up to 20 years. During this period the castings have been subjected to fatigue loading, that is, pressure is repeatedly built up and relieved inside each end cover. The stresses in the covers resulting from this internal pressure are, in fact, quite low. Indeed, a number of covers have not failed even though a crack has extended most of the way through the casting. The few failures resulted when the remaining metal was insufficient to withstand the two-ton internal force under service conditions.

A poorly designed bell-flange junction appeared to be responsible for the initiation of the cracks. New end covers have since been designed with a generous radius at this corner, and it was anticipated that no further problems should result.

V

VALVES

TROUBLES WITH valve spindles, in one instance because of a sharp corner initiating fatigue, and in the other because of a valve arrangement permitting debris to cause galling, are described.

Galvanic action resulted in severe intergranular corrosion of an aluminum alloy usually considered suitable. Several failures by stress-corrosion cracking are explored, particularly in martensitic stainless steel and the aluminum–10% magnesium alloy which shouldn't be in marine service.

Caution in use of 10% aluminum bronzes is suggested following dealuminification of some valve components. The significance of quality in a casting and in a weld repair are clearly demonstrated in failure of bronze valves and of a valve component, respectively.

Development of a longitudinal crack from quenching and of a circumferential crack from a stress-raising corner, both in one spindle, are described. Errors in material choice and in machining caused failure in two other examples.

A-1 Valve spindle fatigue failure problem.

A-3 Galling of stop valve spindles.

B-1 Intergranular corrosion of aluminum–silicon alloy valve body.

B-3 a) Stress corrosion cracking of martensitic stainless steel steam valve components.
 b) Exhaust rejection valve failures by stress corrosion cracking.
 c) Beta brass flooding valve failure by stress corrosion cracking.
 d) Aluminum–magnesium alloy valve failure by stress corrosion.

B-5 Flooding valve failure by crevice corrosion.

B-8 Air cock failure by dealuminification.

C-1 Poor quality casting causes valve failure.

C-3 Monel exhaust valve failure at a weld repair.

C-5 Stainless fuel pump relief valve failure.

C-7 Stainless steel valve seat corrosion problem.

C-8 Valve bonnet failure at machined notch.

A-1 Valve Spindle Fatigue Failure Problem

A number of valve spindles from a steam valve were submitted for examination because others had failed in service. Those that failed had shown the normal fatigue markings originating in each instance from one side, suggesting a bending stress.

Of the particular group submitted for evaluation, only one proved to have a small crack as revealed by magnetic particle examination. It was interesting to note that the other 5 spindles had a generous radius at the corner as in Figure 5-1. The stress alleviating effect of the well-rounded corner allowed the presence of unavoidable bending stresses caused by misalignment with no danger of cracking.

Figure 5-1 Valve spindle (at left) is cracked at the corner (arrow). The other spindle has a generous radius at this location.

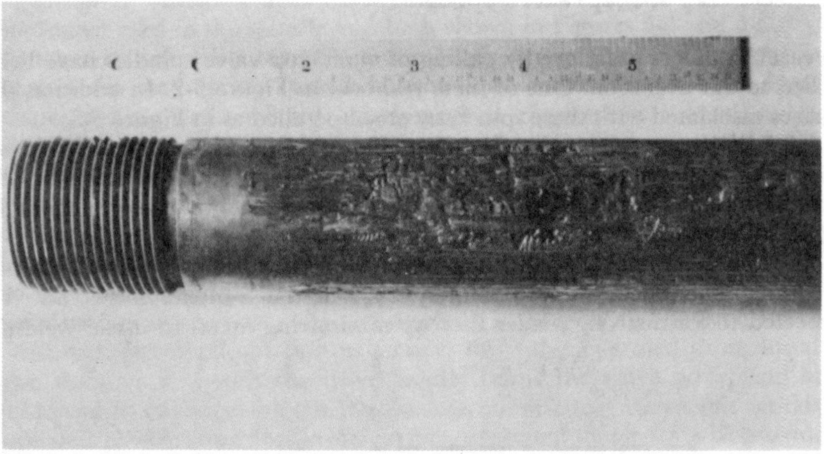

Figure 5-2 Galled spindle from main stop valve.

Figure 5-3 Inside of bush of main stop valve. Note that gouging and scoring most severe at upper end which was furthest from packing end.

A-3 Galling of Stop Valve Spindles

Several instances of failure by galling of main stop valve spindles have been called to our attention. One of these is shown in Figure 5-2. In addition the bushes associated with these spindles were also galled as in Figure 5-3.

Initially it was suggested that the trouble could be related to composition. The specified metals were an 18–8 free machining stainless steel for the spindle, and an S-monel for the bush. The free-machining component in the stainless steel, whether sulphur or selenium, plus approximately $\frac{1}{2}\%$ molybdenum, could be expected to alleviate galling and act as a lubricating medium. The silicon (approximately 2%) in the S-monel might also be expected to form strong oxides thereby minimizing metal to metal welding

Figure 5-4 Failed sanitary sea suction valve body.

or galling. It was soon determined, however, that the specified materials were in fact used in the spindle and bush shown in Figures 5-2 and 5-3.

Further investigation turned up another instance in which the specified material was used for the spindle but the bush was ordinary monel. In this instance galling also occurred but damage to the bush was much lighter. It was apparent that composition of the materials was not basic to the galling problem.

Closer examination of the inside of the damaged bushes, as shown in Figure 5-3, revealed that the damage was confined to the upper zone (which was at the opposite end from the packing). In other words, neither the packing nor its components could be involved. The evidence tended to suggest that some foreign material, such as scale, was working its way in and causing trouble at this end. It turned out that in service, this valve operated in an upside-down position, i.e., with the valve handle below the valve body, and one could readily envisage foreign materials accumulating above this working junction. The operating people had, in fact, suggested that a drain be provided just above the bush to carry away condensed moisture and possibly foreign material. This was recommended as a sensible approach to avoid having material (scale) lodge between spindle and bush.

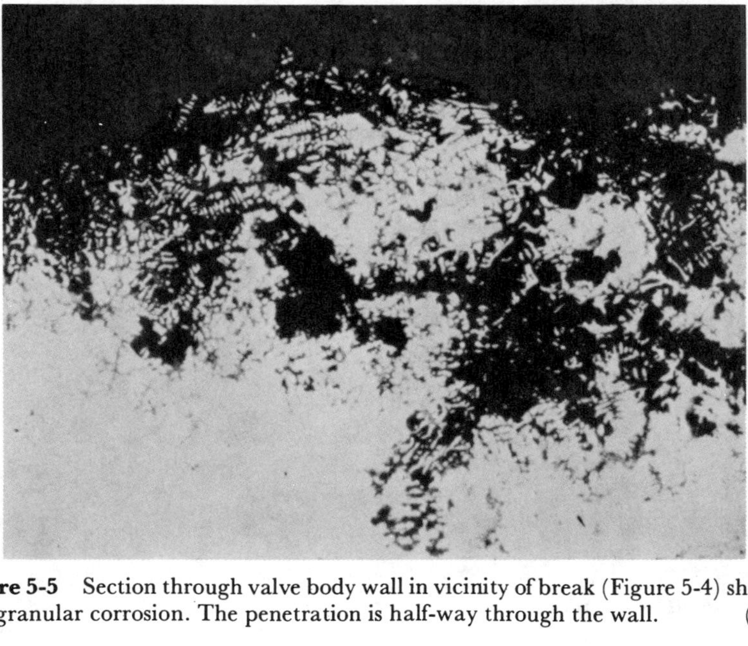

Figure 5-5 Section through valve body wall in vicinity of break (Figure 5-4) showing intergranular corrosion. The penetration is half-way through the wall. (×10)

B-1 Intergranular Corrosion of Aluminum–Silicon Alloy Valve Body

A shipside valve from sea suction was broken during removal at a refit. Figure 5-4 shows the brittle rupture at the flange.

The body of the valve was confirmed spectrographically as being the aluminum–6% silicon alloy specified. The valve trim, however, was in austenitic stainless steel.

Figure 5-6 Stress-corrosion cracking failure of valve feather. Note brittle appearance of fracture (Item 6, Figure 5-8).

Examination under the microscope of samples cut from the rupture area indicated extensive interdendritic corrosion as shown in Figure 5-5. It was also noted that the corrosion penetration at the flange was some 3 times deeper than in the rest of the valve body. In view of this, it was concluded that some factor was involved beside the general galvanic effect of the internal stainless parts. Field investigation revealed that this flange was coupled to another stainless steel fitting with resultant intensified galvanic action.

Of the aluminum alloys normally used in sea water this aluminum–6% silicon alloy is still considered one of the best.

B-3 a) Stress–Corrosion Cracking of Martensitic Stainless Steel Steam Valve Components

An apparent brittle failure of a pilot valve feather shown in Figure 5-6, was brought in for examination.[63] The initial tendency was to consider that design, namely a drastic change in section, was probably responsible for failure. However, there was no readily apparent means to apply the force required to rupture even the small cross section shown.

Accordingly, in search for possible clues, a piece was cut out for metallurgical and chemical examination. Microscopic examination showed that a fine intergranular cracking had occurred in the martensitic structure of

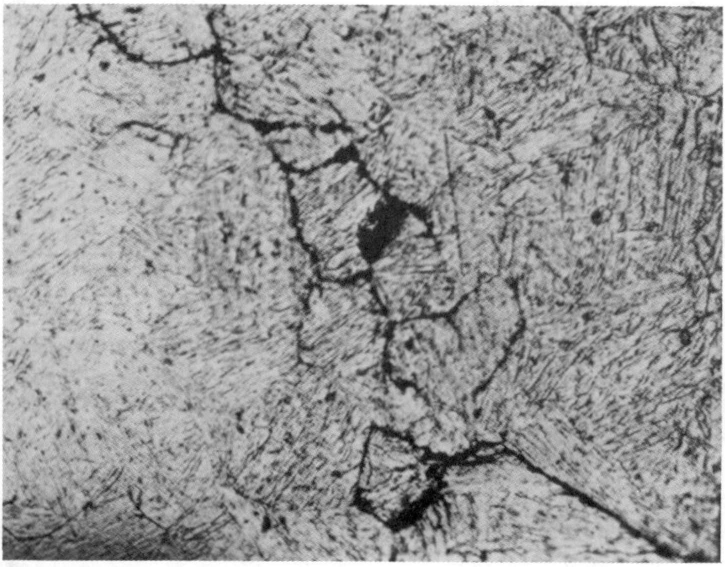

Figure 5-7 Stress-corrosion cracking of martensitic stainless steel valve component (Figure 5-6).

Figure 5-7. With this stress–corrosion cracking, it was evident why this valve component fell apart.

Analysis gave the material as AISI 414, which is a 12% Cr and 2% Ni heat treatable stainless.

In view of the fact that this material could fail by stress corrosion cracking in the environment of the steam valve, and in view of the great potential hazard in steam service, it was considered an urgent matter to examine other valves.

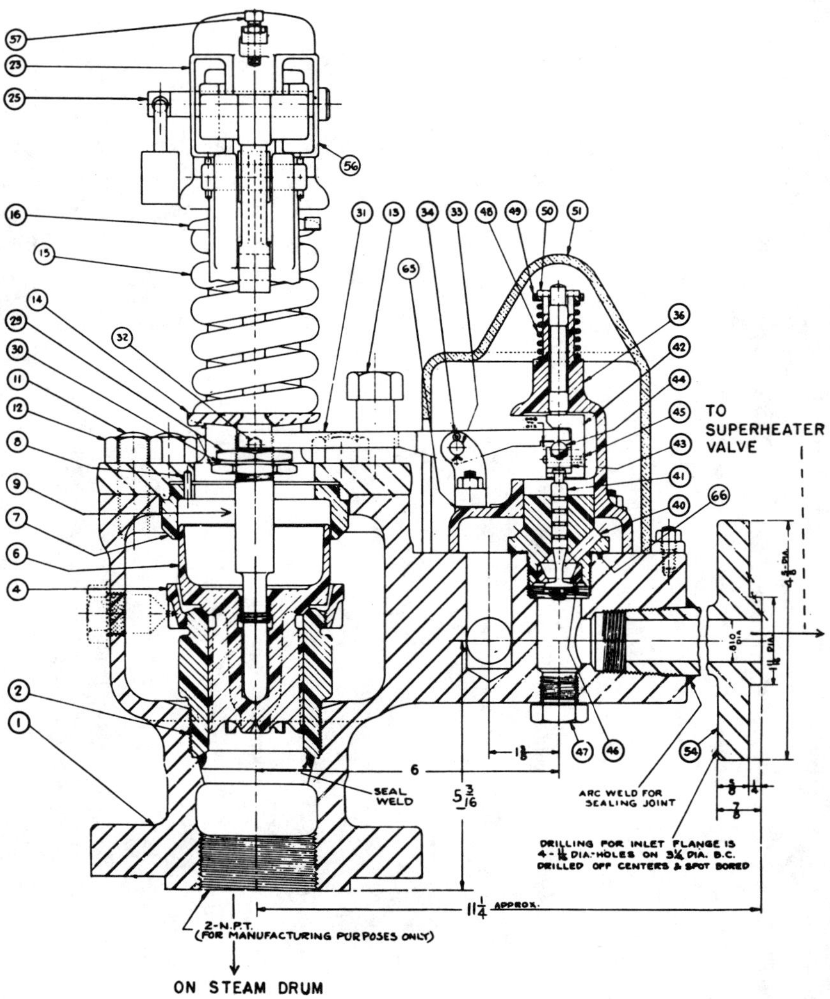

Figure 5-8 Actuator assembly of steam safety valve showing the various components.

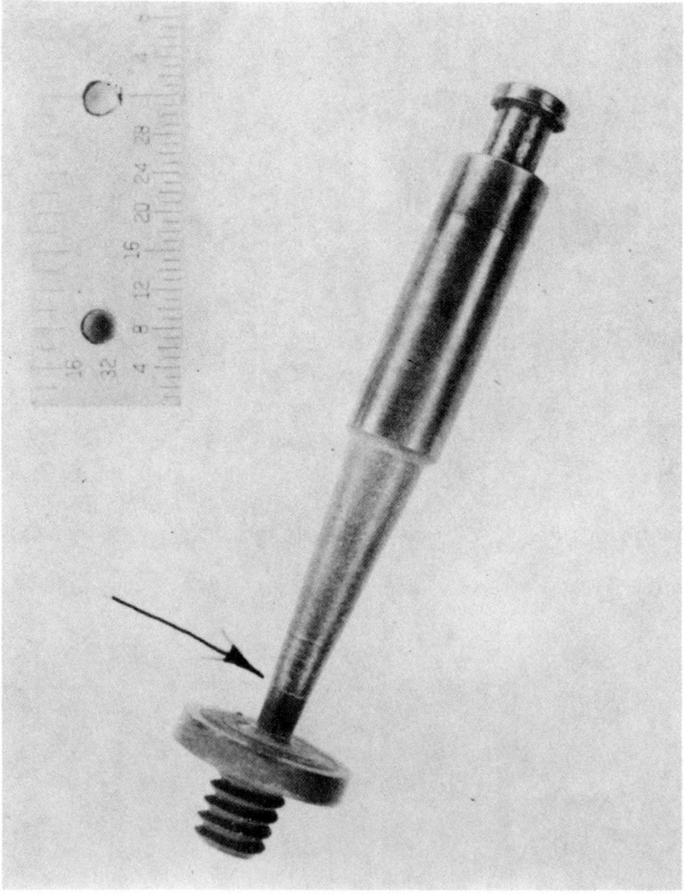

Figure 5-9 Cracked actuator valve feather (Item 41, Figure 5-8). Note location of crack away from radius.

Examination of the relevent drawings as in Figure 5-8, revealed that eight other components in this particular valve were of the same composition. Magnetic particle examination, occasionally supplemented with fluorescent dye penetrant, showed that 3 of these 8 components were cracked:

item 41—actuator valve feather (Figure 5-9)
item 40—actuator seat bushing (shown in section in Figure 5-10)
item 42—actuator (Figure 5-11)

As a matter of interest, it was subsequently learned that several actuator valve feathers (item 41) had failed over the years and had simply been dis-

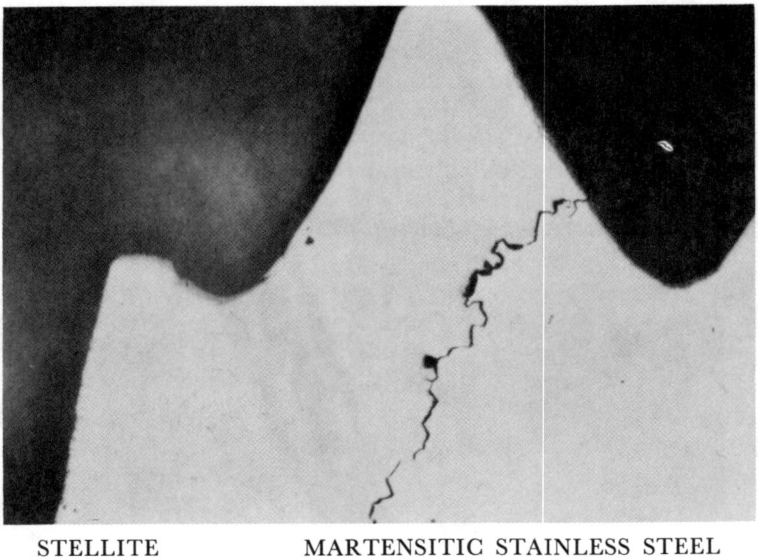

Figure 5-10 Stress corrosion crack in martensitic stainless steel valve component (Item 40, Figure 5-8).

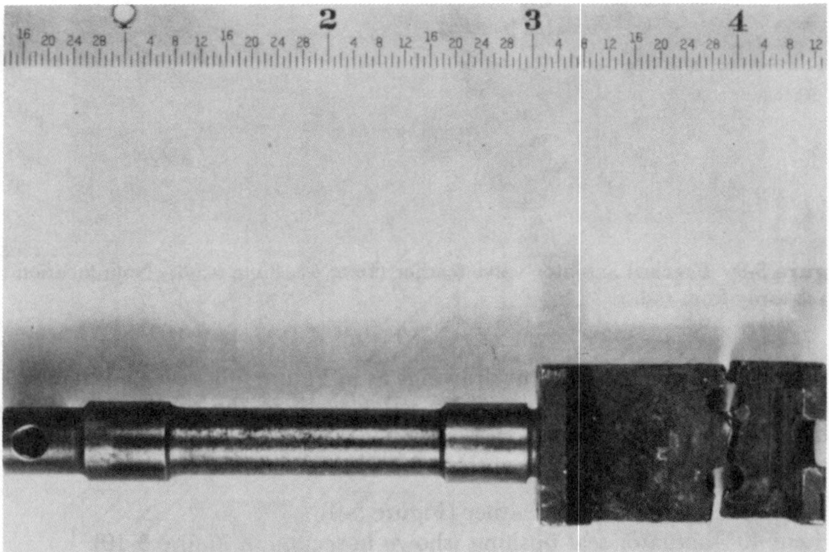

Figure 5-11 Martensitic stainless steel actuator failure at rivet holes (Item 42, Figure 5-8).

carded, attributed to "fair wear and tear" on a small, weak-looking component.

Failures of actuator valve seat bushings (item 40) had also occurred, generally under the hexagonal head at the beginning of the threaded portion—attributed often to excessive torquing in assembly.

At first glance, failure of the actuator (item 42) might be attributed to the rivet holes. Examination under the microscope, Figure 5-12, confirmed that stress corrosion cracking was present.

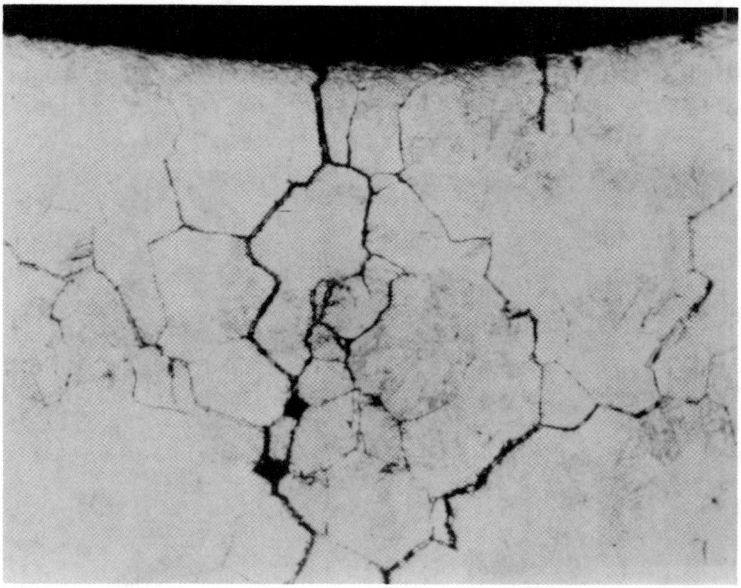

Figure 5-12 Stress corrosion cracks in martensitic stainless steel actuator (Figure 5-11). Rivet hole is at top.

As far as the corrosive environment is concerned, it was first thought that chloride ions, either from chemical cleaning of the boilers or from the atmosphere, were responsible for the corrosion attack. Although a few of the failures could be attributed to the presence of this chemical, it was very unlikely that the chloride could have made its way to some of the components. Hardened martensitic stainless steel parts have been known to fail by stress-corrosion cracking in pure hot water systems* and a source of chloride ions was therefore not required for failures to take place by this mechanism.

* Trozzo P. S. and McCartney R. F. "Relationship of microstructure and stress corrosion cracking of type 410 stainless steel", *Corrosion*, Vol. **16**, March 1960, p. 26.

It was interesting to note that austenitic stainless steel components in these valves had not failed by stress-corrosion cracking in the environment involved. This supported the position that relatively little chloride was likely present because the austenitic stainless steels are known to be subject to stress-corrosion cracking in hot chloride environments.

The working stresses for these valve components were generally quite small; as is the case in most stress-corrosion cracking failures we've experienced,[63] the high stresses were residual. These high locked-in stresses were produced when the martensitic steel parts were hardened.

Since the corrosive environment could not be altered, remedial action had to consist of changing the material. Alternatives suggested included:

a) Heat treating the AISI 414 components at 1150° F. This would result in a greatly increased resistance to stress corrosion cracking but hard facing

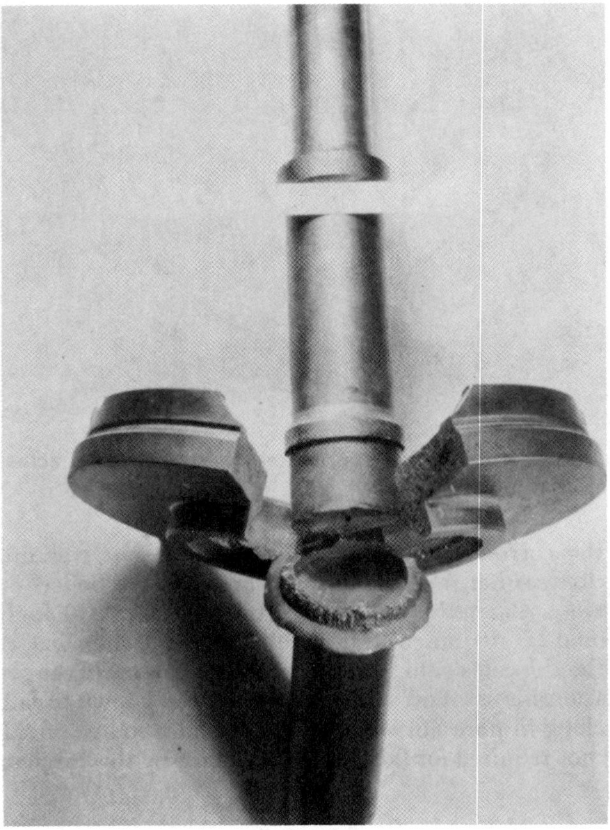

Figure 5-13 Failed martensitic stainless steel exhaust rejection valve.

Figure 5-14 Fractured surfaces of disc of valve of Figure 5-13 showing sequence of failure.

would be needed on valves and seat areas for hardnesses above Rockwell C26.

b) Replace components in Inconel, with suitable hard facing on valve and seat areas if required. It should be noted that a stellite seating surface was provided on the actuator valve seat bushing (item 40).

B-3 b) **Exhaust Rejection Valve Failures by Stress Corrosion Cracking.**

A 6 inch diameter disc (martensitic stainless steel) of an exhaust steam valve broke into 4 pieces in service, as shown in Figure 5-13.

Examination of the fractured surface shown in Figure 5-14 suggested a

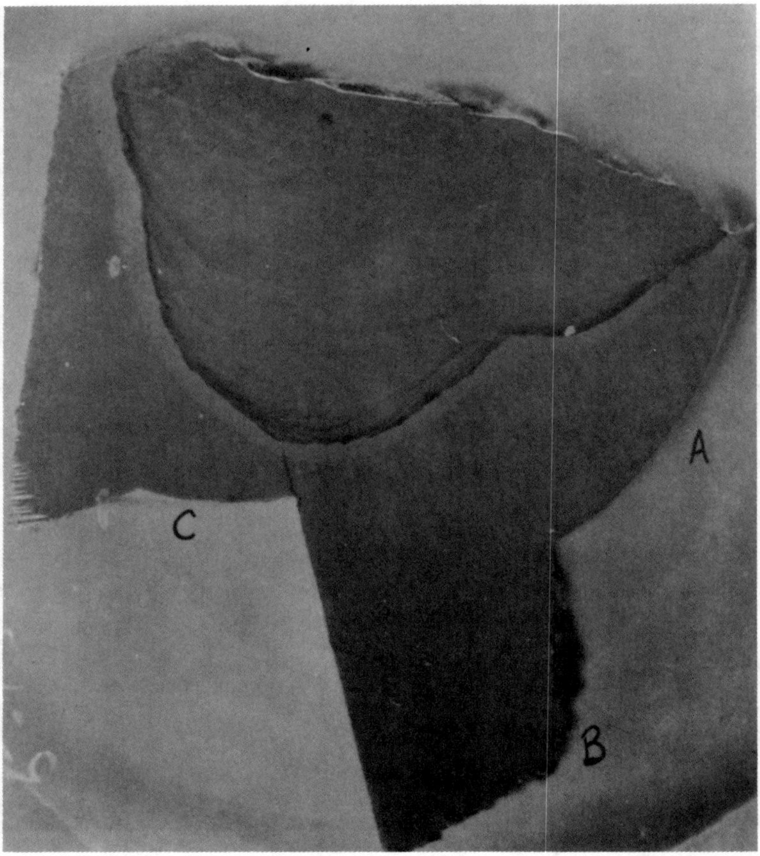

Figure 5-15 Section through weld joint showing in section fracture faces matching those of Figure 5-14. Note that fracture A initiated in heat affected zone.

possible sequence in the separations. The smooth separation (A) apparently initiated the failure. The discontinuity in rupture (B) proves that the transverse crack occurred prior to (B). This sequence was later confirmed with another disc received in a partly cracked condition, i.e., both (A) and the transverse crack were present but (B) had not started. The edge of the weld appeared to be the initiating site.

A section was cut across the weld, as shown in Figure 5-15. Fracture (A) had initiated in the heat affected zone next to the weld and propagated further into the parent metal.

The real answer to the failure mechanism became apparent on examining a micro-section, Figure 5-16. Stress corrosion cracking initiated in the heat affected zone, where the metal was highly stressed (residual) and in a sus-

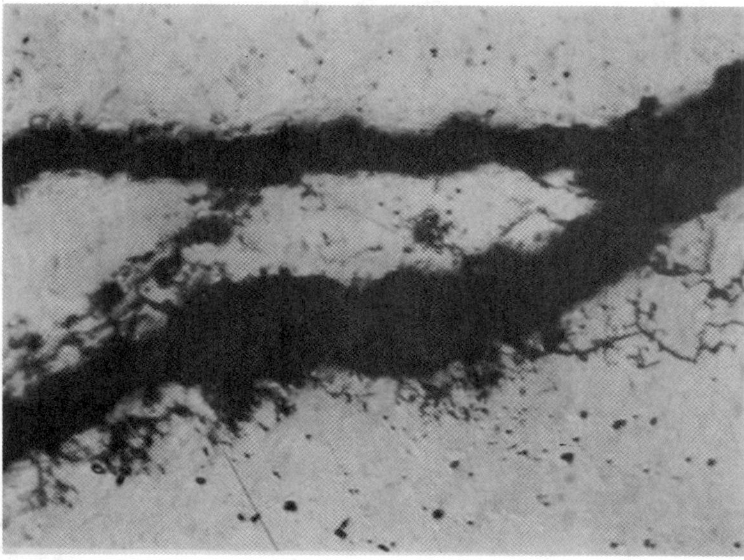

Figure 5-16 Metallurgical structure near fracture (Figure 5-15) showing stress corrosion cracking. (×400)

ceptible metallurgical condition. This type of failure can take place in hardened martensitic stainless steels in a steam environment. Possible remedies are discussed in Section 5.B-3 (a) above.

B-3 c) **Beta Brass Flooding Valve Failure by Stress Corrosion Cracking**

The failure of a high strength brass link from an underwater flooding valve was investigated.

Analysis showed the following composition:

58% copper
2% Al
1% Fe
2% Mn
remainder zinc

A metallurgical examination confirmed that the link was basically all beta phase, and large grained. At lower magnification the microstructure was similar to that shown in Figure 7-19. Branched, trans-granular cracking was evident as shown in Figure 5-17. In another area, there was also some cracking along the intergranularly precipitated alpha-phase.

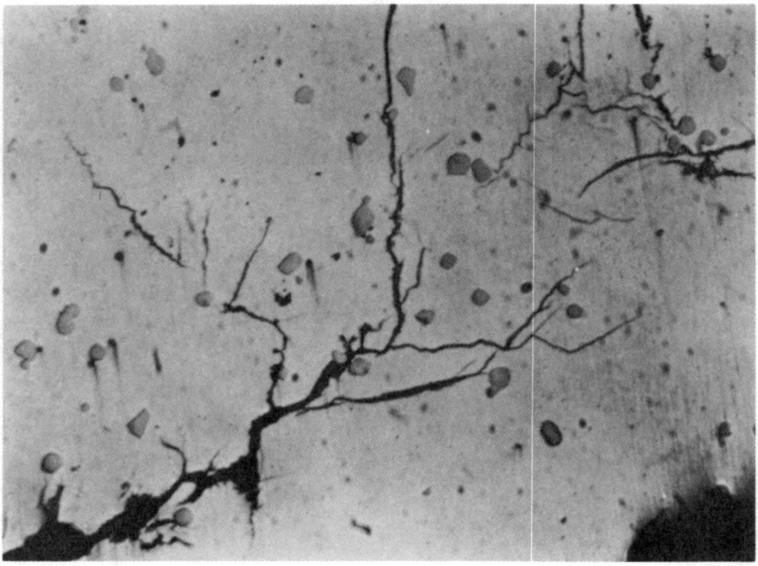

Figure 5-17 Branched transgranular cracks in β-brass sea water flooding valve link. (×400)

High strength beta brass appears to be subject to transgranular stress corrosion cracking in sea water. A safer choice than β-brass in sea water would be tin bronze.

B-3 d) **Aluminum–Magnesium Alloy Valve Failure by Stress Corrosion**

The aluminum–10% magnesium alloy had been specified for various components because of high strength and good corrosion resistance. The high strength was derived from heat treating but unfortunately, the possibility of getting reliable and thorough heat treatment was not always good. And unless the heat treatment is correct, corrosion resistance will be poor.

As an example, here is a valve body, Figure 5-18, which disintegrated in sea water service. Metallurgical examination showed the severe stress corrosion cracking along grain boundaries, as well as along slip planes, resulting from bolting stresses as well as inherent residual stresses. This attack is shown in Figure 5-19.

As a matter of interest, samples of the alloy were cut from another valve that had not been in service, and examined metallurgically. The microstructure shown in Figure 5-20 is comparable with that obtained when this

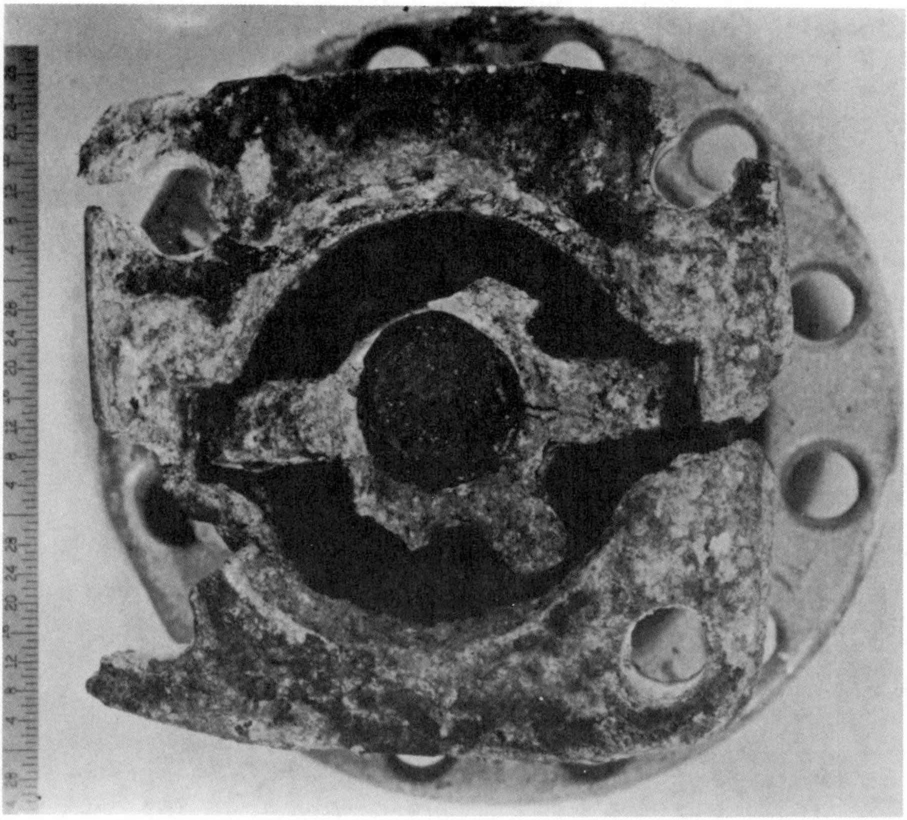

Figure 5-18 Disintegrated aluminum–magnesium alloy sea water valve body due to stress corrosion.

alloy was furnace cooled. In contrast, the microstructure for a correctly heat treated sample would be free of the second phase (the Al–Mg intermetallic compound). In this condition the alloy has maximum corrosion resistance; however, some sub-microscopic precipitation may occur with time and decrease the corrosion resistance. It was shown by Alcoa Laboratories that exposing a 6% magnesium alloy to 212° F for one week is equivalent in precipitation effect to that produced in 24 years at room temperature. And this amount is known to be detrimental.

Chemical analysis of the valve body showed an 11.6% magnesium content, which is greater than the 10% magnesium specified. In addition it would be expected that higher magnesium contents would markedly increase susceptibility to precipitation and resultant stress corrosion cracking.

The aluminum–10% magnesium alloy should be avoided in marine

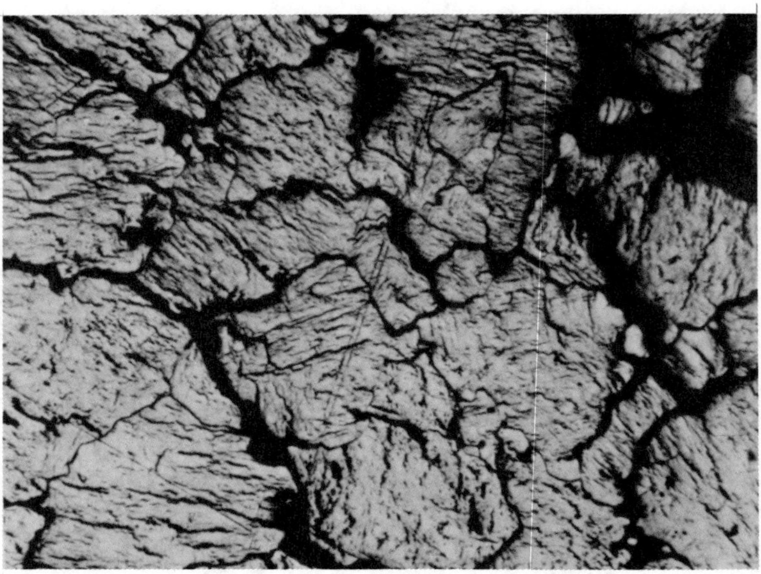

Figure 5-19 Photomicrograph showing severe stress corrosion cracking along grain boundaries and slip planes. (×150)

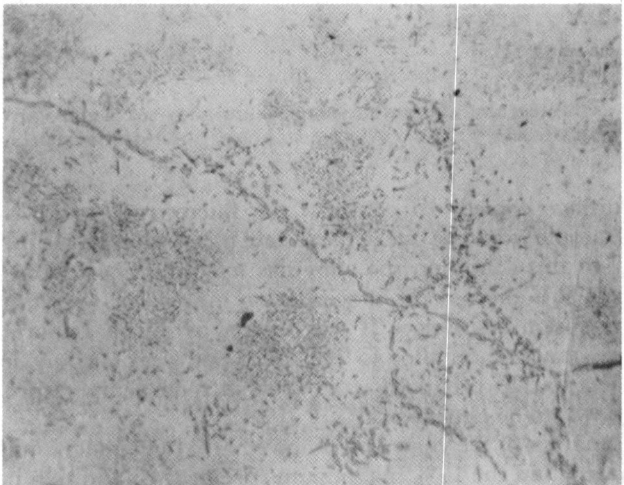

Figure 5-20 Structure of furnace cooled aluminum–10% magnesium alloy showing second phase predominantly along grain boundaries. (×400)

service, in favour of aluminum–silicon alloys which, while not as strong, produce sounder, more reliable castings.

B-5 Flooding Valve Failure by Crevice Corrosion

A leaky valve in salt water service was brought in for examination. In the process of dismantling the valve, it was apparent that the leak was associated with considerable corrosive attack in the seating area as shown in Figure 5-21.

Spectrographic analysis showed that all valve parts were a chromium stainless steel.

Figure 5-21 Corroded and cracked martensitic stainless steel sea water flooding valve.

From the location of the deepest corrosion pits, it was evident that the attack was caused by crevice corrosion—a failing to which stainless steels are particularly susceptible.

Further examination showed that this valve seat was also in the process of cracking by stress corrosion (see arrow in Figure 5-21).

Many other examples of trouble with stainless steels in sea water service suggest that stainless be avoided, and that these components be replaced by tin bronze with monel seating parts.

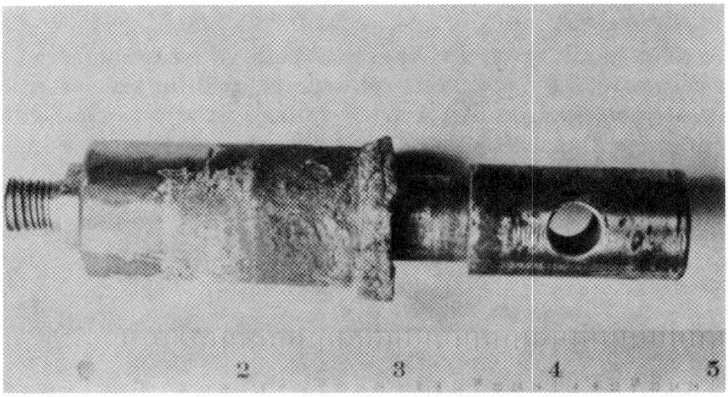

Figure 5-22 Stem from air cock valve showing dealuminification attack on bronze bushing.

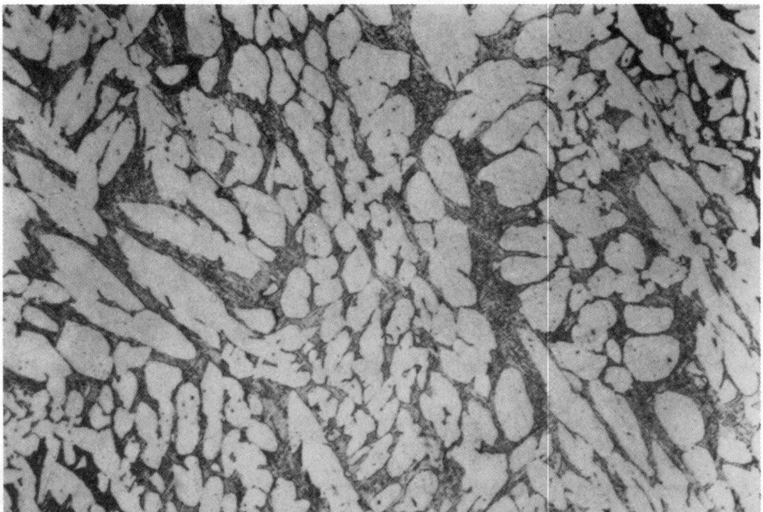

Figure 5-23 Two-phase microstructure of 10% aluminum bronze (ferric chloride etch) showing light copper-rich phase and dark eutectoid. (×200)

B-8 Air Cock Valve Failure by Dealuminification

The stem from a top air cock on a marine boiler began to seize and became inoperable. Both the stem and its bronze bushing were submitted for examination.

During dismantling it was noted that the bronze bushing appeared to be affected by corrosion as evident in Figure 5-22.

Spectrographic analysis showed that the corroded bushing was an aluminum bronze containing about 10% aluminum.

Examination of the surface of the bushing revealed the presence of copper, similar to that found associated with the dezincification of brass. Aluminum bronzes, particularly those containing enough aluminum to produce a two-phase micro-structure, have been known to corrode by a mechanism similar to dezincification; namely, dealuminification. The dealuminification

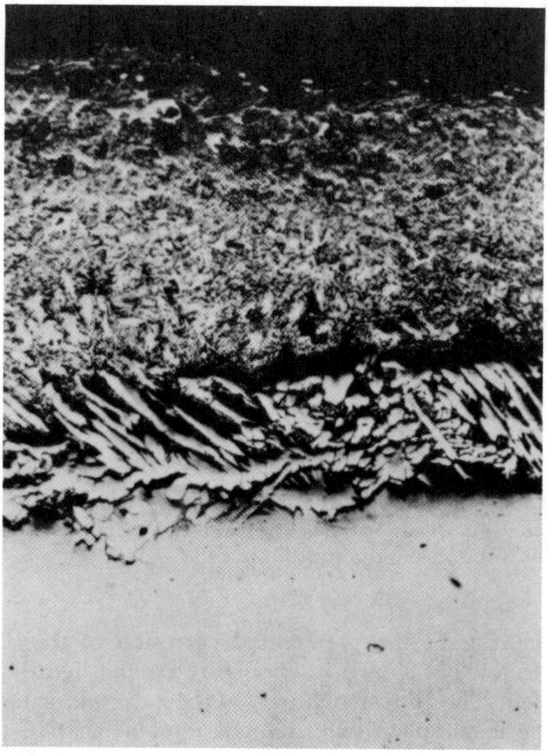

Figure 5-24 Microstructure of corroded aluminum bronze sleeve. Sound metal at bottom, deposited copper at top. Two-phase structure consists of light α-phase and corroding eutectoid. Unetched. (×80)

of aluminum bronze is more common in highly corrosive environments, but also takes place in contaminated steam environments. This particular aluminum bronze was exposed to stagnant condensed steam containing CO_2.

Microscopic examination showed a two-phase structure similar to that of 60–40 brass, as shown in Figure 5-23. Microexamination also confirmed the dealuminification. The sequence of the corrosion attack is indicated by Figure 5-24, a microphotograph of the corroded bushing (unetched). The sound metal is noted at the bottom of the figure. The first stage of the corrosion attack consists of the preferential corrosion of the aluminum-rich part of the eutectoid, as shown in the figure by the lightly attacked area at the bottom left. The copper-rich part of the eutectoid is next to corrode. When this stage is complete two products remain; that is, the original α phase and deposited copper (in the original eutectoid sites). The last stage of the dealuminification consists of the corrosion of the α phase. The large area at the top of the figure is mostly copper which has redeposited at each of the corrosion stages.

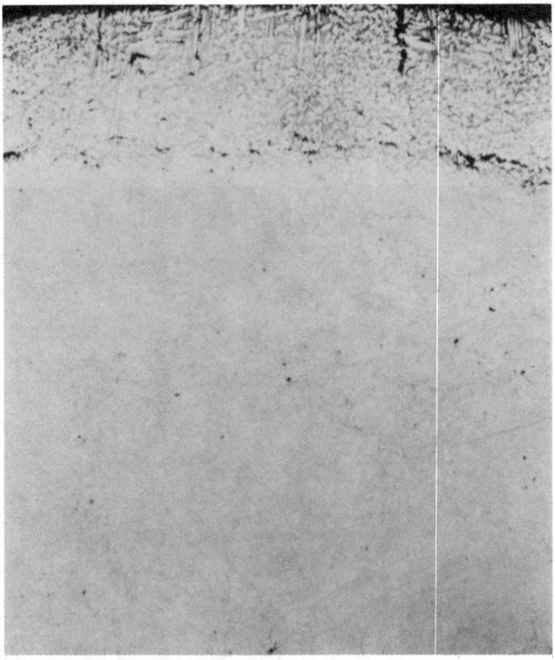

Figure 5-25 Microstructure of corroding aluminum bronze valve stem showing cracked two phase structure above (the eutectoid is corroding) and extensive advance of corrosion attack into the sound metal. Unetched. (×50)

A subsequent failure of a similar cock demonstrated how the first stage of corrosion in the aluminum bronze can advance to considerable depth, as in Figure 5-25, facilitating an apparent brittle failure.

It appears, then, that a 10% aluminum bronze is not a suitable material for cocks and valves which are subject to an environment of stagnant condensed steam. A single phase aluminum bronze, one containing less than 5% aluminum, would serve this purpose.

Figure 5-26 Failed valve body showing dendrite indications on fractured surface.

C-1 **Poor Quality Casting Causes Valve Failure**

Two failed small valves were brought in for examination. The tin bronze castings parted in the threads as shown in Figure 5-26.

Visual examination suggested that dendritic structure had some influence on the fracture—traces of a dendritic pattern could be distinguished under the low-power microscope.

Sectioning and subsequent examination under the microscope indicated, as in Figure 5-27, an excessive amount of shrinkage. This shrinkage, a casting defect related to the last metal solidifying in the inter-dendritic space, would greatly weaken the metal. It may be readily seen then how separation could occur, giving a dendritic pattern.

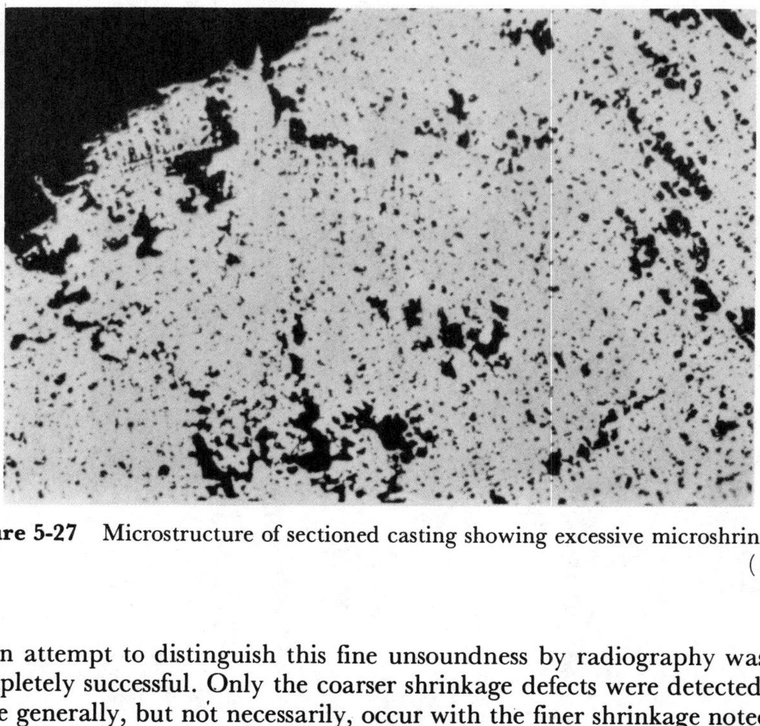

Figure 5-27 Microstructure of sectioned casting showing excessive microshrinkage. (×80)

An attempt to distinguish this fine unsoundness by radiography was not completely successful. Only the coarser shrinkage defects were detected and these generally, but not necessarily, occur with the finer shrinkage noted.

Greater care in foundry techniques on the part of the valve manufacturer was obviously required.

C-3 **Monel Exhaust Valve Failure at a Weld Repair**

An exhaust valve hinge pin which cracked (as shown in Figure 5-28) on being removed, was submitted for examination. It was reported that removal was difficult and extra force was required.

A section was cut adjacent to the crack and examined under the microscope. It immediately became apparent that the fracture was primarily in a coarse, porous weld zone, as shown in Figure 5-29. The evidence suggested

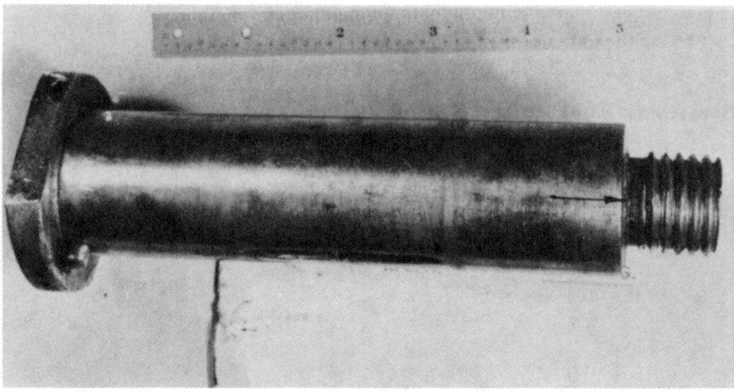

Figure 5-28 Monel exhaust valve hinge pin which failed in the threads. (arrow).

Figure 5-29 Section through failed pin (Figure 5-28) showing weld structure at left containing porosity. Wrought structure shown at right.

that on some previous occasion the threaded portion had been turned down, built up with welding and re-threaded. This would have been satisfactory except that no care was exercised in the welding, giving the weak structure shown.

The sharp change in section between bearing surface and the threaded portion is also not desirable.

Both base metal and weld metal were confirmed as monel spectrographically.

C-5 Stainless Fuel Pump Relief Valve Failure

A one inch relief valve from a fuel oil pump was received for examination. A longitudinal defect extending the entire length of the spindle and terminating at the valve portion was noted as in Figure 5-30.

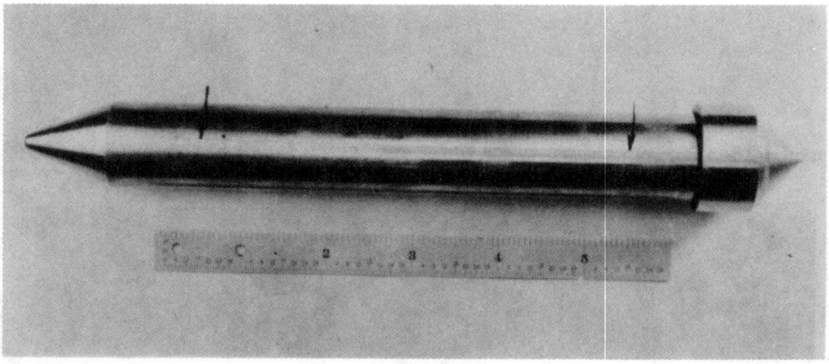

Figure 5-30 Fuel pump relief valve spindle showing longitudinal crack extending full length of spindle and terminating at the valve.

Spectrographic analysis confirmed the material to be Type 431 Stainless Steel. A hardness test indicated the component to be in the fully hardened condition represented by a Rockwell hardness of Rc 46.

A transverse microsection through the spindle showed that the longitudinal defect was a radial crack extending approximately one third of the way through the cross section. In order to establish the reason for the crack terminating at the junction between valve and spindle, a longitudinal section was made at this point. A circumferential crack emanating from the abrupt change of diameters was discovered extending from the notch. This otherwise undetected crack extended almost halfway through the spindle. The location and nature of this crack is illustrated in Figure 5-31. It became apparent that the circumferential crack had acted as a crack arrestor for the longitudinal crack. This is confirmed in Figure 5-32.

Figure 5-31 Circumferential cracking at change of section from spindle to valve originating at sharply machined corner. (×50)

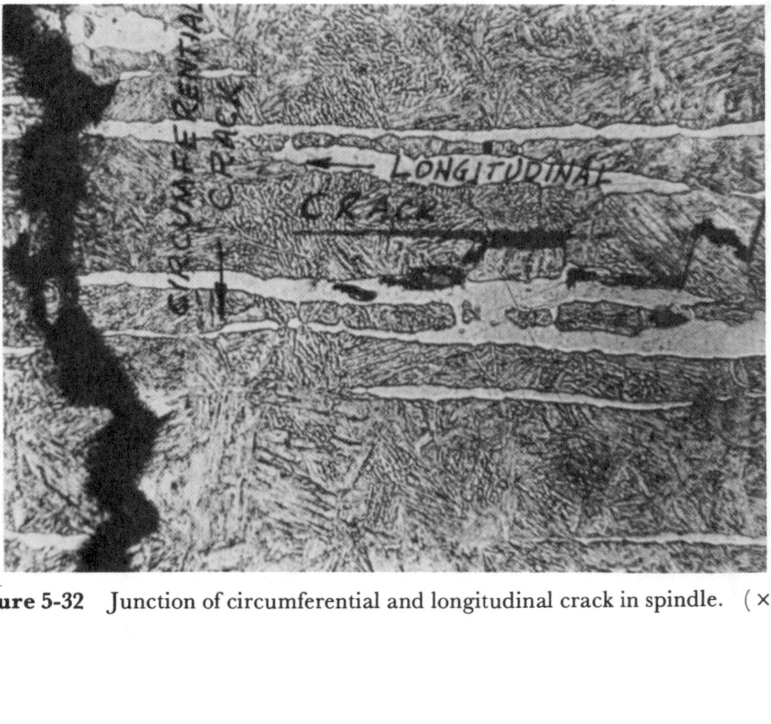

Figure 5-32 Junction of circumferential and longitudinal crack in spindle. (×270)

It will be observed that all the micrographs show the cracks to jig-saw to a remarkable degree, i.e., the protuberances of one side have matching cavities in the other. This crack form, in addition to the high hardness, suggests that the failure resulted from quenching during heat treatment. It should be noted that quench cracks do not necessarily occur while the work is in the quenchant. They may develop after a considerable time lapse if the work is allowed to stand after hardening without tempering. They are unlikely to occur in material which has been subjected to an immediate tempering treatment after hardening.

Hydrogen embrittlement, to which Type 431 stainless is susceptible, may also have been a factor in the failure.

The sharp change of section between the valve and stem provided an undesirable stress raiser for the circumferential crack.

It is recommended that the heat treatment for hardened Type 431 stainless include a tempering process to prevent the formation of quench cracks and to reduce the susceptibility to hydrogen embrittlement. The provision of a radius at the change of section between valve and stem is also imperative.

C-7 **Stainless Steel Valve Seat Corrosion Problem**

The commonly held belief that stainless steel is stainless, or corrosion resistant in almost any environment needs to be constantly dispelled. An example of failure of an 18–8 austenitic stainless steel serving as a valve seat in sea water service is described below. This valve came to us because of doubt whether specified materials had been used in view of the corrosion which resulted.

As shown in Figure 5-33 there has been severe attack of the seat.

Chemical analysis showed that main parts of the valve (except the seat) were leaded red brass i.e., 85 Cu–5 Pb–5 Zn–5 Sn.

Initiation of the corrosion attack was undoubtedly by the stainless steel suffering crevice pitting in the seat area while closed. Subsequently corrosion would continue because the stainless steel, no longer passive, would be anodic to the red brass adjacent.

This failure would not have occurred if the specified disc and seat material, namely monel, had been used.

C-8 **Valve Bonnet Failure at Machined Notch**

A hull stop valve bonnet from a submarine was submitted for examination following discovery of a leak.

Visual examination confirmed that a crack had penetrated to the outer shoulder surface, and this was also shown on sectioning, Figure 5-34. Although not visible in Figure 5-34, a small crack was also present at the diametrically opposite corner; in other words, the crack persisted for most of the way

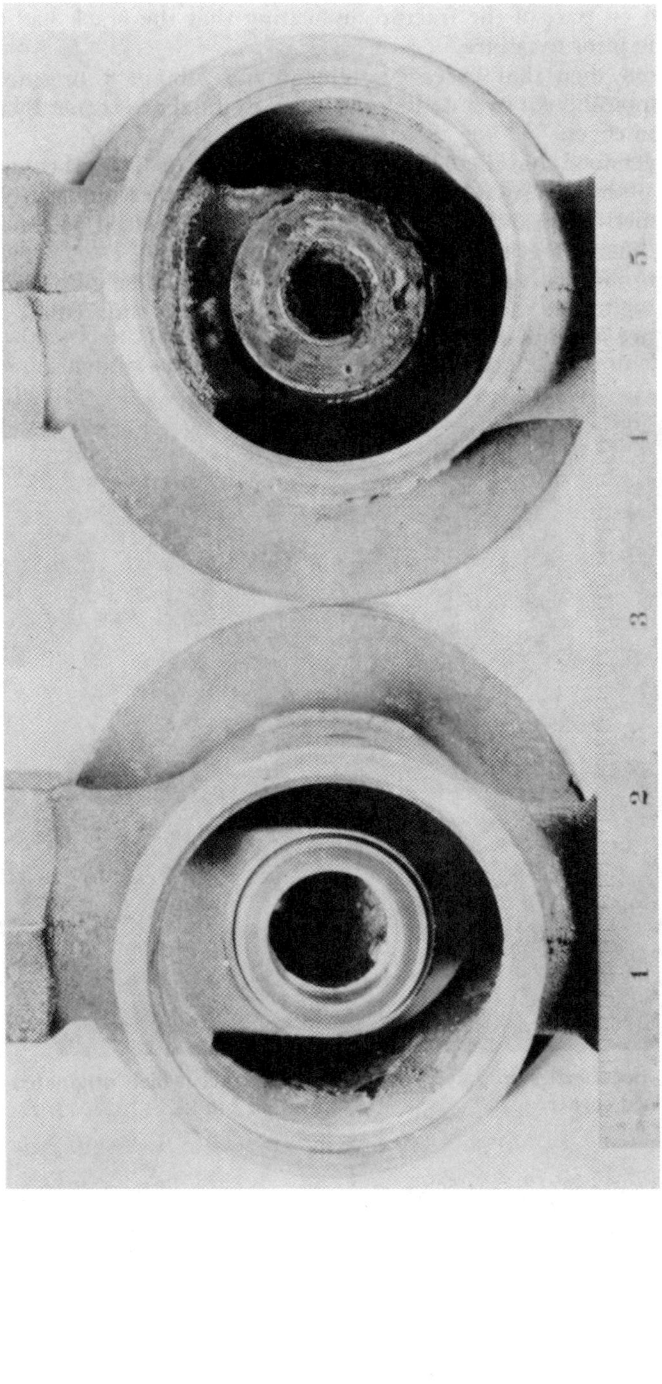

Figure 5-33 New and corroded valve seats of the same type of sea water valve. Seat is stainless steel and valve body is red brass.

around. Severe deformation from hammering was also noted on the hexagonal faces on the bonnet exterior. When the crack was split open, there was a discoloration on part of the fracture indicating that the crack had existed for some time prior to failure.

It appeared then that excessive tightening to obtain a pressure tight joint was responsible for overloading the metal in the sharp corner by a stress concentration effect.

It was understood that these fittings were subjected to 3000 psi compressed air. Because of the hazard, it was recommended that all these units be checked by a dye penetrant examination, and that they be replaced as convenient with fittings having a generous radius at the inner corner.

This failing illustrates a common and widespread tendency by machinists and even designers to instinctively prefer a tidy, neatly squared-off corner. This preference requires education to eliminate.

Figure 5-34 Sectioned stop valve bonnet showing crack which originated at the sharply machined corner.

VI

PIPING AND PIPING SYSTEMS

FATIGUE FAILURE in a hydraulic line initiated in sharp thread roots under the influence of vibration. Ballooning bursts in copper lines indicated that normal service pressures were greatly exceeded. Somewhat similarly, bursting of a sound stainless steel expansion bellows was proof of a sudden and drastic pressure increase because there was evidence of the reversing 45° shear characteristic of face-centred cubic metal exposed near an explosive source. Abuse of a copper–nickel–iron pipe in fabricating and the attempt to disguise the subsequent braze repair were uncovered.

Copper–nickel–iron is not resistant to moisture from funnel gases. Some aluminum alloys, particularly those containing 10% magnesium, disintegrate readily in sea water. Cold worked brass is susceptible to "season cracking" and therefore its exposure to bird droppings can be hazardous. An aluminum alloy hydraulic line failure by corrosion cracking endangered life on an aircraft. Copper lines will not resist impingement from sea water flow at rates for which the copper–nickel–irons would be suitable.

Inadvertent use of an alkaline inhibitor resulted in severe attack of aluminum fresh water lines. Better choice such as a clad aluminum line for salt water cooling plus avoiding a couple to stainless steel would have avoided serious corrosion of another line.

An unusual cracking and peeling of a galvanized layer on a $2\frac{1}{2}$ inch line was traced to bath contamination with iron. Craze cracking of a copper line during bending was attributed to inclusions of intergranular oxide.

A-1 Fatigue failure of steel pipe at the threads.

A-4 a) Copper tube failure by overloading.
b) The "hammer" failure of an 8-inch expansion bellows.

A-5 Failure of 95–5 copper–nickel–iron line by overheating.

B-1 a) Corrosion of copper–nickel–iron drain lines.
b) Aluminum sea strainer corrosion failure problem.

B-3 Failure of cold worked brass tube by "season cracking".

B-4 Aircraft hydraulic line failure.

B-7 Impingement attack of copper water line.

B-8 Failure of aluminum cooling lines by corrosion—an inhibitor problem.

C-7 Corrosion of aluminum tubing by salt water.

C-9 a) Cracking of the galvanized layer on pipe when bending.
b) The cracking of copper pipe on hot bending.

A-1 Fatigue Failure of Steel Pipe at the Threads

The hydraulic line powering the steering mechanism on a destroyer failed and was submitted for examination. Fracture of the steel pipe, shown in Figure 6-1, was in a threaded joint.

It was apparent from the fatigue markings that failure initiated in a thread root and progressed with time under the influence of stresses associated with the vibration of the steering gear machinery. The sharpness of the thread roots may be noted in Figure 6-2.

Figure 6-1 Failed end of pipe showing fatigue markings.

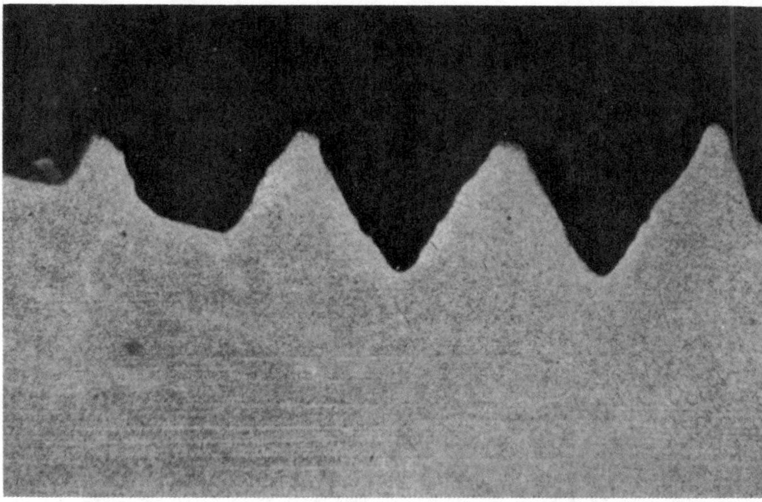

Figure 6-2 Threads of failed pipe section. Note the relatively sharp root radii.

Because of the seriousness of a failure in the steering gear on a destroyer, two recommendations were made:

a) that a more generously radiused thread root be used and if possible thread roots should be rolled (see T-8), and

b) that the line should be better supported to lessen stresses caused by vibration.

A-4 a) **Copper Tube Failure by Overloading**

Rupture of a $1\frac{1}{2}$ inch diameter copper tube, serving as an indicator pipe on a ship, was reported. The failure, shown in Figure 6-3 took place at the end of the pipe adjacent to the main engine.

It is understood that the indicator pipe is subjected to 225 psi, 390° F steam. The hoop stress in the copper tube at this pressure would be about 2000 psi, much lower than the tensile strength of copper at 390° F.

The ballooning of the tube and the thinning at the rupture indicate that the failure was not associated with an unsoundness of the tube. Analysis showed that the tube was of annealed copper.

Two possible failure mechanisms are suggested. If, for some reason, the pressure and temperature of the steam rose to a level considerably higher than the values shown, a failure of the above type could be envisaged. Also, copper has rather poor resistance to creep at temperatures above 400° F.

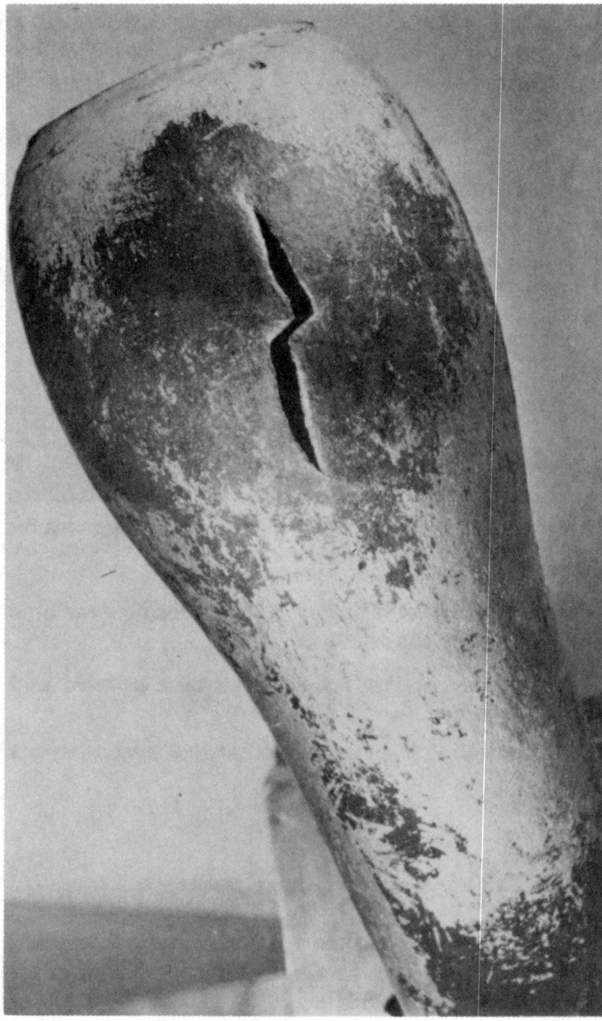

Figure 6-3 Rupture of $1\frac{1}{2}$ inch copper pipe by overloading. Note stretching from normal diameter at bottom.

Possibly the tube became enlarged by creep unnoticed over a long period. A number of other pipes in similar service were subsequently examined, and it was concluded that this failure was an isolated case.

At a later date, another burst in copper was experienced. Again the reported pressures in the tube could not possibly have caused the burst. The evidence was, however, that normal pressures were drastically exceeded.

A-4 b) The "Hammer" Failure of an 8-inch Expansion Bellows

The failure of a buried 8 inch main steam line resulted in the release of a tremendous amount of steam which lifted and carried to a distance of two feet a slab of concrete of about 10 feet by 10 feet. It was fortunate that the failure took place on a Saturday when the yard was not in operation.

The ruptured bellows is shown in Figure 6-4. It may be noted that there was a considerable outward bulging of the stainless steel bellows, sufficient to flatten out corrugations in this area. Another view, of the inside of the bellows is shown in Figure 6-5. Here may be noted the many "tidal" marks of many condensate levels which existed during the service period of the bellows. The ruptures were directly above and centred over the "tidal"

Figure 6-4 Ruptured 8-inch stainless steel expansion bellows for steam heating plant.

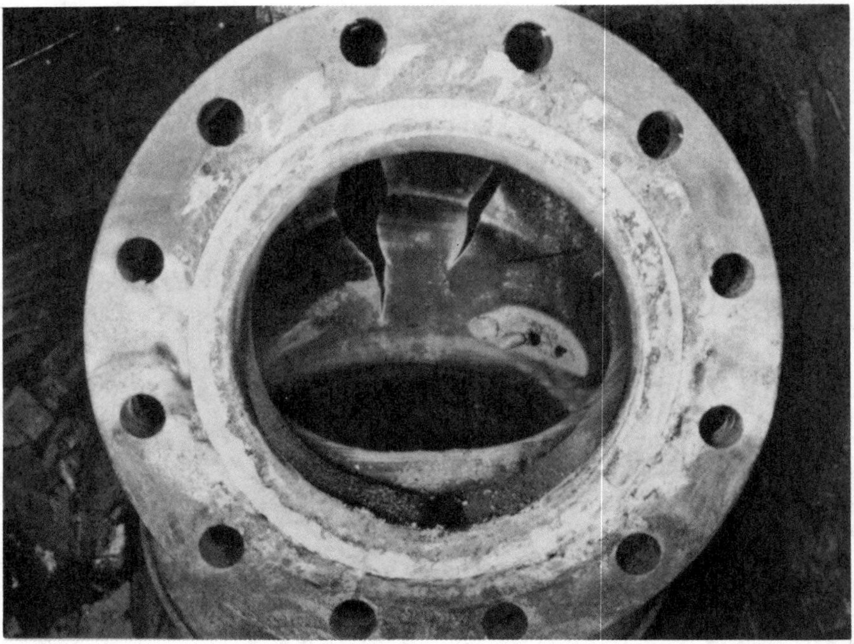

Figure 6-5 Inside the bellows showing "tidal" marks of many condensate levels. Note also the weld (arrow).

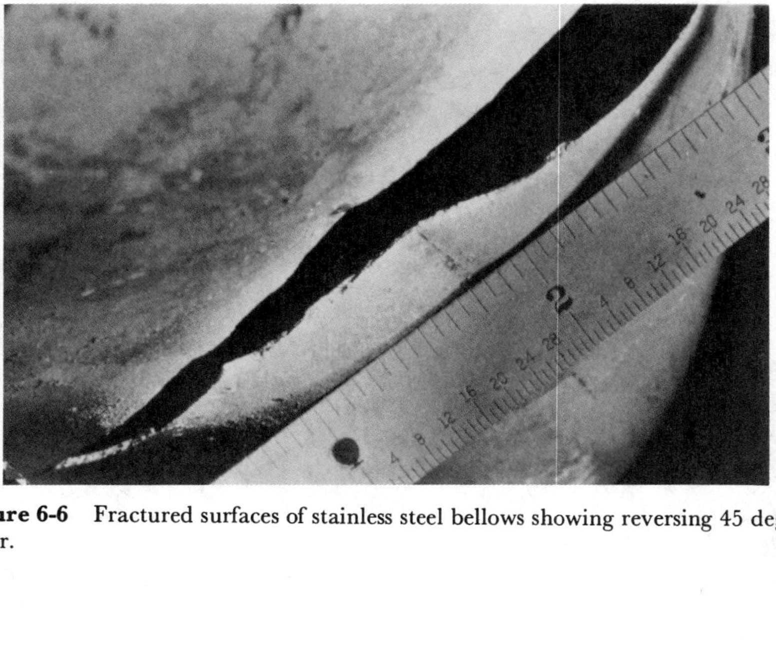

Figure 6-6 Fractured surfaces of stainless steel bellows showing reversing 45 degree shear.

Figure 6-7 Fractured surface of aluminum sheet ruptured by a high explosive. Note similarity to that of Figure 6-6.

marks. It appeared that there was some connection between the condensate accumulation and the centering of the rupture in this area.

A close-up photograph of the actual rupture surfaces, Figure 6-6 shows a reversing 45° shear characteristic of aluminum or stainless steel exposed fairly near to an explosion centre. A similar fracture is shown, Figure 6-7, in a piece of aluminum which was ruptured by a high explosive.* The almost complete absence of stretching in the bellows rupture eliminates the possibility of a slow and steady pressure increase.

There was no evidence of fatigue cracking. This type of cracking would likely have been transverse rather than longitudinal. No other evidence of prior weakness or unsoundness was found.

It was, in fact, of interest that a butt weld ran parallel to the main rupture and 7 inches away (Figure 6-5). The heat affected zone adjacent to the weld would have been a likely area for separation because of the more ductile larger grains. That this area did not fail was probably due to the suddenness and abnormal pressure applied.

Also, the fact that there were two ruptures (one small and one large) rather than one, confirms that the applied pressure was sudden—otherwise there would have been time for pressure release with one opening.

Calculations based on an 8-inch cylinder of 0·04 inch wall thickness suggested that pressures of over 400 psi would have been required to stretch

* Courtesy of T. S. Sterling of the Defence Research Establishment Valcartier, Valcartier, Quebec.

and split the stainless steel. Evaluation of actual explosive pressures was outside the authors' experience.

The failure of the bellows appears to be due to an extremely drastic and sudden increase in pressure of a type which might occur adjacent to an explosion centre. It is believed that a "hammer" mechanism must have been responsible.

A bypass valve to equalize pressure should have been installed in this large steam line. This would eliminate any large steam fluctuations during the opening up periods.

A-5 **Failure of 95–5 Copper–Nickel–Iron Line by Overheating**

An unexpected failure in a copper–nickel–iron cooling line was called to our attention. This copper–nickel–iron, in service for a relatively short period, was considered capable of resisting corrosion by sea water at moderate flow rates.

Visual examination of the inner surface of the cooling line revealed pitting in a sporadic fashion at, and near, the bend as shown in Figure 6-8, of the attached photographs. Examination at low power showed the presence of checks or cracks in the vicinity of these pits. Some impingement attack was noted but the mode of formation of the pits was not clear.

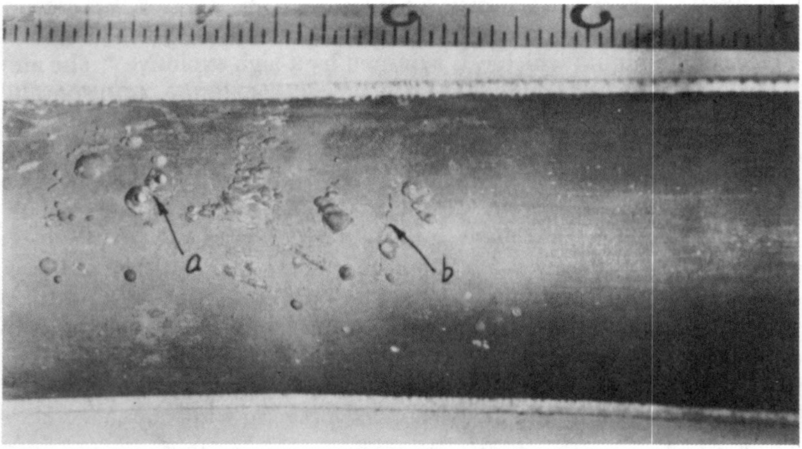

Figure 6-8 Inside of copper–nickel–iron cooling line showing pits at "a" and a crack at "b".

Chemical analysis showed these lines to be 94 copper–5 nickel containing 1.23% iron.

Sections were cut from the line for micro-examination. The line showed

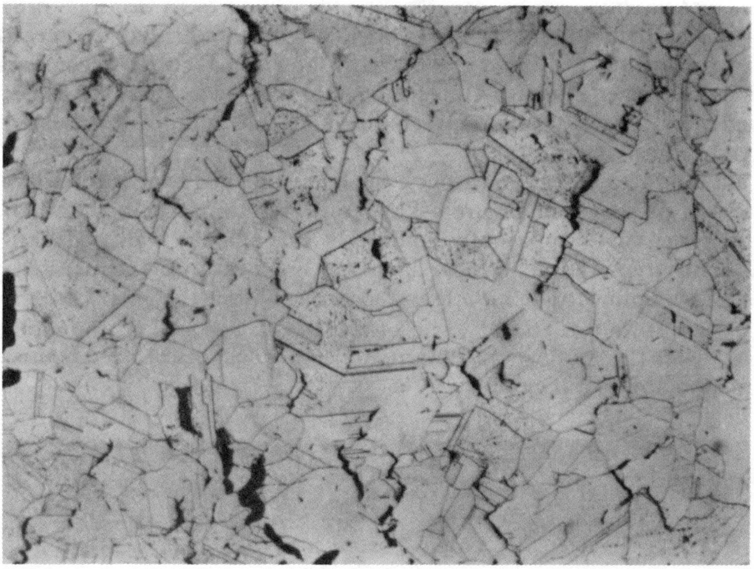

Figure 6-9 Section cut through copper–nickel–iron line (Figure 6-8) near impingement attack. The large grain size compared to Figure 6-10 confirms exposure to high temperature. Note cracks. (×150)

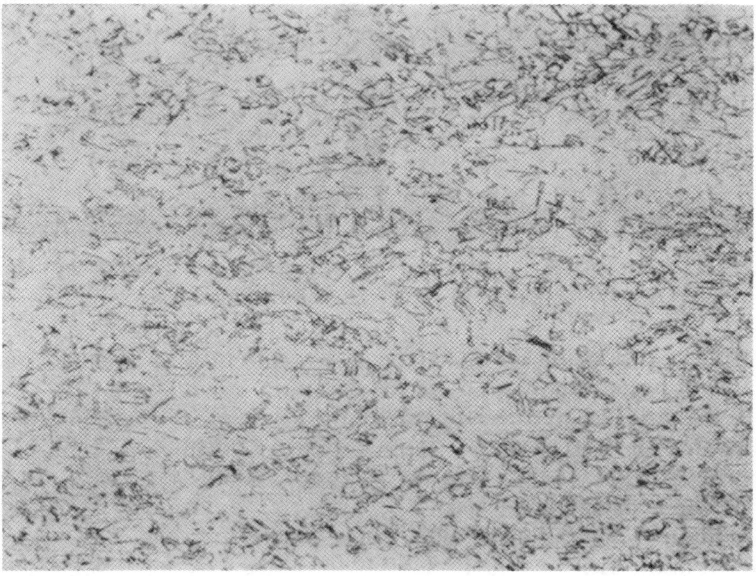

Figure 6-10 Microstructure of copper–nickel–iron line at some distance from the heated area. (×150)

evidence of having been overheated at some period in its history—probably when it was bent to shape. The evidence which indicated overheating is the creep cracking and large grain size as shown in Figure 6-9 compared to the fine grain size (Figure 6-10) of the metal well away from the bend. The impingement attack was largely confined to this zone of overheated and enlarged grains.

In another area, a section through the finely cracked zone was particularly interesting. As shown in Figure 6-11 there were many cracks at the surface, and the largest had apparently gone right through the wall and had been subsequently repaired. The braze had been smoothed off and painted over. At higher magnification the apparent checks within the metal adjacent to the large rupture turned out to be intergranular fusion, i.e., the metal had been heated to the point where some melting took place at the grain boundaries. It indicated abuse of the metal.

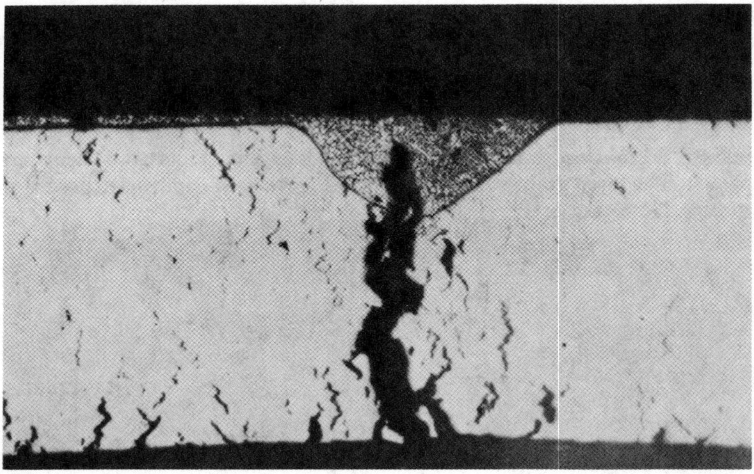

Figure 6-11 Section cut through cracked cooling line ("b" Figure 6-8) showing that the crack had been repaired by brazing by the manufacturer. (×28)

The 95–5 copper–nickel–iron pipe is susceptible to a loss of corrosion resistance if heated unduly, i.e., at temperatures much above 1000° F for a period of time. It was therefore recommended that bending operations be performed with controlled heat input, or where possible, with no heat. It was suggested that Tempilsticks should prove suitable for indicating metal temperature on the pipe surface.

In summary, the failure of this copper–nickel–iron line does not reflect adversely on the metal, rather it points to the abuse to which it was subjected. The manufacturer was clearly to blame for allowing damaged material to pass his quality control checks.

B-1 a) **Corrosion of Copper–Nickel–Iron Drain Lines**

Deterioration of two lengths of funnel drain line from a ship was called to our attention. These drain lines, whose inner surfaces are shown in Figure 6-12, served to lead off moisture collected in the funnel area.

Figure 6-12 Internal deposits and perforations of 90–10 copper–nickel–iron drain lines.

Part of the drain line had been repaired and one was a replacement for original equipment—noted by a change in composition to a slightly richer alloy. Original installation specified copper–nickel–iron (95–5); the replacement was copper–nickel–iron (90–10).

The inner surface of the drain lines (Figure 6-12) was heavily coated with deposits, the thickest in the region of perforation. Analysis of these deposits showed that a water soluble fraction was strongly acidic, and that 80% of the material was hydrated basic iron sulphate. This confirmed the likely contaminants in this area namely, fuel residues, fresh and salt water, and iron corrosion products from deterioration of the inner funnel.

Copper–nickel–iron piping is susceptible to attack by sulphurous and sulphuric acid. According to Uhlig[51] the presence of an oxidizing agent such as ferric sulphate may accelerate the corrosion rate in sulphuric acid by three hundredfold. It is not expected that the 90–10 alloy will perform significantly better than the 95–5 which it has replaced. Corrosion attack in the form of perforation was noted in the 90–10 portion of the pipe.

It was suggested, because of the variety of corrodents involved, that it would require experimental evaluation of promising materials, including lead-lined steel, copper, and some plastics.

B-1 b) **Aluminum Sea Strainer Corrosion Failure Problem**

An aluminum strainer in a salt water suction line was submitted for examination because the flange part shown in Figure 6-13, had essentially disintegrated. The basket part of the strainer was relatively sound.

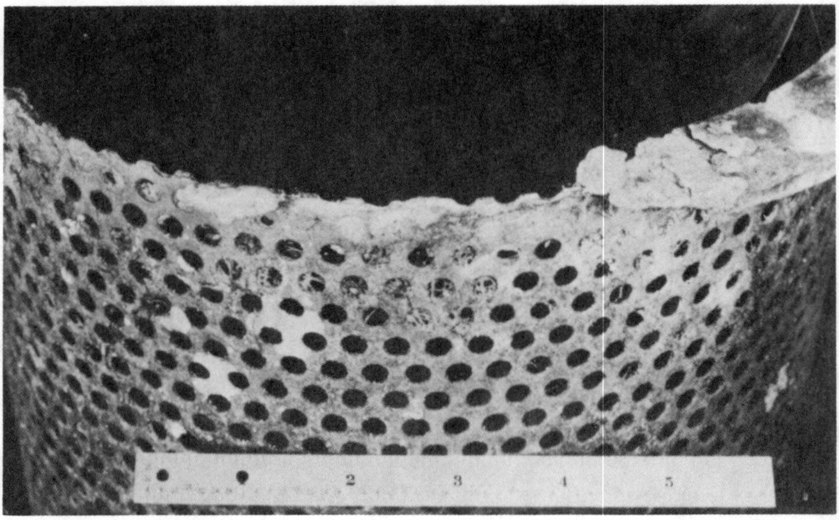

Figure 6-13 Corroded flange of aluminum sea water strainer. Flange is aluminum 10% magnesium and basket is a $2\frac{1}{2}$% zinc, $\frac{1}{2}$% chromium, aluminum alloy.

Spectrographic analysis showed that the flange was an aluminum–10% magnesium alloy, and that the basket was an aluminum–$2\frac{1}{2}$ zinc–0.25 chromium alloy.

Metallurgical examination of the corroded flange indicated that disintegration was by stress-corrosion. This casting was also quite porous.

The microstructure suggested that the metal had been suitably heat treated (5. B-3 (d)) making it relatively resistant to corrosion; however microstructural changes (precipitation of the second phase at the grain boundaries) had probably taken place over the years, resulting in susceptibility to stress-corrosion cracking.

The good condition of the basket was probably the result of a cathodic protection effect by the corroding flange.

It was also noted that a handle had been welded onto the flange of a second strainer examined and that the metal had disintegrated near the weld. The heat resulting from welding had an obviously detrimental effect on corrosion resistance of the aluminum–10% magnesium alloy.

None of the aluminum alloys resist sea water corrosion particularly well, although a better choice in aluminum than the present combination is possible.

B-3 Failure of Cold Worked Brass Tube by "Season Cracking"

Splitting of a length of brass tubing while stored in an open shelter was reported.[63] The split tube is shown in Figure 6-14.

Analysis showed that the brass was a standard 70 copper–30 zinc.

Microstructural examination of the metal at a fine branch of the main rupture, revealed intergranular cracks and evidence that the metal was cold worked, Figure 6-15.

Stress-corrosion cracking or "season cracking" was the obvious mechanism for the cracking. From the size of the separation at the split it could be estimated that the residual stresses were high, approaching the yield strength.

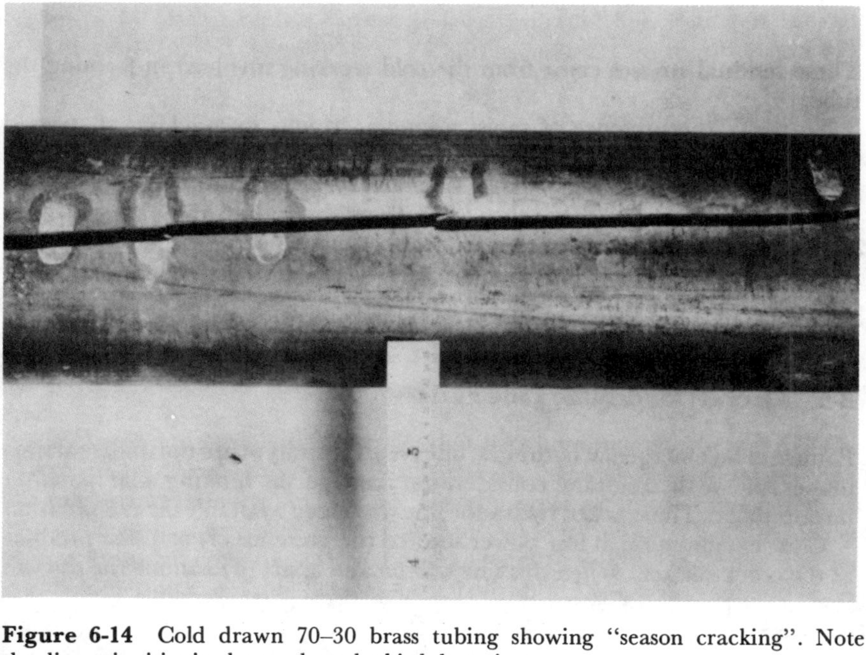

Figure 6-14 Cold drawn 70–30 brass tubing showing "season cracking". Note the discontinuities in the crack at the bird droppings.

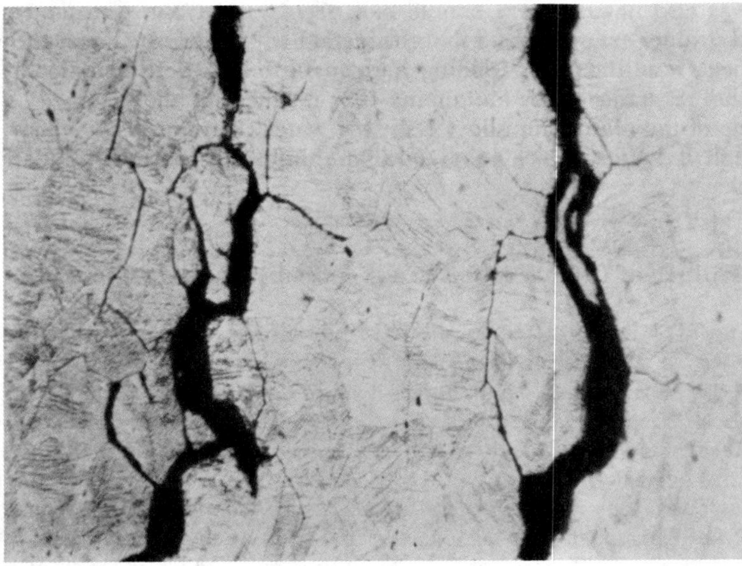

Figure 6-15 Intergranular stress corrosion cracks in cold drawn 70–30 brass tubing of Figure 6-14.

These residual stresses came from the cold working involved in forming the tube.

Normally the presence of moist ammonia results in cracking of stressed brass—in this instance the source of ammonia could be surmised from the presence of whitish excreta, as may be noted on closer examination of Figure 6-14. It is noteworthy that there were discontinuities of the main crack at each of the dropping marks.

This is a rather interesting example of the hazards associated with storage of metals and metal parts.

B-4 Aircraft Hydraulic Line Failure

Failure of an emergency hydraulic line on an aircraft made the undercarriage inoperable with resultant considerable damage on landing and potential hazard to life. The cracked hydraulic line was made available for examination.

Close examination at low power showed the presence of corrosion products at the crack surface. When this line was broken apart to examine the fracture surfaces, a portion of this surface was dark, suggesting corrosive action as in Figure 6-16.

A metallurgical section cut through one of these fracture surfaces and at

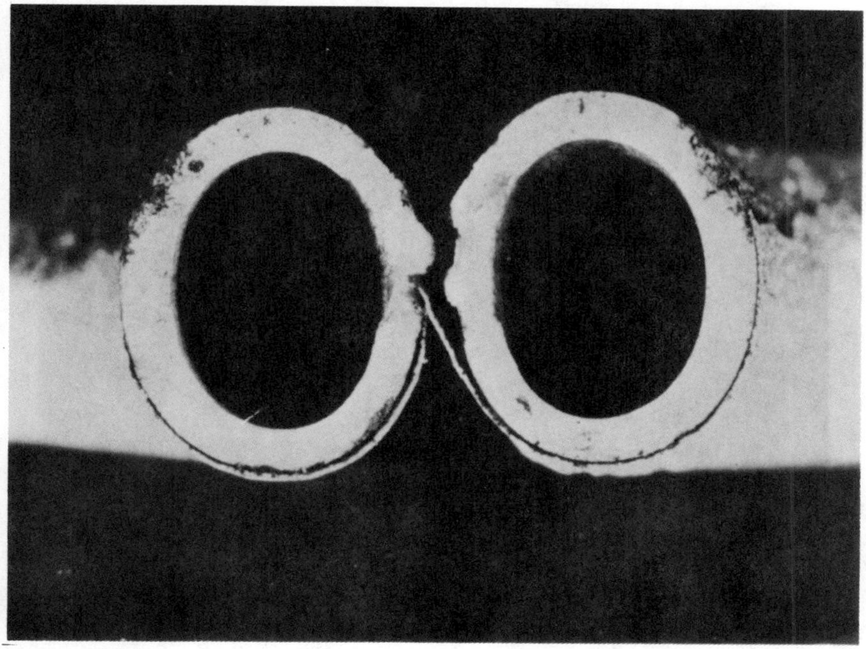

Figure 6-16 Mating rupture surfaces of failed hydraulic line. Note dark zone at 10 o'clock and 2 o'clock respectively. (×7)

right angles to it gave the result shown in Figure 6-17. The intergranular corrosion had penetrated approximately 70% through the tube wall. In addition, at least one additional corrosion pit was detected approximately $\frac{1}{4}$ inch from the rupture zone.

Following these findings, the equivalent hydraulic line on the other side of the aircraft was made available. On this line also, similar corrosion was detected in at least three locations at and close to the same location which failed on the other hydraulic line. Subsequently the same intergranular corrosion to 30% wall penetration was found on other aircraft of the same make.

Spectrochemical analysis confirmed that the aluminum alloy was AA 6061.

The evidence suggested that this particular area of the emergency line was exposed to moisture and sea air, thereby resulting in corrosive action at any weakness in the protective coating. This AA 6061 material has a tendency to intergranular corrosion when exposed to salt-carrying moisture and its use in something as critical as aircraft hydraulic lines was condemned. A far safer choice would be an appropriate clad tubing (clad on the outside); alternatively this emergency line could be replaced at frequent intervals.

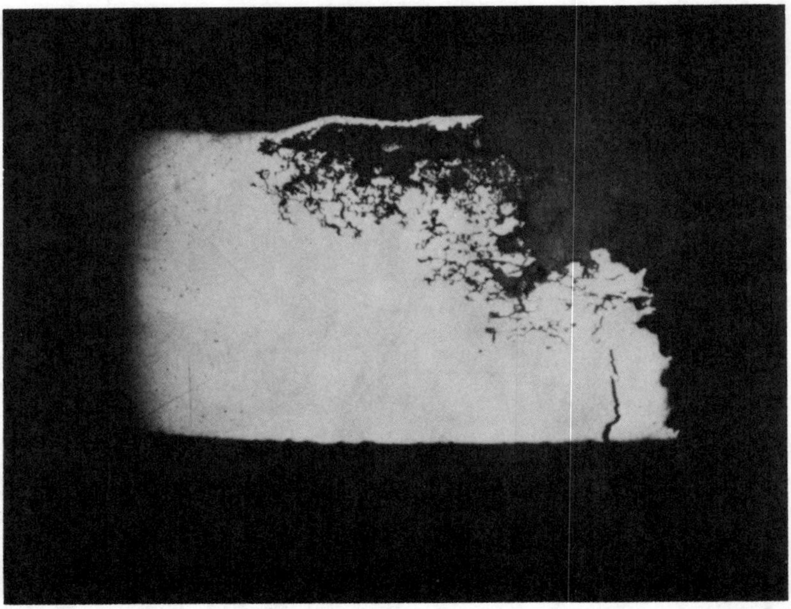

Figure 6-17 Severe intergranular corrosion associated with the fracture, Figure 6-16. The corrosion extends in from the outside. (× 50)

B-7 Impingement Attack of Copper Water Line

Leaking of a sea water supply line on a ship was reported. This copper line carried pre-heated sea water at rates up to 6 feet per second.

The pipe was sectioned and almost the entire inner surface had corroded. The absence of corrosion products was noted and was an indication of a scouring type of action. Closer examination of the corroded surface showed horseshoe shaped pits, Figure 6-18, a classical identification of impingement attack. A section cut through this area showed (as in Figure 3-9) the undercut profile contributed by the erosion component of the corrosion-erosion attack.

In this instance, the specified material was 95–5 copper–nickel–iron which would have adequate resistance under these flow conditions.

B-8 Failure of Aluminum Cooling Lines by Corrosion—An Inhibitor Problem

Perforation of several aluminum fresh water cooling lines on a ship resulted in great inconvenience. The cooling water was normally inhibited with the

Figure 6-18 Corrosion-erosion (impingement) attack in copper water line. Note the horseshoe shaped pits.

usual chromate inhibitor and no trouble had been experienced in the past. It was, therefore, surprising when a rash of failures occurred.

Analysis of the inhibitor soon showed that there had been an inadvertent substitution of a highly alkaline material, normally used for degreasing in boilers.

It was, however, interesting to examine some of the corrosion damage. Sectioning of one of the many pits was rather revealing because of an unusual attack through the clad layer, Figure 6-19. Normally the clad layer of an anodic aluminum alloy, inside a stronger aluminum alloy pipe, corrodes and thereby protects the underlying metal. Under the present conditions, however, the cladding proved to be cathodic.

C-7 **Corrosion of Aluminum Tubing by Salt Water**

An aluminum salt water cooling line was submitted for examination after leaking in service. It had been noted that leaking occurred in a cracked area adjacent to the flange. The cracked area is shown in Figure 6-20.

The pipe was split open, Figure 6-21, showing that there was pitting attack generally down the length of the pipe. A close-up, Figure 6-22, of the flange end showed the severe deterioration of the pipe and the inside of the flange.

A section from the pipe at some distance from the badly corroded area,

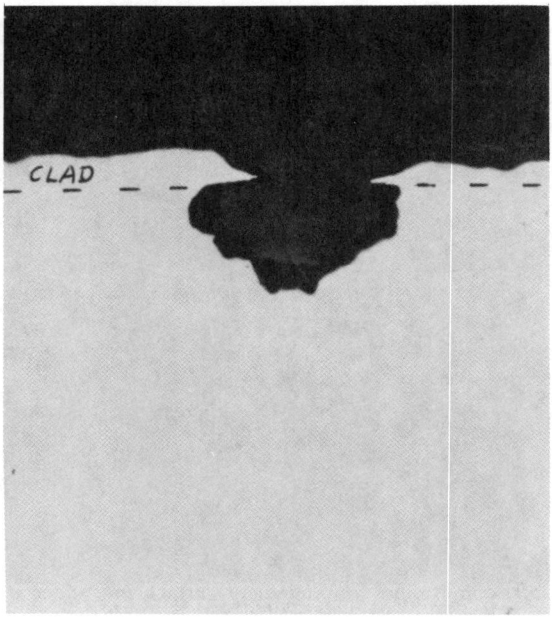

Figure 6-19 Pit in sectioned aluminum pipe. Note that the cladding is cathodic to the tube material.

Figure 6-20 Cracks in aluminum pipe near flange caused by internal corrosion.

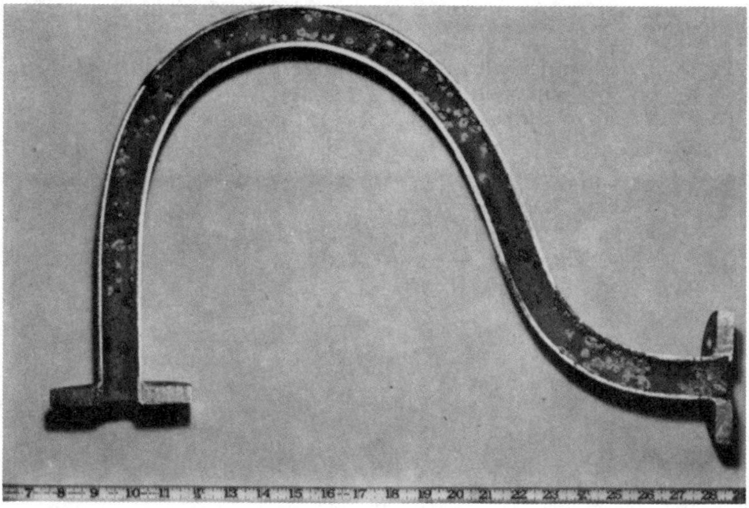

Figure 6-21 Sectioned aluminum tubing showing pitting over the length of the tube and severe corrosion at right flange.

Figure 6-22 Section at flange showing extent of galvanic corrosion attack at the cracked zone (Figure 6-20).

was examined under the microscope, Figure 6-23. Extensive intergranular attack was obvious.

Analysis showed that both pipe and flange were AA 6061. This material is known to be subject to corrosion in sea water.

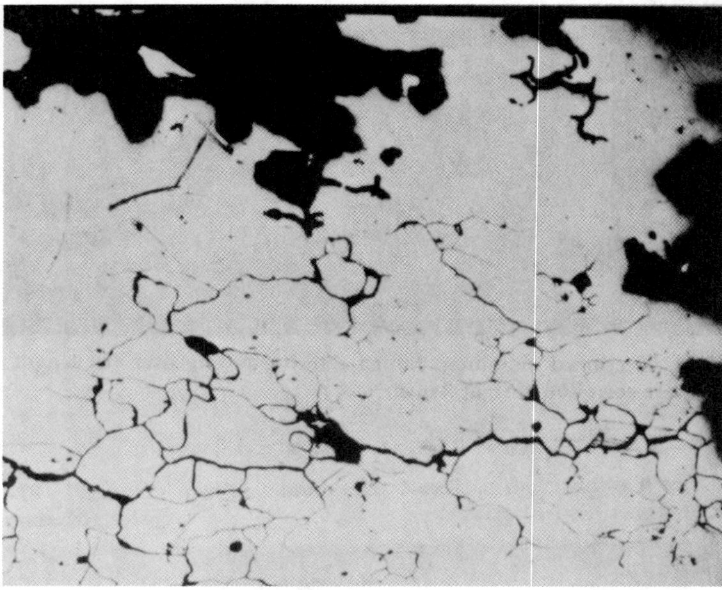

Figure 6-23 Intergranular corrosion at galvanically corroded area of aluminum pipe. (×400)

The severe corrosion at the flange end was attributed to galvanic action: the flange was bolted to a stainless steel adapter. The galvanic problem could be eliminated by the use of a suitable aluminum alloy (if remainder of system aluminum) or by isolating with a non-metallic fitting.

To avoid serious pitting, the best remedy is the use of AA 3003 clad with AA 7072.

C-9 a) **Cracking of the Galvanized Layer on Pipe when Bending**

On several lengths of $2\frac{1}{2}$ inch galvanized pipe which were being bent the coating broke up in the manner shown in Figure 6-24. Concern was expressed as to whether the bond was weak over straight lengths of the same pipe already in service.

Analysis of a large piece of zinc coating and superficial filings from the galvanized layer of the pipe revealed that there was an unusually high percentage of iron (approximately 4%) distributed evenly throughout the

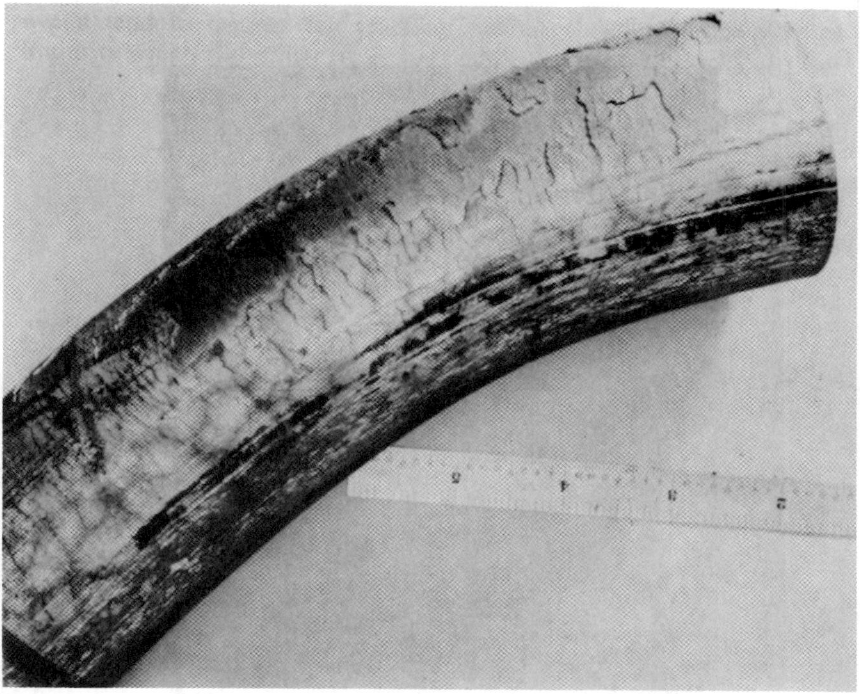

Figure 6-24 Bent section of steel pipe showing cracked and peeling galvanized coating.

galvanized layer. Apart from about 0.3% lead and 0.1% copper, no other metals were present in significant amounts. The aluminum and cadmium contents, for example, were essentially nil.

The even distribution of the iron–zinc phase (Figure 6-25) indicates that the iron was present in the zinc bath and not the result of the formation of an intermetallic compound between the iron pipe and the zinc bath during the dipping process. The heavy contamination of iron in the zinc has resulted in a very brittle coating which has broken away during bending.

Microscopic examination of sections from the straight length suggested that the steel–zinc bond was otherwise satisfactory.

C-9 b) The Cracking of Copper Pipe on Hot Bending

The problem of severe cracking of 1 inch copper piping during hot bending was brought to our attention.

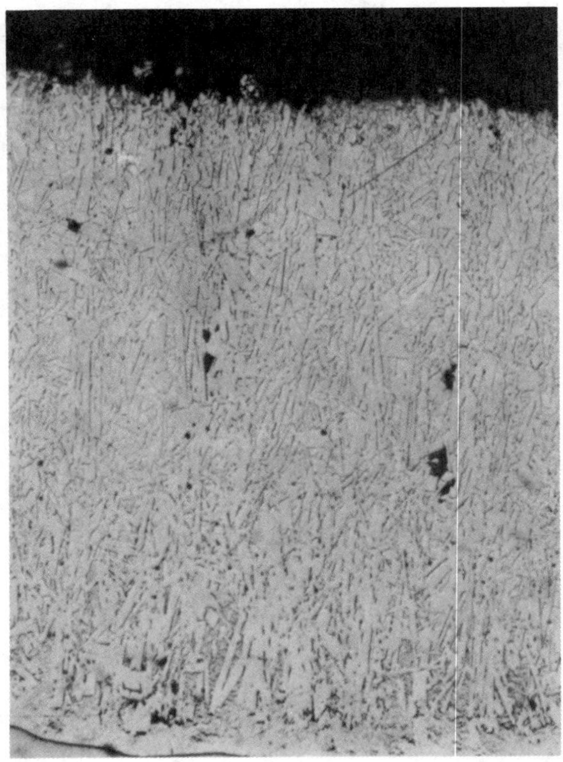

Figure 6-25 Section through galvanized coating showing even distribution of large amounts of the iron–zinc phase.

This cracking took place on only one piece of four being bent at the time. The remote possibility that this one length was overheated was checked by carefully applying the minimum amount of heat required to just bend the pipe. Cracking was still noted.

Visual examination (and examination at low magnification) showed the presence of fine cracks on both inner and outer surfaces of the copper pipe at the bend region. An example of the type of cracking is shown in Figure 6-26.

Metallurgical examination confirmed the intergranular nature of the cracks. This is shown in Figure 6-27. It is of interest to note that these cracks were apparently limited in depth, approximately 0.016 inch on the outer surface and approximately 0.012 inch on the inner surface. This feature would tend to suggest that cracking was associated with some defect or impurity which also extended to that depth.

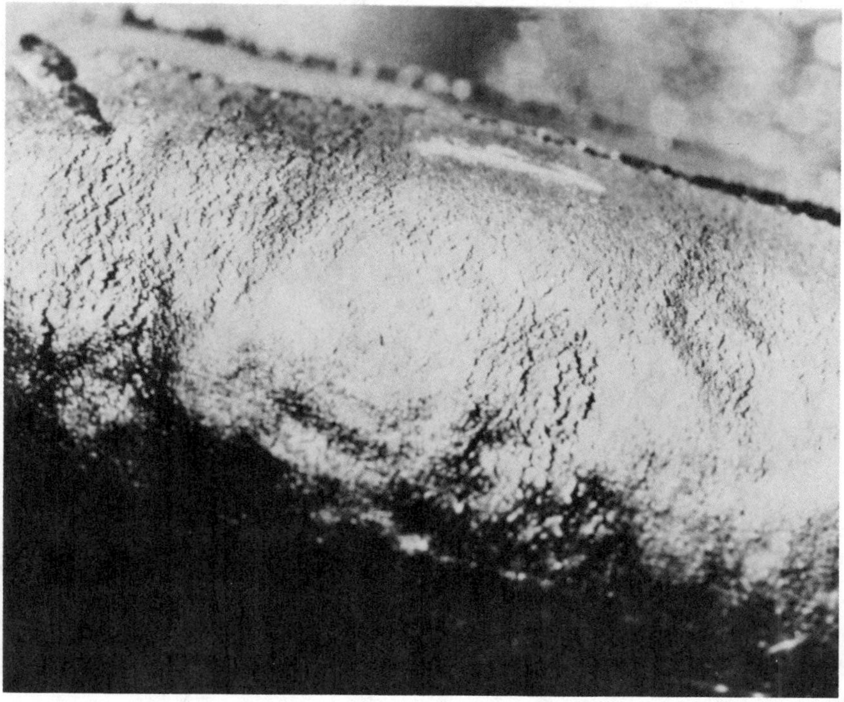

Figure 6-26 Cracking on surface of bent area of copper pipe. (×8)

Micro-examination of areas some distance removed from the bends showed the presence of intergranular oxide networks. These are shown in Figure 6-28 for an outer surface. It might be pointed out that depth of this oxide was not uniform down the length of the pipe. Generally, it was shallower on the inner surface than on the outer.

It is a known fact that hydrogen gas can penetrate heated copper and reduce any oxide particles which may be present. This reaction produces steam which is at a high enough pressure to rupture the metal.

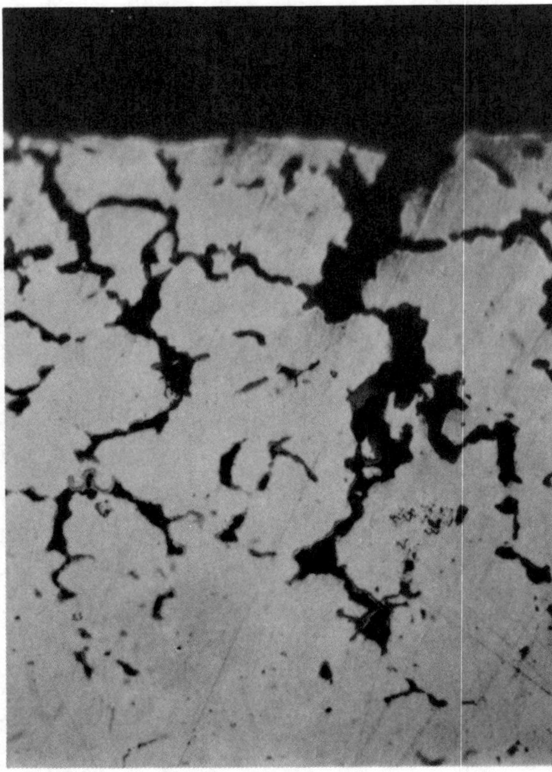

Figure 6-27 Section through copper pipe at cracked area. (×150)

Experimental hot bending in hydrogen-containing flames has resulted in cracking similar to that experienced by the Plumbing Shop; bend tests after heating in an electric furnace produced a different type of cracking—the difference between the two pointed strongly to an effect such as would be expected from internal pressure acting locally to assist the separation at grain boundaries.

It is concluded that the cracking of copper pipe as experienced locally was attributable to inferior metal, i.e., containing an injurious impurity.

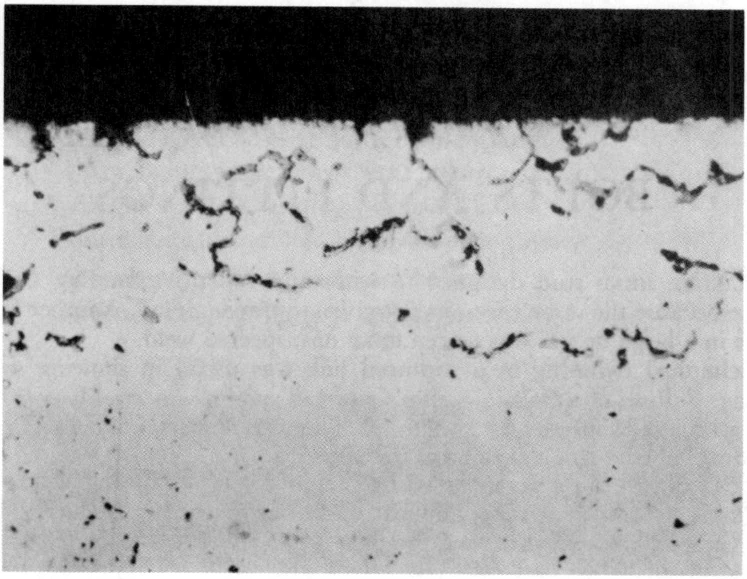

Figure 6-28 Section some distance from cracked area showing oxide network responsible for the cracking. (×150)

VII

BOLTS AND FITTINGS

A CYLINDER HEAD stud designed to withstand repetitive loading failed in fatigue because the same care was not given to thread detail. Another fatigue failure in a large tie rod was traced to an unsuspected weld.

Mechanical twinning in a ruptured link was useful in showing sudden loading. Failure of a bolt in a ship's steering system was readily attributed to poor impact properties. Seizure from inadequate lubrication led to shearing of hydraulic wing jack bolts on an aircraft.

Studs exposed in sea water failed by corrosion fatigue through inadequate protection against corrosion. Lifeboat davit pins fell apart because stressed beta brass is not resistant to ammonia type compounds. An aluminum scaffold clamp failed and others had their reliability seriously impaired by stress corrosion cracking in marine atmosphere. Many other failures by stress corrosion cracking of bolts and fittings have been experienced including those of brass and also the copper-containing aluminum alloys. Crevice corrosion is also a problem with aluminum bolts not properly inhibited and where galvanic effects can intensify the action.

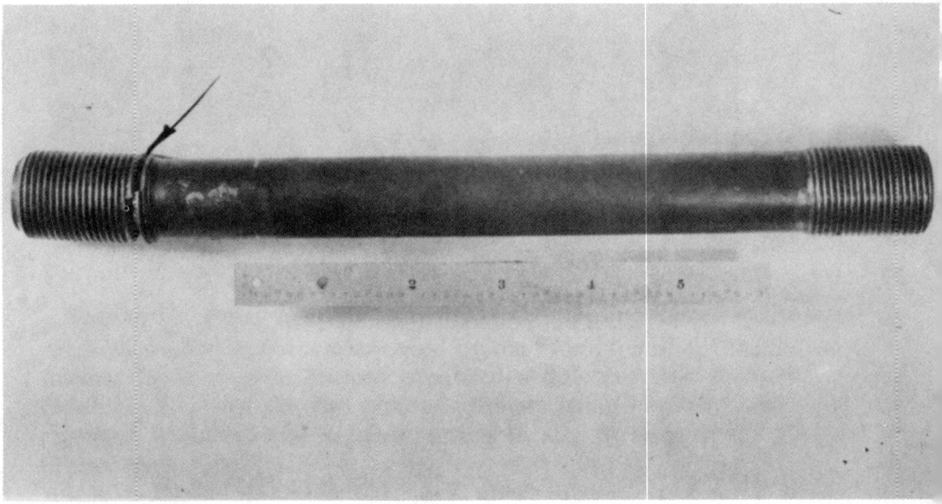

Figure 7-1 Steel stud which failed by fatigue at the first thread.

Improper heat treatment resulted in failure on impact loading of recoil cylinder bolts, and in another instance, in failure of helicopter securing bolts by "delayed fracture". Changes in weld fabrication were suggested in order to avoid loss of sonar targets through failure of the towing eye.

A-1 a) Failure of diesel holding down bolts by fatigue.
b) Fatigue failure of tie rod at weld repair.

A-2 a) Wrought iron link failure by impact loading.
b) Failure of quadrant bolt in ship's steering system.

A-4 Shearing of hydraulic wing jack bolts.

B-2 Corrosion fatigue failure of underwater studs.

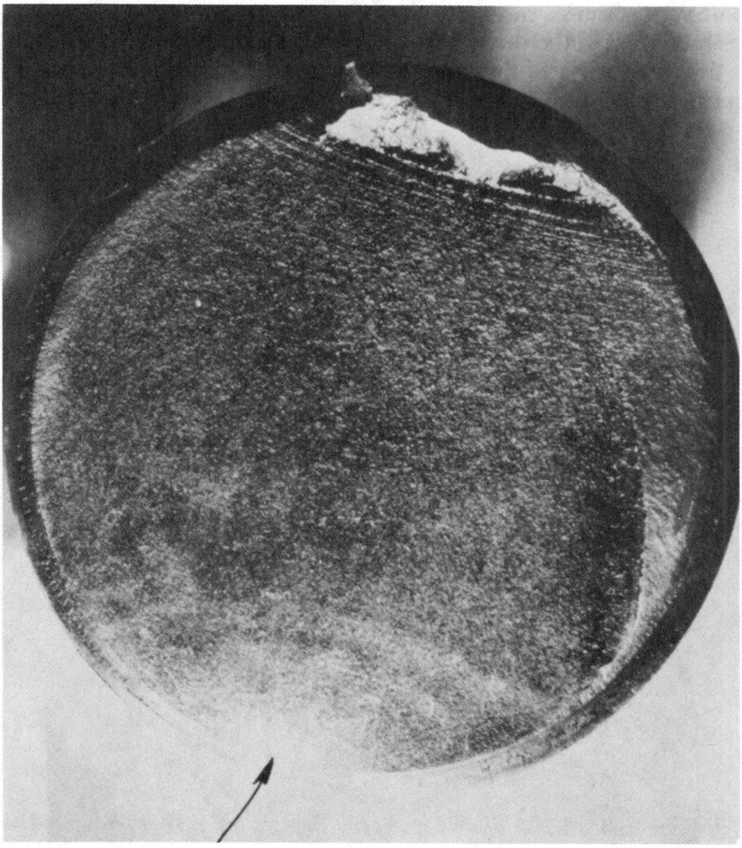

Figure 7-2 Fractured end of cylinder head stud showing fatigue markings. Fatigue initiated at arrow.

B-3 a) Lifeboat heel pin failure by stress corrosion cracking.
b) Aluminum scaffold clamp failure by stress corrosion cracking.
c) Brass securing bolt failure by stress corrosion cracking.
d) Air compressor unit adapter failure by stress corrosion cracking.

B-5 Crevice corrosion of aluminum bolts in wooden hulls.

C-5 a) Impact failure of recoil cylinder securing bolts.
b) Delayed fracture of Type 431 stainless steel bolts.

C-8 Towing eye failure of sonar target.

A-1 a) **Failure of Diesel Holding Down Bolts by Fatigue**

A cylinder head stud from a large diesel generator was submitted for examination. The stud is shown in Figure 7-1.

The stud failed in the threads as a result of metal fatigue. Figure 7-2 shows, on the fractured surface, the tell-tale markings associated with metal fatigue.

Spectrographic and metallurgical examination showed that the stud was

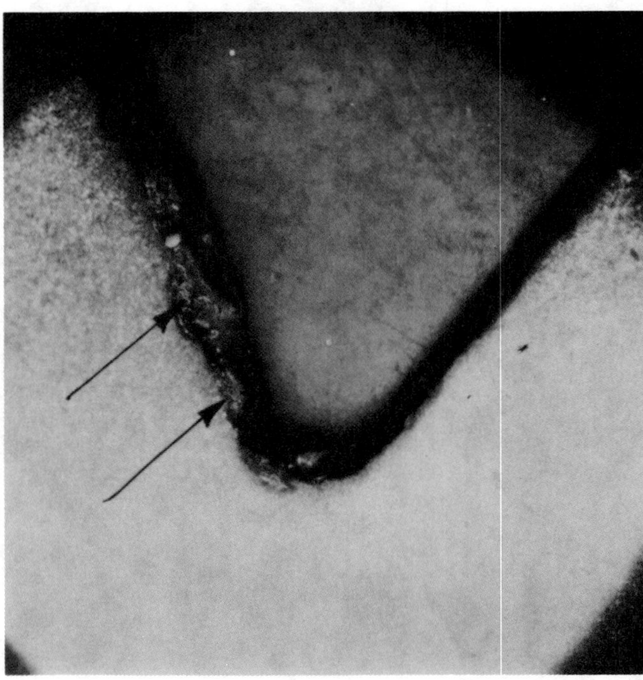

Figure 7-3 Root of thread of cylinder head stud. Note the torn metal, from machining, at arrows. (×200)

fabricated from a good quality high strength steel, AISI 4140 of approximate tensile strength of 150,000 psi.

The overall design of the stud appeared to be good, as consideration had apparently been given to the stress-raising effect of the first thread. However, examination of the thread roots indicated that the machining was rough, as shown in Figure 7-3. Also, the radius of the thread root was very small.

This was an interesting failure in that so much care was apparently taken to design a bolt to minimize fatigue effects, and yet the thread detail and machining quality were ignored.

A-1 b) **Fatigue Failure of Tie Rod at Weld Repair**

Failure of a turbo blower tie rod was brought to our attention. The failed end of the tie rod is shown in Figure 7-4. The corroded character of the surface is evident.

Figure 7-4 Failed part of turbo blower tie rod showing external corrosion attack.

Additional detail of the fracture is shown in Figure 7-5. From this the fatigue markings may be noted, and these indicate that cracking proceeded slowly from some defect on the surface.

A small piece was cut out of the rod at the rupture and examination under

Figure 7-5 View of fractured surface of tie rod showing fatigue markings.

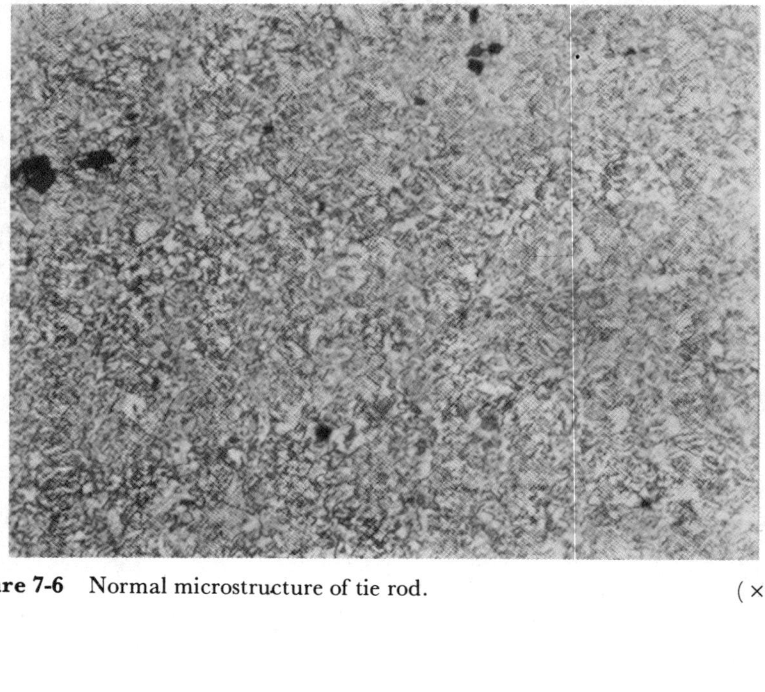

Figure 7-6 Normal microstructure of tie rod. (×230)

the microscope showed, surprisingly, that some welding had been done on the outer surface in the region of the break. The normal microstructure of the tie rod is shown in Figure 7-6 and the microstructure of the weld area is shown in Figure 7-7.

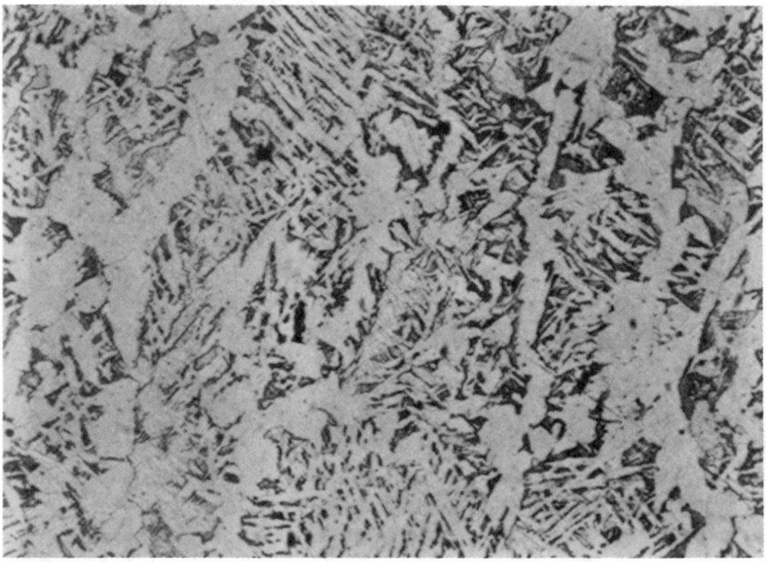

Figure 7-7 Microstructure of area at fracture. This metal is a weld deposit. (×230)

Chemical analysis confirmed the presence of weld metal of a composition containing less alloy than the rod itself. The reason for this small amount of welding was not known but it was very likely that some welding defect or crack served to initiate the fatigure failure.

The external corrosion shown in Figure 7-4 is undesirable as the resulting corrosion pits could also initiate failure. Some protective coating or plating would appear desirable.

A-2 a) **Wrought Iron Link Failure by Impact Loading**

The failure of a swivel piece from an anchor chain resulted in loss of the anchor. A part of the fractured swivel piece was recovered and was submitted for investigation.

A close-up of the fractured surface is shown in Figure 7-8. The probable initiating flaw may be noted, as well as the magnetic particle indication of the forged weld junction. A small rusted area suggested that there was a transverse crack for a time prior to failure. A cross-section cut throught the defect, Figure 7-9, showed the extent of the defect.

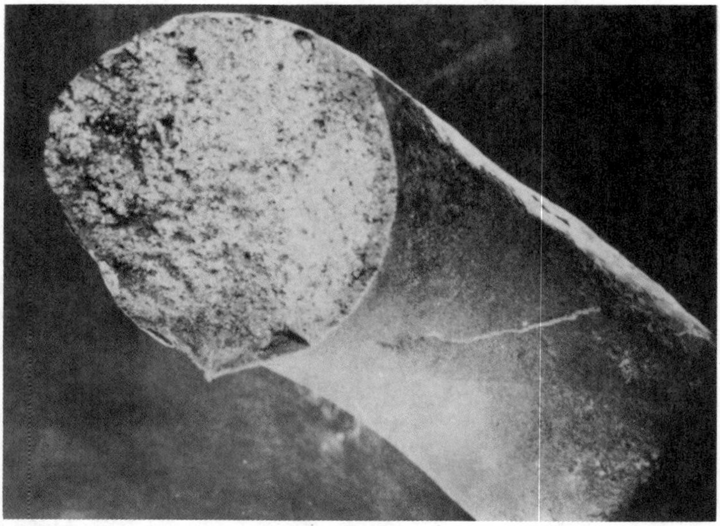

Figure 7-8 Fractured face of wrought iron link. Note the magnetic particle indication outlining the forge weld. Fracture initiated at bottom in forge weld.

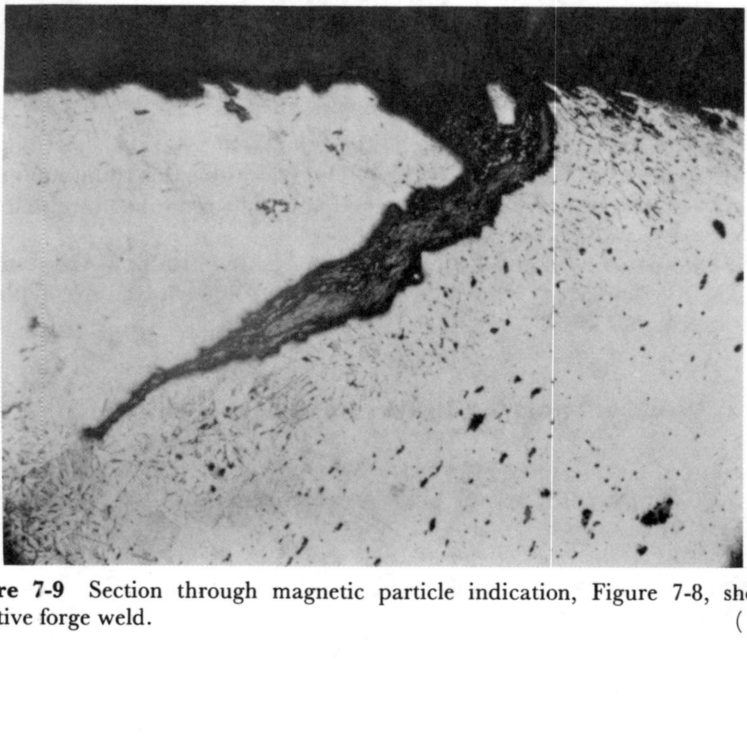

Figure 7-9 Section through magnetic particle indication, Figure 7-8, showing defective forge weld. (× 26)

Micro-examination showed the structure to be essentially a wrought iron containing an assortment of elongated slag patches. An interesting finding was the presence, close to the break, of "mechanical twins" (Figure 7-10). These represent a slight change in atomic arrangement within a grain and are caused in iron by suddenly applied loads or blows.

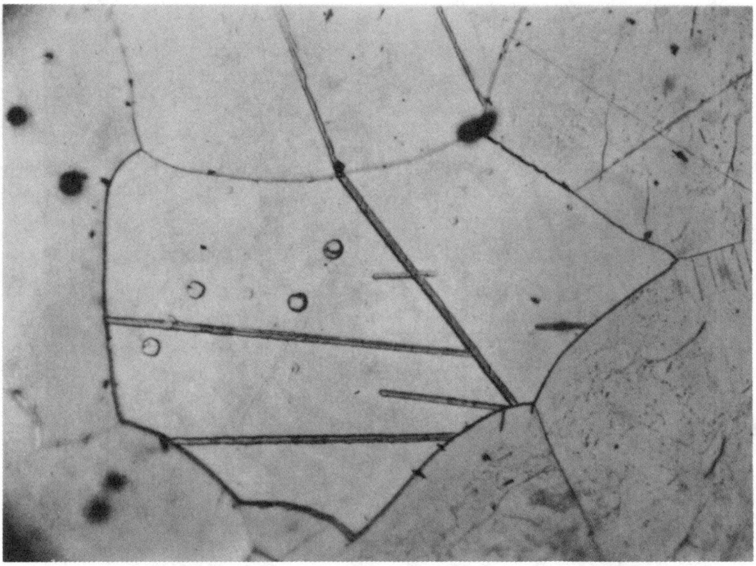

Figure 7-10 Microstructure near fracture showing mechanical twins in iron grains. (×400)

The evidence tends strongly to support the picture of a brittle failure resulting from sudden loading in the presence of a notch (the defective weld or the rusted zone).

In order to avoid this type of failure in future, it was recommended that swivel piece links be checked by magnetic particle inspection (after removal of the paint) with two objectives in mind:

a) to detect minute cracks which may have developed in excessively cold worked metal on the inner surface, and

b) to check quality of the welds.

Shallow defects could be ground out; deeper ones would result in rejection. The practice of annealing wrought iron or mild steel chain would be worthwhile to minimize the possibility of cracks being initiated in the cold worked area of a link.

A-2 b) **Failure of Quadrant Bolt in Ship's Steering System**

Failure of a bolt in the steering system almost resulted in loss of a ship in heavy seas. The failed bolt, Figure 7-11, was submitted for examination.

This fracture surface was cleaned of rust and dirt to permit visual examination. The coarse grained structure revealed after cleaning, strongly suggested brittle failure.

Figure 7-11 Failed quadrant anchor bolt from ship's steering system. Impact failure took place during storm.

As a matter of interest, the fracture may be compared with that from a winch part which failed in simple tension, i.e. overload. This tension failure is shown in Figure 7-12.

Figure 7-12 Stud, from a winch, which failed in simple tension by overloading.

The brittle steering bolt was sectioned and examined under the microscope. Examination showed an undesirable microstructure consisting of dark unresolved "pearlite" and a deleterious ferrite network, as shown in Figure 7-13.

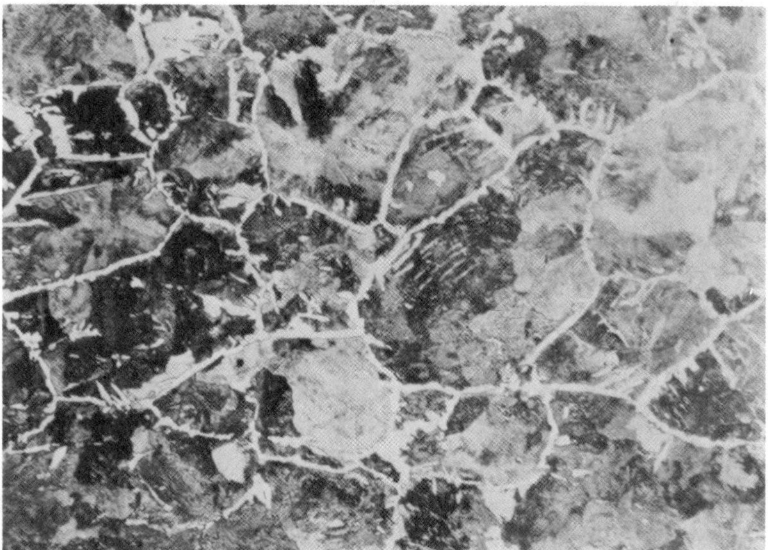

Figure 7-13 Microstructure of failed AISI 1045 anchor bolt (Figure 7-11). Ferrite network surrounds unresolved pearlite.

This steel was in fact AISI 1045, confirming the unsatisfactory intermediate quench (very little ferrite present).

In view of the nature of the fracture and the poor metallurgical structure, Charpy V-notch specimens were machined from the failed bolt. Impact strengths of 4 and 7 foot-pounds were recorded at 32°F, confirming that the steel had very poor impact properties. A standard good quality steel would have impact properties, determined by Charpy V-notch of at least 30 to 40 foot-pounds.

It was recommended that future replacements be made in a better steel, heat treated to give adequate toughness.

A-4 Shearing of Hydraulic Wing Jack Bolts

Two failed bolts from a hydraulic wing jack on an aircraft were evaluated. The failure took place when the aircraft's wings were being folded.

Both bolts failed at the first thread as shown in Figure 7-14. The fractured surfaces suggested shear under torsional loading conditions.

Figure 7-14 Sheared hydraulic wing jack bolts.

Investigation of the arrangement on the aircraft showed that a bushing surrounded the central portion of the bolt. Unless the bolt turned freely in the bushing, a torsional load was placed on it when the wings of the aircraft were folded. Inadequate lubrication appeared to be the contributing factor resulting in seizure.

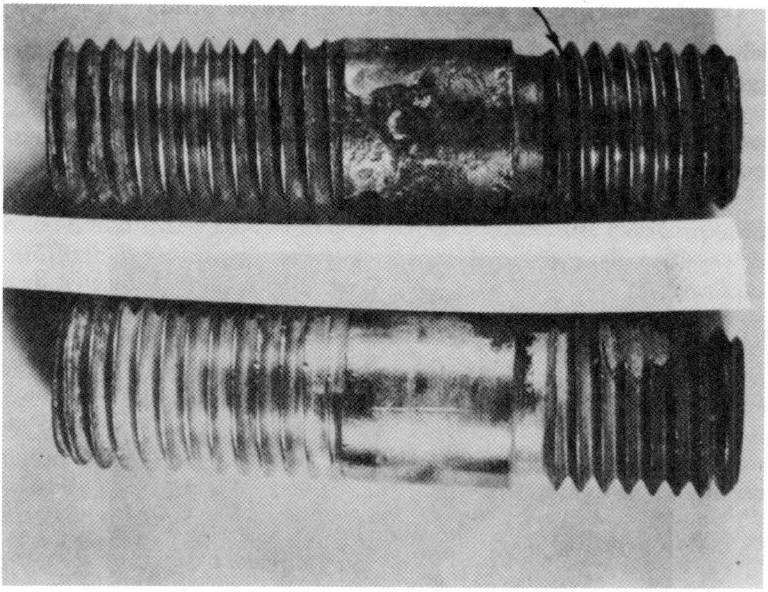

Figure 7-15 Corroded stud (upper) with arrow marking location of crack in those which failed. The lower stud was cadmium plated.

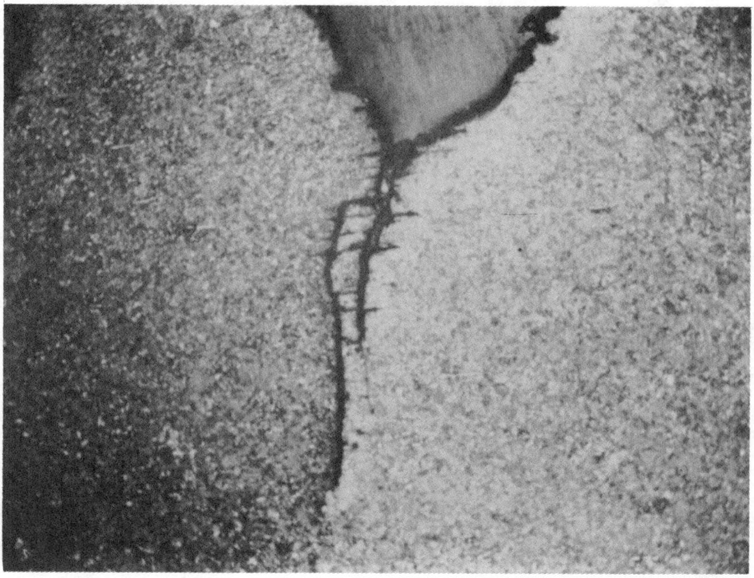

Figure 7-16 Corrosion fatigue crack at thread root. This crack was detected by magnetic particle inspection. (×22)

B-2 Corrosion Fatigue Failure of Underwater Studs

Failures of several studs which fastened an asdic dome to a warship were brought to our attention. Two of these studs which had not failed are shown in Figure 7-15.

One of these studs was brighter and free from corrosion. Analysis confirmed that it was cadmium plated, and that both studs were plain carbon

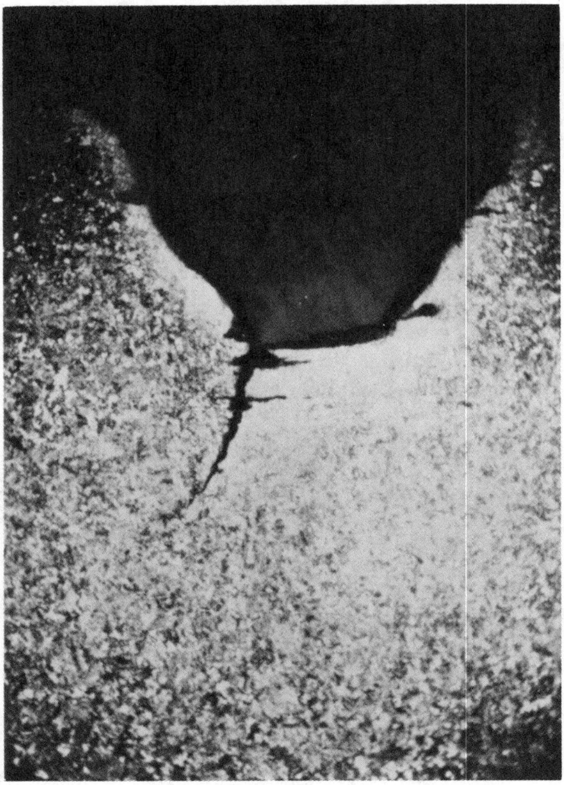

Figure 7-17 Corrosion-fatigue crack at root of thread not detected by magnetic particle inspection. (×80)

steel. The other stud was somewhat corroded externally. Magnetic particle examination revealed the presence of a fine crack at the first thread root (arrow in Figure 7–15).

Sectioning through this fine crack and examination under the microscope showed that cracking was by corrosion fatigue, Figure 7–16. It is of interest to note that only the corrosion component of the joint corrosion-mechanical action was extending along the bolt in the rolling direction whereas the mechanical component is predominant in the transverse cracking.

Examination of other thread roots indicated that additional fine cracks were present, Figure 7–17. This crack would not be readily shown by magnetic particle techniques.

It was noted that a stud with only a minor amount of cadium plating left was still free from corrosion fatigue cracks. It appears, then, that cracking took place in those studs which had not been plated. However, it was possible that, in the cases of the cracked studs, cadmium plating was originally present but had already corroded off.

Cadmium plating is usually recommended for atmospheric corrosion protection rather than underwater service. A hot dipped galvanized steel bolt might serve the purpose more satisfactorily. It should be added that higher strength alloy steel studs would have been expected for this service; however, the absence of protection against corrosion rather than low strength was the cause of the stud failures.

Figure 7-18 Two pieces of a broken heel pin. Note the diagonal cracks at right of smaller piece.

B-3 a) **Lifeboat Heel Pin Failure by Stress Corrosion Cracking**

Parts of two heel pins from a destroyer's lifeboat davits were submitted for examination. As these would be relied upon to function under extremely hazardous conditions, the apparent brittle failure of these heel pins shown in Figure 7-18 was viewed with great concern.

The failure was apparently associated with seizure resulting from periods of inactivity. Visual examination showed that no provision had been made for lubrication—there were no grease grooves and the clearance, 0.003″, was too fine. The smaller piece of Figure 7–18 shows diagonal cracks, proof of torsional stresses.

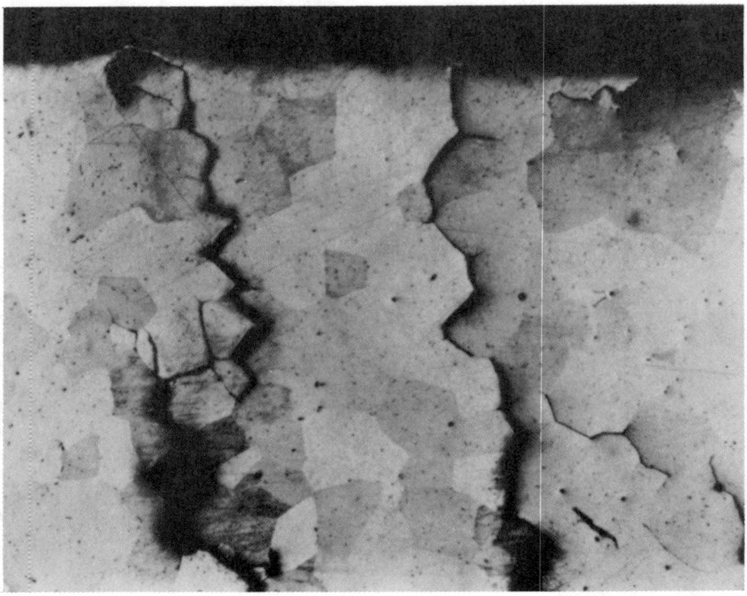

Figure 7-19 Intergranular stress corrosion cracks in high strength beta brass heel pin. (×80)

Micro-examination of pieces cut from near the fracture showed the clear cut intergranular cracking of stress corrosion, Figure 7-19.

Spectrochemical analysis of the cracked pin indicated it to be a high strength brass containing 55–60% copper, 1% aluminum, 0.4% manganese, remainder zinc.

The material was a beta brass which is one of the brasses most susceptible to stress corrosion cracking in marine environments. A marine environment includes many possible contaminants or ammonia type compounds; for example, bird droppings and decaying organic matter.

In conclusion, failures resulted when the pins seized through lack of lubrication, and were subject to torsional loads in handling of the lifeboats in the corrosive environment. Measures to ensure a constant grease supply, thereby excluding moisture and contaminants, were recommended. A better material such as a wrought tin bronze would also be an additional safeguard.

B-3 b) Aluminum Scaffold Clamp Failure by Stress Corrosion Cracking

Scaffolds are commonly used during refit work on ships for access to high areas. These scaffolds consist of aluminum pipe fastened with forged aluminum clamps. If one of these clamps parted with no prior warning lives could be lost. One such parting occurred, but fortunately only when a scaffold was being dismantled. The failed scaffold clamp is shown in Figure 7–20 and the fractured surface gave no indication as to the cause of the failure.

Metallurgical sectioning of the failed hinge of the clamp indicated that the fracture was associated with numerous cracks in the hinge. These are shown at low magnification in Figure 7–21. A higher magnification view, Figure 7–22, shows a network of fine intergranular cracks—typical of stress-corrosion.

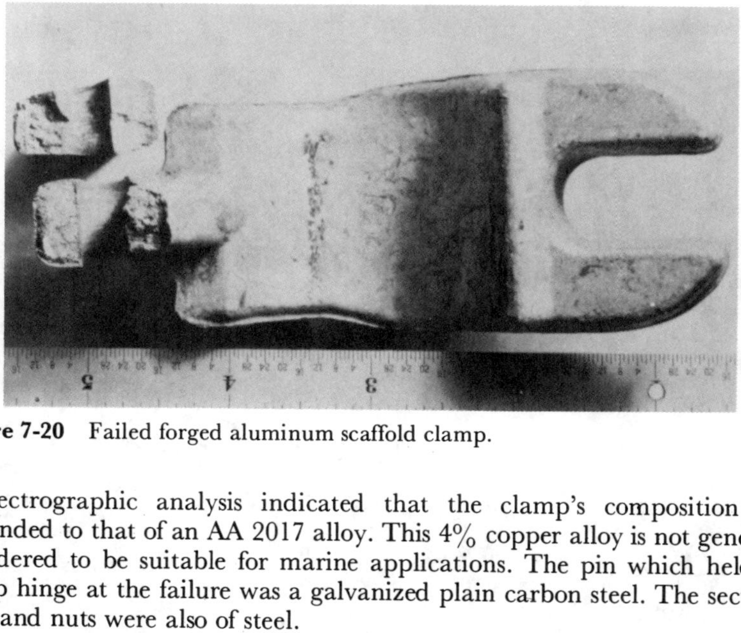

Figure 7-20 Failed forged aluminum scaffold clamp.

Spectrographic analysis indicated that the clamp's composition corresponded to that of an AA 2017 alloy. This 4% copper alloy is not generally considered to be suitable for marine applications. The pin which held the clamp hinge at the failure was a galvanized plain carbon steel. The securing bolts and nuts were also of steel.

Six additional aluminum clamps were subsequently submitted for testing purposes. An examination revealed no evidence of corrosion in any of the

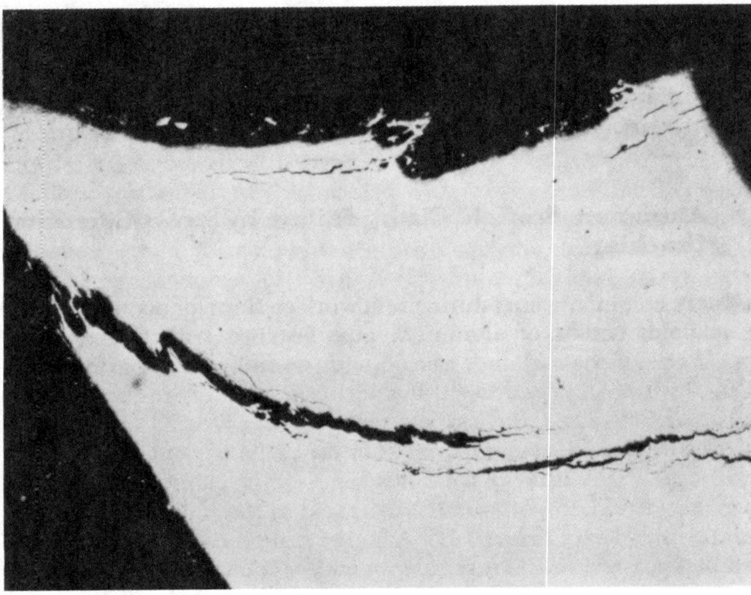

Figure 7-21 Section through failed aluminum scaffold clamp showing stress corrosion cracking. (×10)

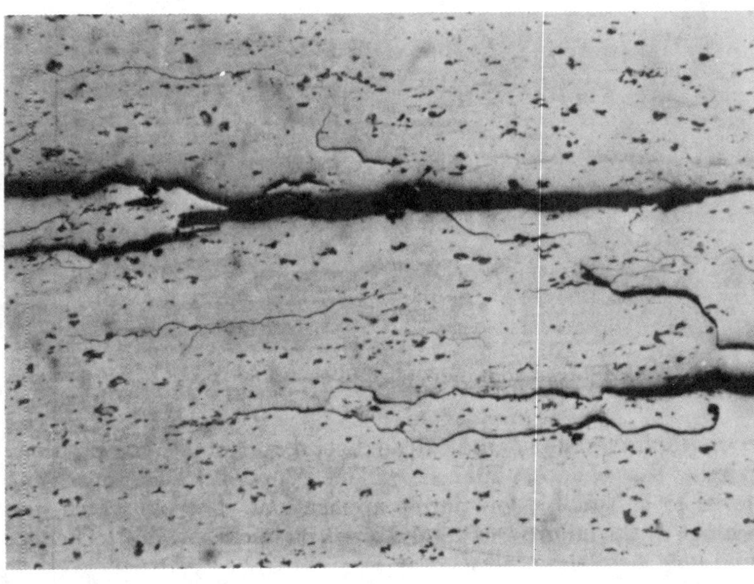

Figure 7-22 Microstructure of aluminum scaffold clamp at failure showing intergranular stress corrosion cracks. (×80)

clamps. The samples were subsequently clamped onto a jetty, where they were subjected to salt spray and occasional splashing by the sea water.

Two clamps were removed after six months. The surfaces of the clamps were weathered and the once-galvanized steel pins and bolts showed signs of rusting. (No rusting was noted in the case of the original failure). Sectioning of the clamp hinge indicated stress-corrosion cracking from the inside of the hinge. The cracks shown in Figure 7–23 extended half way through the forging and are identical to those of the original failure.

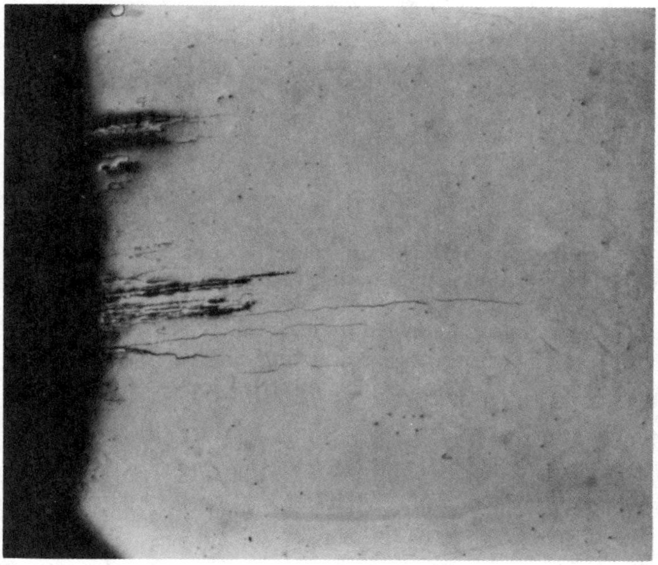

Figure 7-23 Stress corrosion cracks in scaffold clamp hinge after six months exposure to sea spray. (× 10)

It was recommended that the aluminum clamps be replaced with the more reliable, less expensive, galvanized forged steel clamps.

B-3 c) **Brass Securing Bolts Failure by Stress Corrosion Cracking**

The failure of four $\frac{1}{2}$ inch diameter radar antenna securing bolts was brought to our attention. Two of the bolts shown in Figure 7–24, were examined.

Spectrographic analysis showed that the bolts and nuts were manufactured from a free cutting brass (61% Cu, 2.7% Pb, remainder Zn). Chemical analysis confirmed that the bolts were in contact with aluminum—the white product in the bolt threads was a hydrated aluminum oxide.

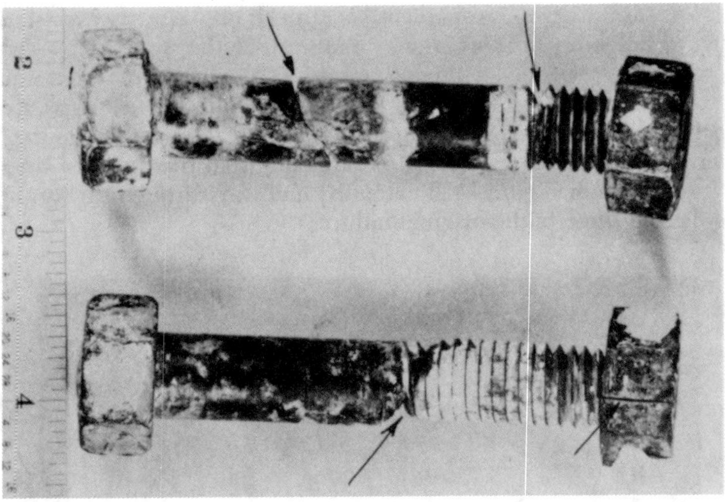

Figure 7-24 Stress-corrosion cracking failure of brass securing bolts. The white material on the bolts is aluminum corrosion product.

Microscopic examination showed a crack extending into the bolt from the fractured surface. The intergranular cracking nature of the attack as shown in Figure 7–25 confirmed that failure was by stress corrosion cracking.

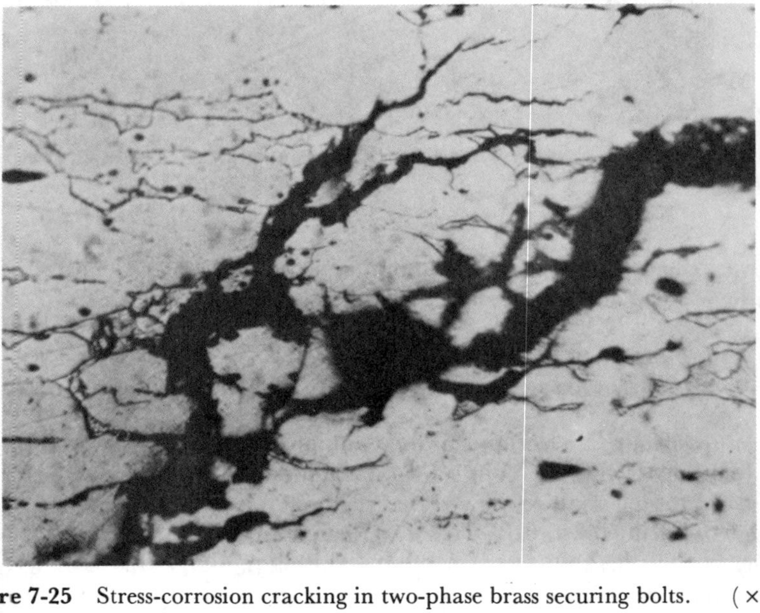

Figure 7-25 Stress-corrosion cracking in two-phase brass securing bolts. (×200)

While it might have been expected that aluminum would cathodically protect the brass bolts—in fact, the aluminum corrosion products would soon electrically isolate the two metals. Galvanized steel bolts would have been a better choice for this service.

B-3 d) Air Compressor Unit Adapter Failure by Stress Corrosion Cracking

An examination of an adapter nut from an air compressor was undertaken when cracks were noticed. The cracked aluminum nut is shown in Figure 7–26.

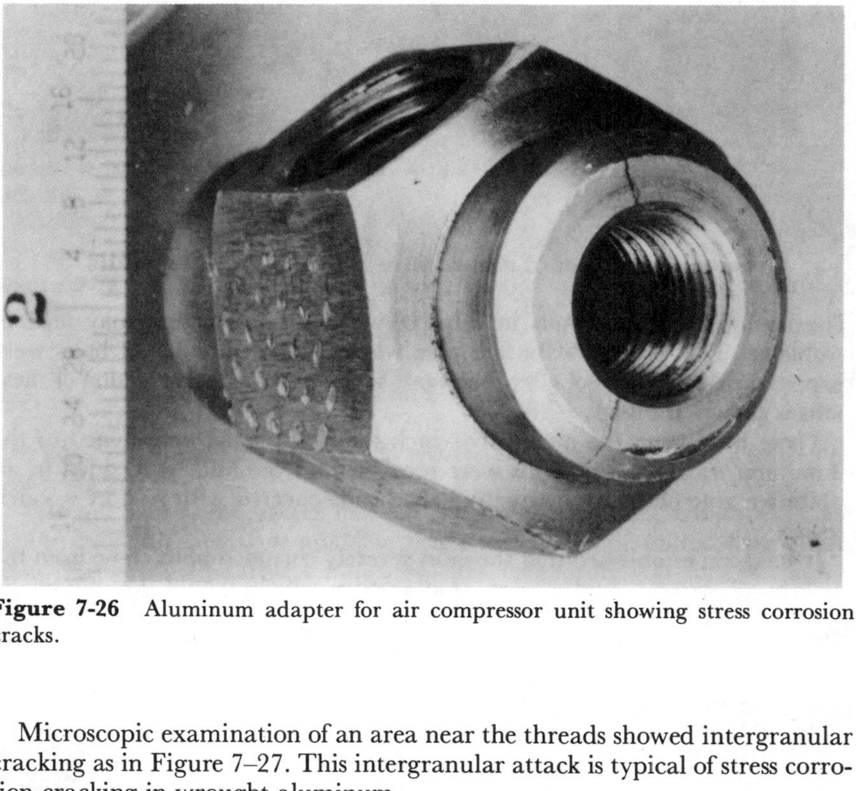

Figure 7-26 Aluminum adapter for air compressor unit showing stress corrosion cracks.

Microscopic examination of an area near the threads showed intergranular cracking as in Figure 7–27. This intergranular attack is typical of stress corrosion cracking in wrought aluminum.

Spectrographic analysis indicated that the alloy was AA 2024. This 4% copper alloy is subject to stress corrosion cracking in a marine environment and should be avoided.

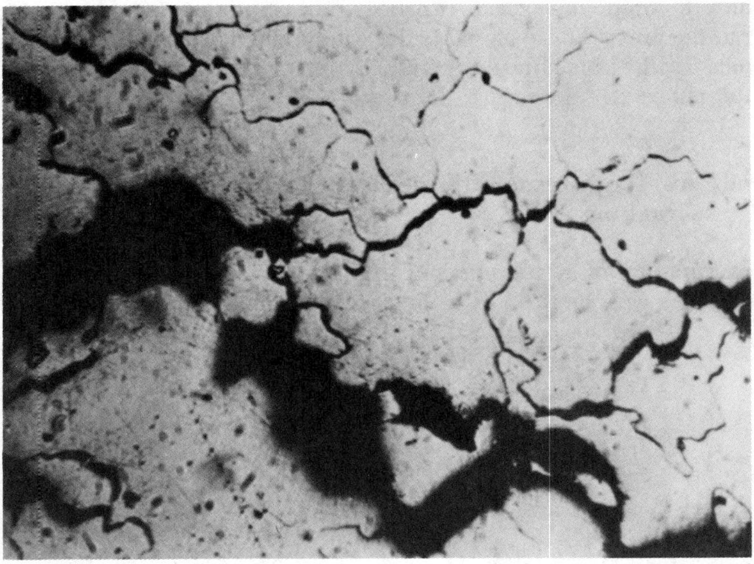

Figure 7-27 Stress corrosion cracking in AA 2024 aluminum alloy compressor adapter. (× 250)

B-5 Crevice Corrosion of Aluminum Bolts in Wooden Hulls

The use of aluminum bolts in ships of wooden construction may lead to problems. This proved to be the case when a large number of bolts were inspected on such a vessel after four years service. A random selection of these bolts is shown in Figure 7–28.

These bolts were the means by which the hull planks were joined to the aluminum framing. The bolts were recessed into the hull, embedded in an inhibitive zinc chromate paste, and the heads covered with a short wooden dowel.

It was soon established that the most severely corroded bolts came from the vicinity of bronze underwater components which would be physically coupled to these bolts.

These bolts were made of alloy AA 6061 which is not particularly resistant to sea water, but it was assumed that the inhibitive paste would serve to protect them. Unfortunately, except for an occasional bolt, the inhibitive material dried out, cracked and allowed moisture to penetrate. The hardened material was apparently insufficiently active to provide the inhibitive function.

It was recommended that for replacement and for future construction the bolts should be bedded in a non-hardening type sealant. In addition, efforts were subsequently made to provide a more durable coating on the bronze underwater components, to minimize galvanic effects.

Figure 7-28 Aluminum hull-bolts showing various amounts of crevice corrosion attack.

C-5 a) Impact Failure of Recoil Cylinder Securing Bolts

The failure of one of four bolts which secured the recoil cylinder of a large gun was reported.

The gun fires 90 rounds per minute, each round resulting in a tensile stress on the bolts. Although one might suspect a fatigue failure under these conditions, the ruptured surfaces indicated that the failure was due to sudden loading. Figure 7–29, a photograph of the ruptured surfaces of one of the bolts, shows that the brittle failure initiated at the root of a thread. Note that fatigue markings are absent.

Time did not permit a study of the design arrangement nor of the stress configuration in the part of the gun concerned with the these bolts. However, a metallurgical examination was made on the failed bolts and on three additional unused bolts. Microscopic examination indicated that the failed bolts were a tempered martensite, Rc 40, and that the roots of the threads were well rounded.

The bolts were shown, by spectrographic analysis, to be a zinc plated low alloy steel, containing about 0.3% Cr, 0.4% Ni and 0.15% Mo.

In view of the type of loading to which these bolts were subject: that is, sudden or impact, it was believed that the ability of the bolts to resist the loading could best be determined by testing their Charpy V-notch impact strength.

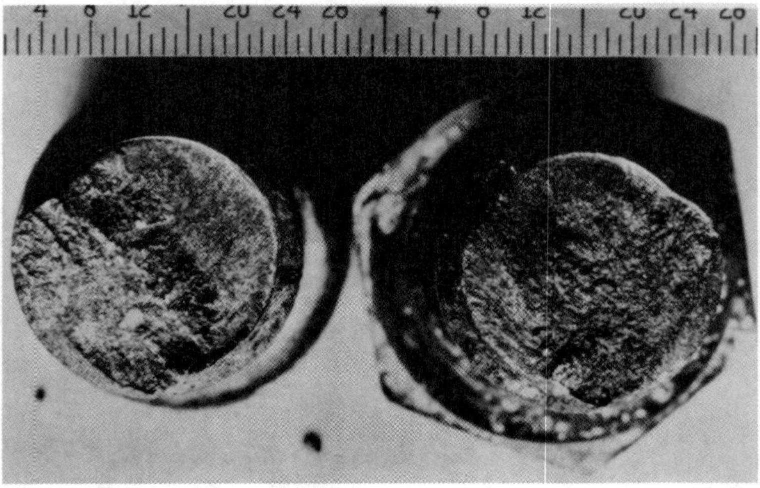

Figure 7-29 Brittle failure of recoil cylinder bolt, which initiated at the root of a thread.

Tabulated below are the results of hardness and impact tests* made on specimens machined from bolts in the as received and heat treated conditions. The tensile strengths are approximate, based on the hardness of the samples.

Specimen	Treatment	Hardness	Tensile Strength	Charpy Impact Strength
1	As received	Rc 40	180,000 psi	17 ft.lb.
2	400°F bake (5 hr)	Rc 40	180,000 psi	17 ft.lb.
3	1000°F temper (1 hr)	Rc 33	150,000 psi	47 ft.lb.

Specimen 1 shows that the Charpy impact strength of the as received bolt was relatively low for the type of service to which the bolt was subjected. The second specimen was baked at 400 °F for 5 hours. If the specimens had been significantly embrittled by hydrogen, this treatment would have removed the hydrogen and increased the impact strength of the bolt.† No

* Not suited for low level hydrogen embrittlement.

† E. R. Parker, *Brittle behavior of Engineering Structures*, John Wiley and Sons Inc., 1957 p. 194.

increase in impact strength was noted. The third specimen, tempered to Rc 33, showed a small decrease in tensile strength; however, the impact strength of the bolt was increased markedly.

Since the securing bolts failed in a brittle manner under sudden loading, it appeared that bolts of higher impact strength should be used. This could be achieved by tempering the present bolts at 1000°F for one hour followed by an air cool. This treatment would destroy the zinc coating on the bolts. These could, however, be galvanized again or metal sprayed.

It was suggested that future replacement bolts be manufactured from a high strength steel, such as AISI 4340, properly heat treated to produce toughness, that is, one with a Charpy impact strength of about 50 ft.lbs.

C-5 b) **Delayed Fracture of Type 431 Stainless Steel Bolts**

The failure of a number of $\frac{7}{8}$ inch diameter helicopter tract securing bolts was brought to our attention. It was understood that the bolts were satisfactorily torqued in place and were later found ruptured even though additional stresses were not applied.

Spectrographic analysis confirmed that the bolts were manufactured from a Type 431 stainless steel as specified.

Metallurgical examination showed that the bolt heads were forged and that they had been hardened by the subsequent air cool. The structure in the

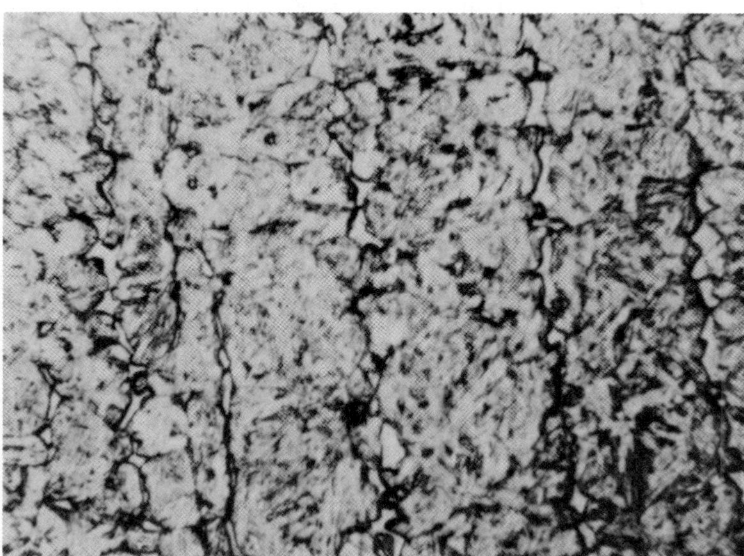

Figure 7-30 Hardened structure of forged steel securing bolt head and neck. (× 500)

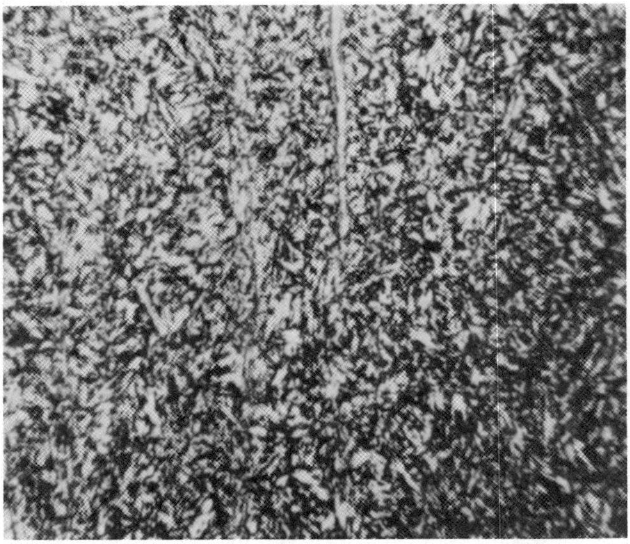

Figure 7-31 Tempered martensite structure of bolt shank away from hardened head. (×500)

bolt head (Figure 7–30) was hard, Rockwell C48, whereas the original structure of the bolt shank (Figure 7–31) was considerably softer, Rockwell C21. It was in the hard brittle metal that failure took place.

The drawing specified bolt strength of 125,000 psi, and the following heat treatment:

"The steel shall be oil quenched from a temperature of 1800° to 1850°F after a soaking period of a half hour at temperature and shall be tempered at not less than 1100°F."

It was apparent that the bolts were not heat treated in this way after forging.

Examination of the bolts also showed that a sharp corner had been machined into the neck of the bolt. Although a generous radius was shown at this location in the drawing, a radius was not actually specified. It is believed that the stress-raising effect of the corner also contributed toward the failure.

The failure mechanism appears to be that of "delayed fracture" which is associated with the hydrogen embrittlement of high strength steels. The heat treatment indicated above should result in the elimination of this problem. A generous radius at the neck of the bolts should also be specified for new bolts.

C-8 **Towing Eye Failure of Sonar Target**

The loss of a sonar target because of the failure of the towing eye shaft was reported by the operators. The bolt had broken about four inches from the eye with the subsequent loss of the target. Only the eye was therefore available for examination.

The shaft failure appeared to have initiated at one edge as shown in Figure 7–32. Microscopic examination of the steel at this edge showed that the microstructure of the parent metal had been heat affected (Figure 7–33). This heat affected metal was associated with a weld (the joint of the shaft and a support bracket) about four inches from the eye.

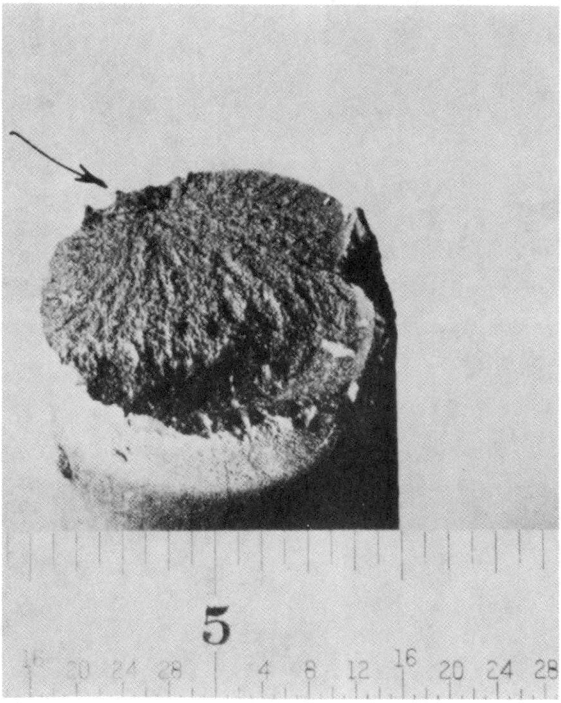

Figure 7-32 Fractured surface of failed shaft. Arrow points to where fracture initiated.

Subsequent examination of a complete sonar target confirmed the above indications. Figure 7–34 shows the beginning of a crack at a support bracket weld (the identical location of the other failure). It is noteworthy that the towing shaft had been subjected to loads large enough to bend the shaft. Under these conditions the maximum stress is found at the edge of the support bracket welds.

The authorities were warned that future sonar target losses could be expected unless the design or nature of fabrication of the towing shaft assembly is

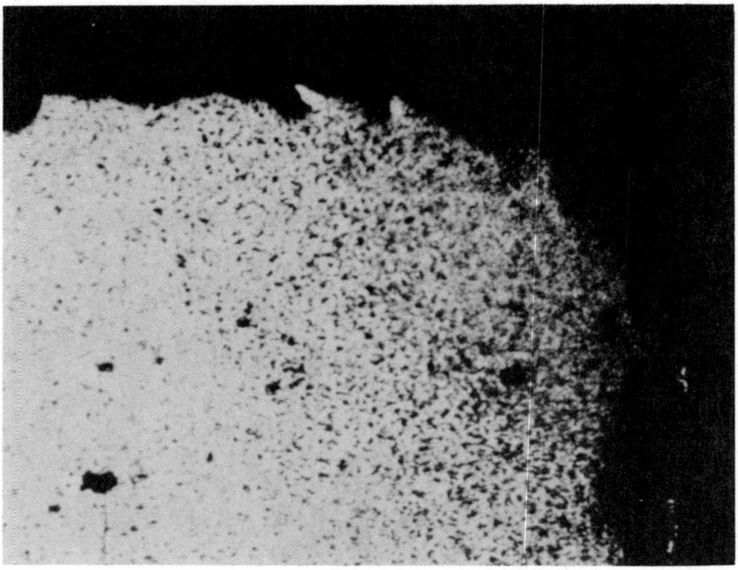

Figure 7-33 Section through initiating site of shaft failure (Figure 7-32) showing heat affected metal at right. (×50)

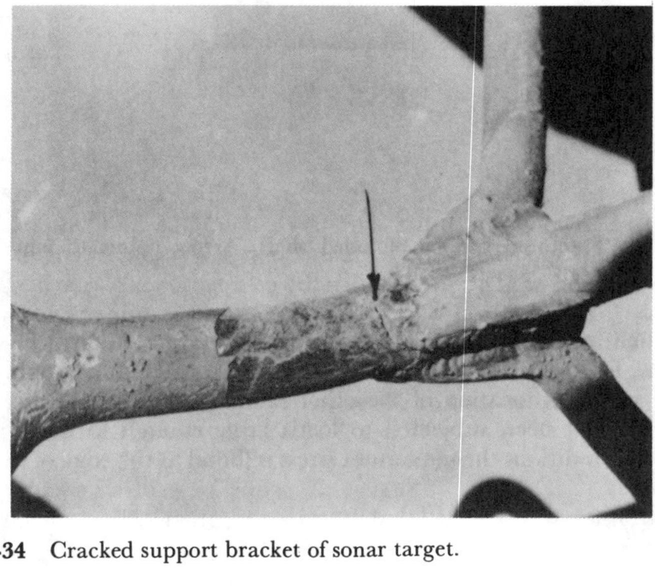

Figure 7-34 Cracked support bracket of sonar target.

altered. For example, the stress level at the edge of the weld under bending loads could be reduced by eliminating the four inch shaft length between the supports and the eye. Also, offset welding of supports onto the shaft would minimize deleterious effects of welding.

VIII

WIRE ROPES

WE HAVE fortunately not been exposed to many problems with wire ropes. The few here described demonstrate specific problems but with wider application. Inadvertent kinking can seriously weaken a wire rope and impair its reliability. Aluminum wire rope, plastic coated, is sensitive to weakening by crevice action if moisture enters at a coating break. Defects in brazing may propagate into sound steel cable giving unexpected failure.

A-6 Steel wire rope failure at kink.

B-5 Corrosion problem with polythene covered guard rails.

C-3 Failure of wire cable at braze.

A-6 **Steel Wire Rope Failure at Kink**

Unexpected failure of a $\frac{5}{16}$ inch galvanized wire rope used on a minesweeping winch was brought to our attention. The rope reportedly failed with a load estimated at approximately 3000 lbs (rated capacity over 5000 lbs), barely missing a deckhand.

An examination of the failed ends of the rope revealed nothing which indicated an inherent weakness in the rope. The individual wires had broken in a normal ductile manner. No signs of wear, corrosion or fatigue were observed at any location on the length of rope submitted.

Two tensile specimens, made from the rope supplied, showed strengths of 5200 and 5300 lbs., the strength expected for this rope when in good condition.

There is, however, evidence that the rope broke at a kink. Instead of all the parted wires being stretched in the direction of the axis of the rope, some of the wires shown in Figure 8–1, have been permanently deformed across the rope. As a result of this observation, a sample of rope was clamped into a loop and tested in tension. A number of failed wires were again noted to have permanently deformed across the rope in a manner similar to those seen in the photograph.

From the available evidence it appeared, then, that the rope failed not because it was corroded or worn, but rather because it had been kinked previous to or at the time of actual failure.

Figure 8-1 Parted end of 5/16 inch wire rope. Note that one strand has been permanently deformed across the rope.

B-5 Corrosion Problem with Polythene-covered Guard Rails

As a possible alternative to galvanized steel wire rope for use as guard rails on ships, a plastic coated aluminum wire rope was used for a period of time. These proved to be reasonably effective except that where the guard rail coating was damaged or worn, a potential weakening of the aluminum developed.

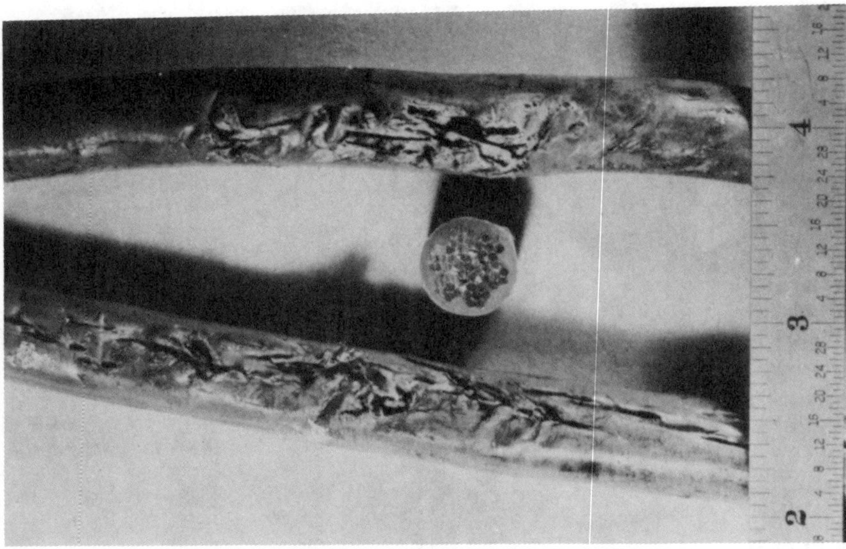

Figure 8-2 Damaged polythene-covered aluminum guard rail wire rope.

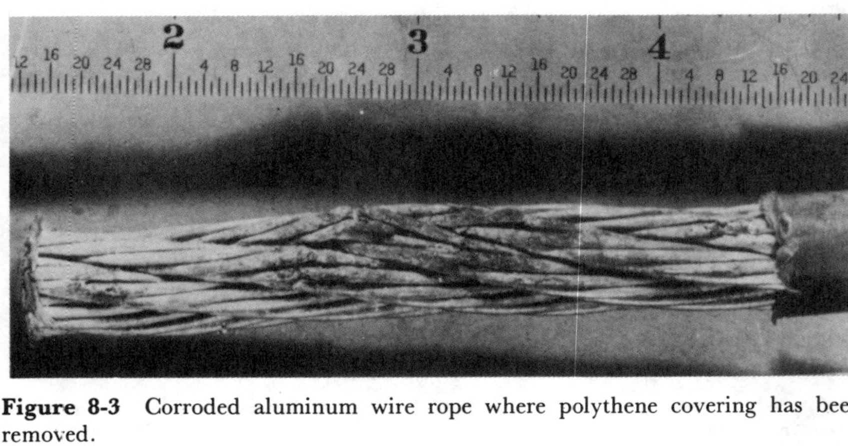

Figure 8-3 Corroded aluminum wire rope where polythene covering has been removed.

The ready separation of the polythene covering may be noted in Figure 8–2 as well as the breaks in some of the wires within. This is shown more clearly in Figure 8–3 with the covering removed from the same area as in Figure 8–2. The beginnings of corrosion attack at the wear-through areas may also be noted in Figure 8–4. Moisture can enter and remains trapped, permitting corrosive action to continue by a crevice mechanism.

As these photographs show, corrosion to the extent of rupture of individual

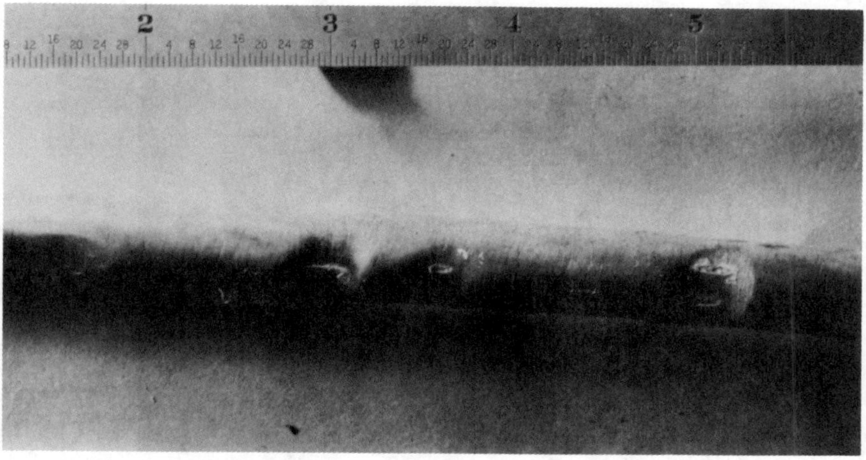

Figure 8-4 Abrasion of polythene covering at stanchions.

wires appears to occur once the polythene cover is broached. The potential weakening, once this occurs, is apparent and reliability is endangered.

Among the various alternatives, the following were suggested:

a) Present system but subject to periodic rigorous inspection. Openings found, regardless of how small they may be, must be opened up, dried and repaired with plastic tape.

b) Galvanized steel wire rope coated with polythene. When corrosion has persisted to the point where the steel is deteriorating, ample warning would be given by the rust staining. The strength of the steel would permit a larger margin of safety and it is less likely that a group of wires would corrode through.

C-3 Failure of Wire Cable at Braze

A valuable piece of deep-sea electronic apparatus was lost because of failure of a steel cable. The end of this cable was available for examination, Figure 8–5.

It may be noted that fractured ends of individual wires showed no evidence of necking, i.e. no ductility. The apparent brittleness of the strands was explained by a microscopic examination which showed that cracks in the brazing alloy had propagated into the steel wire at a number of locations (Figure 8–6).

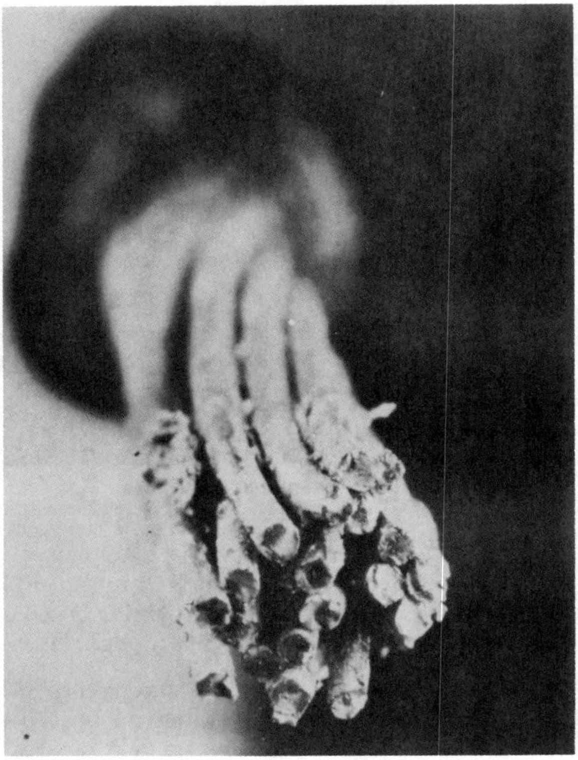

Figure 8-5 Failed end of steel cable. Note that the wires have shown little ductility in failing.

Subsequent examination of a brazed but not stressed cable confirmed that the cracks were not introduced during fabrication. It appears that the cracks in the brazing, which ultimately produced cracks in the wire, resulted from flexural stresses. Stresses due to flexing were particularly bad because the silver brazing was at the junction between the flexible wire and the rigid cable connector.

It was interesting that the tensile strength of the brazed joint as fabricated was over 1000 pounds and the lost apparatus weighed less than 100 pounds.

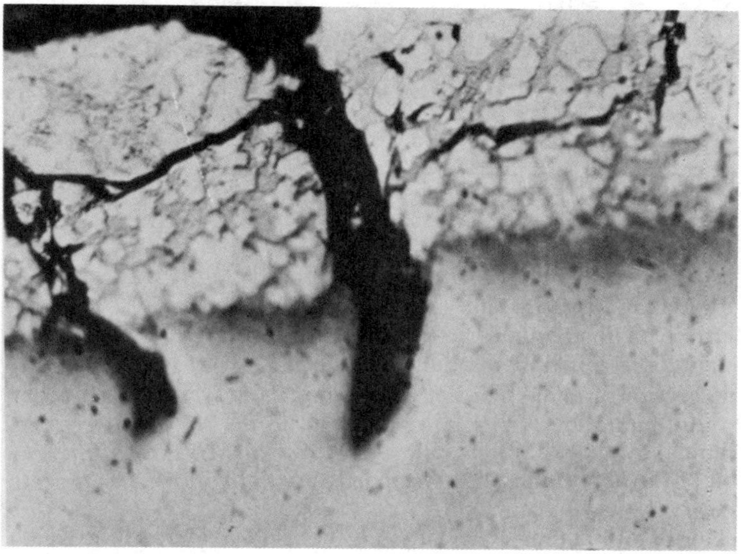

Figure 8-6 Section through wire rope near failure. The cracks in the silver solder (top) have propagated into the steel wire. (×400)

IX

STRUCTURAL AND FITTINGS

An unusually poor quality weld readily proved that it was inadequate for a ship's radar mast. Corrosion fatigue cracking developed in a small ship's hull because of inadequate structural strength. Internal corrosion led to corrosion fatigue cracking in a destroyer's keel.

Disintegration of a stanchion and of a conduit cable was the result of exposure of two aluminum alloys which are not suited to marine exposure: the 10% magnesium alloy and 4% copper alloy. Failure of an antenna by intergranular corrosion of the aluminum alloy was partly attributed to sea water retention by a sleeve.

Cracking of bronze flanges was attributed to negligence by the manufacturer in accepting castings with considerable shrinkage.

A-1 Ship's aluminum mast failure by fatigue at a weld.

B-2 a) Corrosion fatigue crack in ship's hull.
b) Corrosion fatigue cracking of ship's keel.
c) Corrosion fatigue failure of sailing craft centreboard.

B-3 a) Aluminum stanchion disintegration by stress corrosion cracking.
b) Electrical conduit failure by stress corrosion cracking.

B-4 Whip antenna failure by intergranular corrosion.

C-1 Waveguide flanges—a casting problem.

A-1 Ship's Aluminum Mast Failure by Fatigue at a Weld

Failure of the upper 20 feet of a ship's radar mast occurred in catastrophic fashion in moderate seas. It was fortunate that no one was killed considering the size of the mast and associated fittings.

The mast was returned and the failure was examined. Examination revealed, as in Figure 9–1 several features:

a) Severe incomplete penetration. The weld was not much more than two-thirds of the wall thickness.

b) Severe and extensive porosity.

c) Fatigue markings.

Figure 9-1 Fractured face of failed aluminum mast showing incomplete penetration of weld, porosity in weld, and fatigue markings.

There was no question, then, of a poor weld because of porosity and the more serious incomplete penetration which provided the stress concentrating notch for the initiation of the fatigue failure.

Subsequent to this, several other masts of similar construction were radiographed. These also showed porosity and incomplete penetration. Timely repairs prevented this dangerous failure from recurring.

B-2 a) **Corrosion Fatigue Crack in Ship's Hull**

During a hull survey of a small riveted steel vessel an unusual crack indication in the paint film was noted. The laboratory was called in to establish the nature of the cracking.

Visual examination after removal of the paint film showed a fine cracking between rivets. This was confirmed by magnetic particle examination as shown in Figure 9-2.

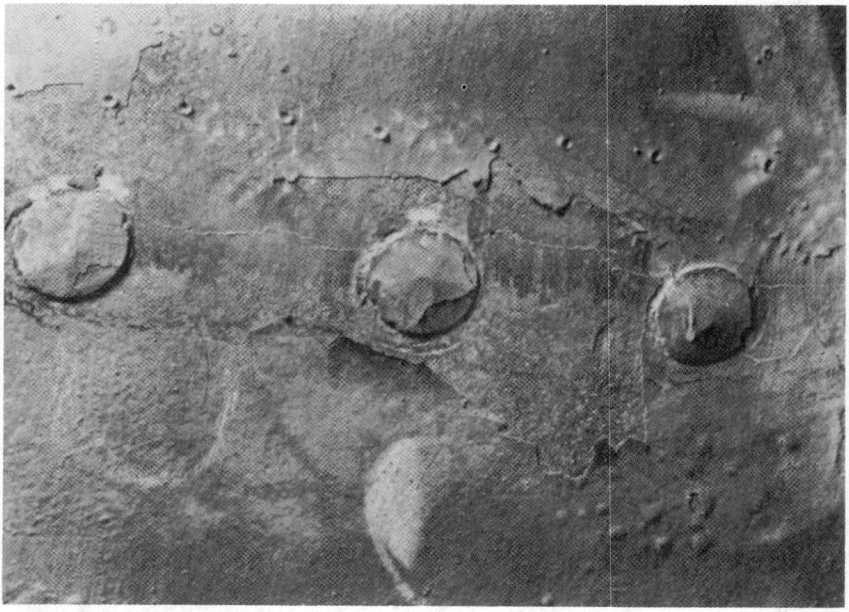

Figure 9-2 Corrosion fatigue cracks which initiated in the ship's bilge, in steel hull. Cracks are outlined by a magnetic particle inspection.

Although it was not possible at the time to obtain a section for closer examination, it was suspected that these cracks were the result of corrosion acting together with mechanical stressing.

Two remedial measures were suggested:

a) The use of cathodic protection to limit the corrosive effect of brackish water in the bilge, and

b) stiffening of this area, as it was known to be inadequately supported.

B-2 b) **Corrosion Fatigue Cracking of Ship's Keel**

Cracking of the keel near the fore-peak on a destroyer was reported. A section of the keel was subsequently submitted for examination. Figure 9-3 shows the crack as it appeared on the outside.

Figure 9-3 Bottom of ship's keel showing corrosion fatigue crack which initiated inside.

The inside of the section shown in Figure 9-4, indicated that the crack extended through to the inside and was alongside a weld. The interior was also severely corroded.

Metallurgical examination showed that the crack initiated inside the keel

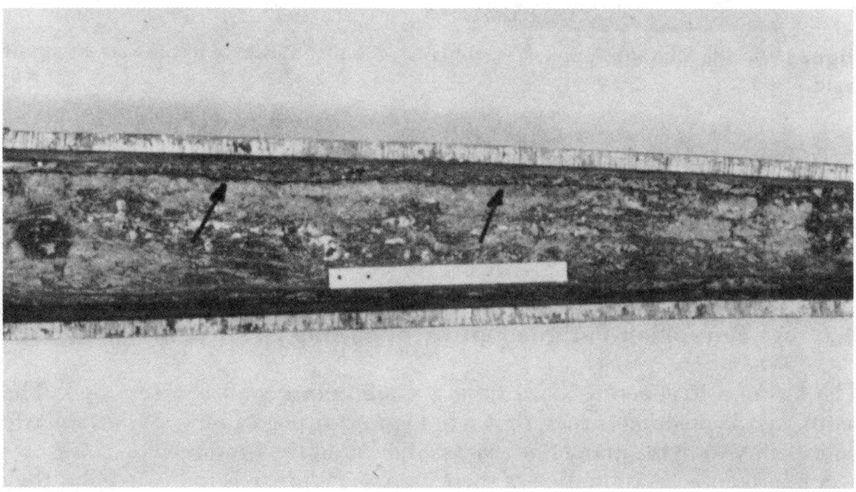

Figure 9-4 Inside of ship's keel, Figure 9-3, showing corrosion fatigue crack alongside weld.

at the weld (Figure 9-5). Corrosion products in the crack confirmed that the failure mechanism was corrosion-fatigue.

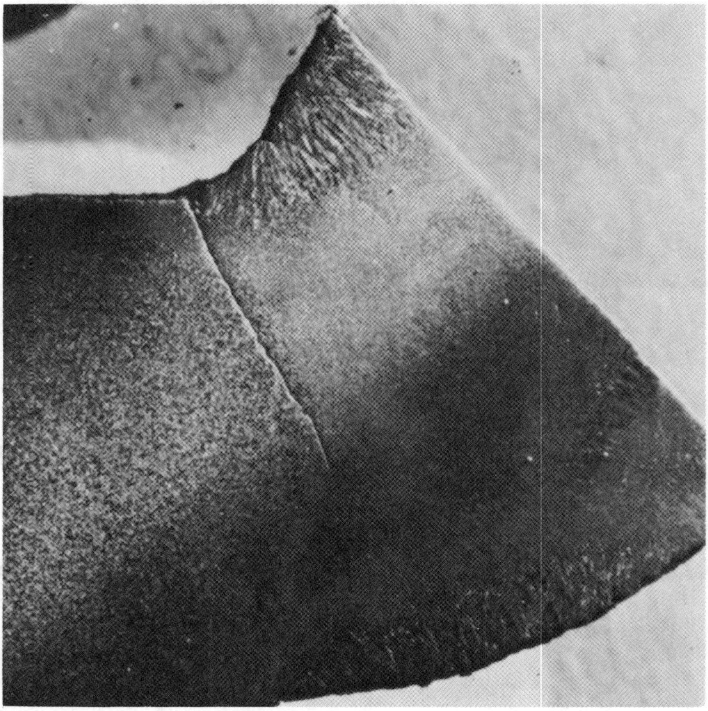

Figure 9-5 Section through keel showing crack which initiated inside keel alongside weld. (×5)

To minimize the recurrence of this type of failure, several measures were recommended. These included stiffening structurally to decrease the stresses associated with "panting", and an immediate remedial step using a spray-on rust-preventative material to displace the moisture in this area.

B-2 c) **Corrosion-Fatigue Failure of Sailing Craft Centreboard**

The loss of a steel centreboard from a small sailing craft was reported. The remnant, the pivoting corner, shown in Figure 9-6, was examined. Particularly noteworthy were the many fine cracks adjacent to the rupture.

A microscopic examination of these cracks (Figure 9-7) indicated that they had formed over a period of time by corrosion and fatigue. A corrosion pit acted as a stress raiser for a fatigue crack. During periods of inactivity the

Figure 9-6 Failed centreboard pivoting corner of sailboat. Note the many cracks near the fractured surface (bottom).

Figure 9-7 Section through a crack in centreboard of sailing ship showing action of corrosion as corrosion fatigue crack propagated.

corrosion attack in the opening often extended laterally, possibly along non-metallic stringers. The steel centreboard finally fractured at one of these cracks.

To prevent recurrence, it was considered wise to examine these centre-boards periodically with magnetic particle inspection.

B-3 a) **Aluminum Stanchion Disintegration by Stress Corrosion Cracking**

Practically complete distintegration of a cast aluminum stanchion base, as shown in Figure 9–8, was brought to our attention for examination. These

Figure 9-8 Aluminum—10% magnesium stanchion which has nearly disintegrated by stress corrosion cracking.

bases support forged aluminum stanchions which, in turn, carry the guard rails.

Spectrographic analysis showed that this casting was an aluminum–10% magnesium alloy. Metallurgical examination showed evidence of typical stress-corrosion cracking, as previously discussed for this alloy (5 B-3(d)).

The aluminum–6% silicon alloy originally specified for this application would have served satisfactorily.

B-3 b) **Electrical Conduit Failure by Stress Corrosion Cracking**

We were asked to examine some examples of severe corrosion of aluminum alloy fittings on electrical conduits in an exposed bridge position on a ship.

Figure 9-9 The aluminum alloy electrical conduit is cracked open at right and nearly disintegrated at left.

Figure 9-10 Section of failed aluminum alloy gland in metallurgical mount. Note an exfoliation-type of attack.

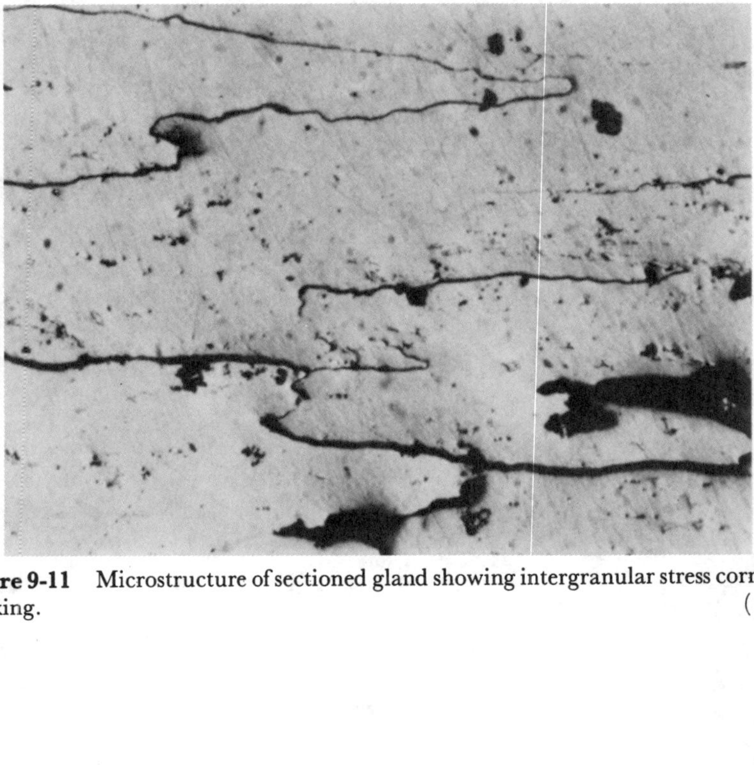

Figure 9-11 Microstructure of sectioned gland showing intergranular stress corrosion cracking. (×80)

Figure 9-9 shows typical examples of gland couplings which are partially disintegrated and appeared swollen. Most of the failures were of couplings, but gland nuts were also attacked. The conduits themselves and the deck glands did not appear to be corroding significantly.

Figure 9-10 shows a section of a failed gland nut illustrating the deep attack along the elongated metal structure. Figure 9-11 is an area on this same section at higher magnification showing the intergranular type of cracking which, as it progresses, results in an exfoliation type of failure. The attack is a form of stress corrosion cracking resulting from the "wedging" action of the aluminum corrosion products.[59]

Analysis showed that the corroded fittings were aluminum alloys of high copper content (AA 2024 variety) which are known to have poor corrosion resistance in marine applications and should not be used on ships.

It is recommended that aluminum alloy materials for such applications be restricted to alloys AA 5254 (or their equivalent) for wrought, and the aluminum–6% silicon alloy.

B-4 **Whip Antenna Failure by Intergranular Corrosion**

The failure of the top section of a 28 foot aluminum whip antenna was brought to our attention. The aluminum tube parted in the sleeve assembly as the antenna was being lowered for inspection.

Examination of the failed antenna shown in Figure 9-12 indicated that the fracture took place above the sleeve (which had produced the circumferential indentation). Metallurgical examination proved that intergranular corrosion was the major factor in the failure of the antenna. Figure 9-13 showed that the aluminum tube had corroded nearly half-way through at the failure. The corrosion took place in the crevice at the junction of the aluminum tube and another fitting.

Spectrographic examination showed that the tube was an AA 6151 alloy.

It appears that the present sleeve assembly allows sea water to enter the crevices between the aluminum components. Corrosion which takes place in the crevices ultimately results in a weakening of the antenna tubing. It was suggested that future failures of this type could be prevented by sealing the crevices, particularly those at the location of maximum stress.

C-1 **Waveguide Flanges—A Casting Problem**

The problem of cracking of waveguide flanges was brought to our attention. A close-up of these cracks is shown in the Figure 9-14. Other flanges were made available for comparison, and were also similarly cracked.

Chemical analysis indicated that all were of essentially similar composition: a type of leaded gun metal containing 4.5–5.5% tin, 4–6% zinc, 4.5–5.5%

Figure 9-12 Failed end of aluminum whip antenna. The cause of the failure, intergranular corrosion, is not apparent.

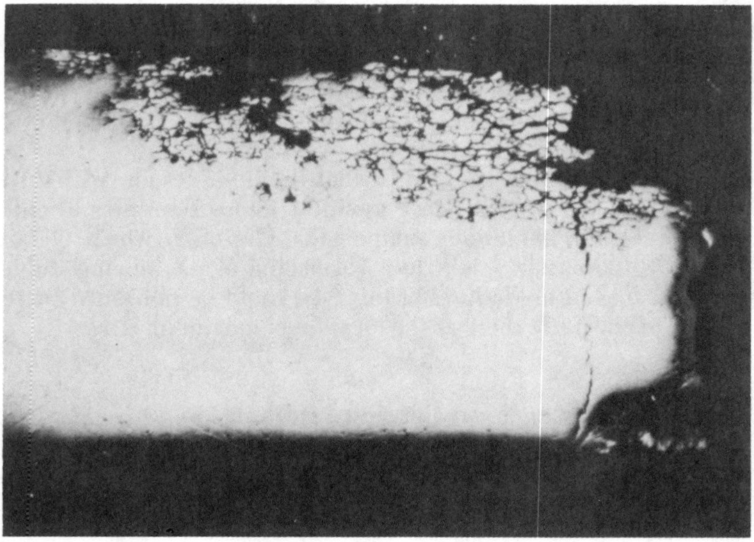

Figure 9-13 Sectioned aluminum whip atenna wall showing penetration of intergranular corrosion. (×60)

lead, often called Ounce Metal (ASTM B62-52). This composition should not of itself lead to cracking (as some brasses might).

Radiographic examination revealed considerable unsoundness (shrinkage) in the cracked flange. The others had considerably less unsoundness. There

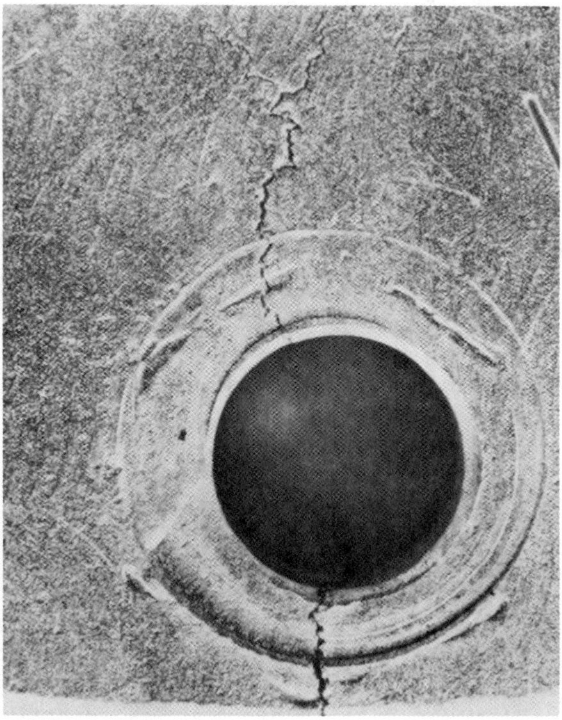

Figure 9-14 Cracks at bolt hole of wave guide flange. These are original shrinkage cracks.

is little doubt that the cracking was related to the defects subsequently shown under the microscope. Figure 9-15 shows a fine crack which was formed by a number of the shrinkage voids linking together. Figure 9-16 shows the excessive coarse shrinkage associated with the visible crack of Figure 9-14.

It was obvious that the cracks were the result of excessive shrinkage, produced by inadequate foundry practice. The manufacturer was negligent in allowing these castings out of his plant.

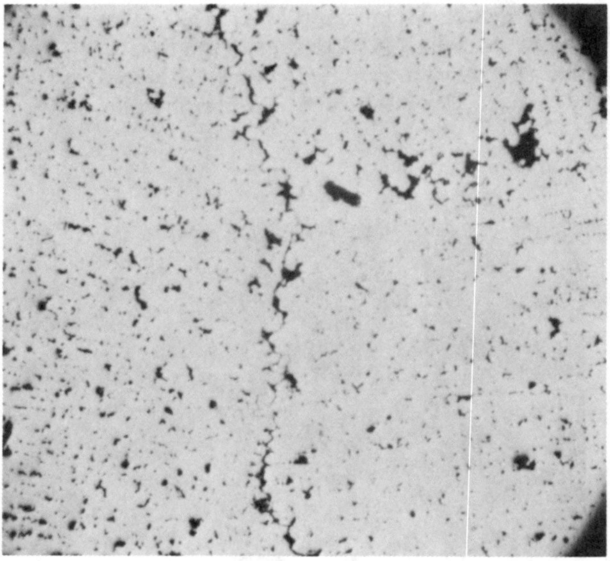

Figure 9-15 Microstructure of relatively sound area of waveguide flange showing microcracks. (×23)

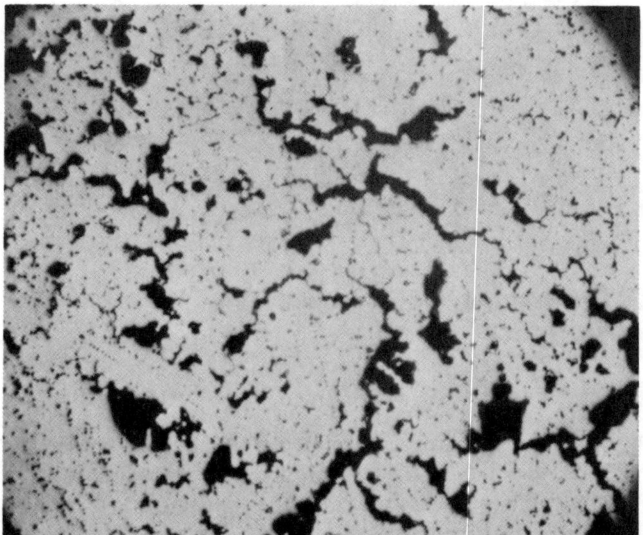

Figure 9-16 Microstructure near cracks in waveguide flange showing gross shrinkage voids. (×23)

X

PUMPS AND FITTINGS

A SHEARED PUMP shaft proved to have been partly ruptured by fatigue initiating in sharp corners of the splines. An unusual intergranular attack in a monel pump shaft was revealed, but the environment responsible was unknown. Stress-corrosion cracking in an aluminum–10% magnesium alloy confirmed again the unsuitability of this alloy for sea water service.

Use of a non-specified bronze and porosity contributed to the impingement of impellers and sealing rings. Another impeller of aluminum bronze, produced "nuggets" after disintegration in service.

Cause of failure of a double-ended piston was shown by radiography to be related to foundry practice. Original quenching cracks led to failure of a large crosshead pin from a pump.

A-1 Scavenger pump shaft failure by fatigue.

B-3 a) Monel shaft failure by intergranular attack.
b) Stress corrosion cracking of an aluminum–10% magnesium pump cover.

B-7 Failures of impellers and sealing rings by impingement.

B-8 Disintegration of aluminum bronze impeller.

C-1 Core shift in pump piston rod.

C-5 Heat treating quench cracks in crosshead pin.

A-1 Scavenger Pump Shaft Failure by Fatigue

A sheared off pump shaft was brought to the laboratory for examination.

Visual examination of the fracture shown in Figure 10-1, revealed that the failure had originated at sharp corners of one of the splines. In fact, a fatigue crack had propagated to the bore at the time of the final shearing.

Metallurgical examination showed that the steel was in a poorly heat treated condition (partially quenched and no temper). Considerable improvement in fatigue strength could be expected from the same steel properly heat treated.

The major improvement lay in eliminating the sharp corners of the splines.

Actually one would not normally expect shear to occur with such a large

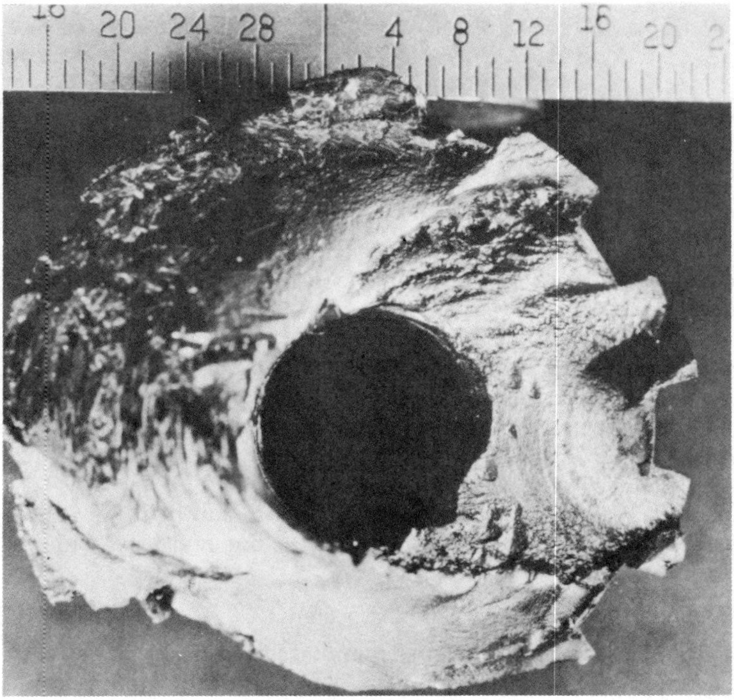

Figure 10-1 Failed end of pump shaft showing fatigue markings from spline at right.

cross-section remaining unless some abnormal overload (i.e. partial seizure) was experienced.

B-3 a) **Monel Shaft Failure by Intergranular Attack**

We were asked to examine a monel shaft from a ship's pump. The shaft had been submitted to the machine shop to be turned down and metal sprayed. It was noted that the surface of the turned down shaft showed many fine cracks.

The cracks extended over part of a bearing surface as shown in Figure 10-2. The fact that the cracks were longitudinal suggests that they were not associated with torsional stresses. A section through the shaft showed that they extended into the shaft radially and also surprisingly, were uniformly spaced at 60° intervals. It appeared likely that the surface of the shaft contained internal circumferential tensile stresses.

Metallurgical examination of the cracks in the shaft indicated that they

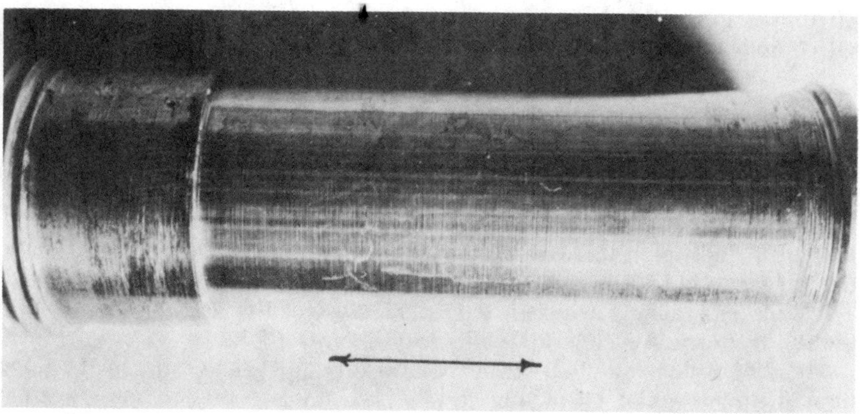

Figure 10-2 Cracked 1 inch diameter monel shaft from service pump. The extent of cracks indicated by arrow.

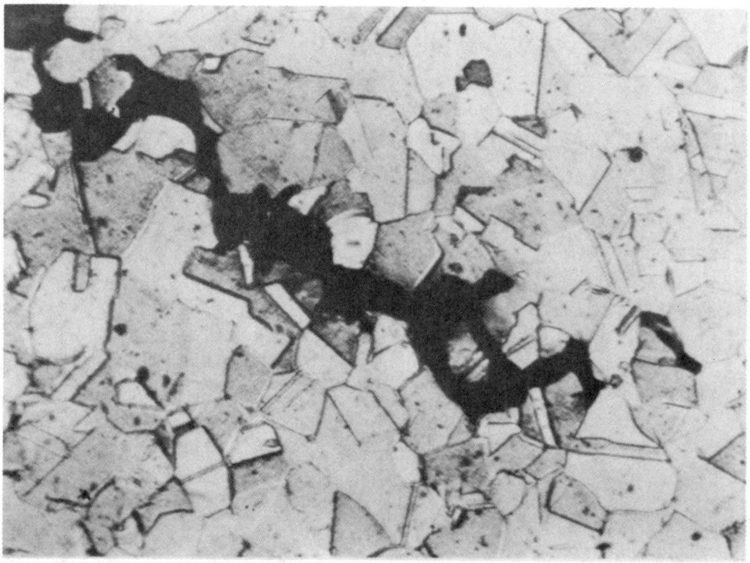

Figure 10-3 Intergranular corrosion of monel shaft. (×650)

were intergranular, apparently resulting from corrosion. Figure 10-3 shows the end of one of the cracks at high magnification.

The cause of the intergranular corrosion of a short length of the monel shaft has not been determined. However, it was not believed that the cracks formed in sea water service. Intergranular corrosion of monel is usually attributed to one of a few exotic corrosive conditions (mercury or mercury salts), none of which, to our knowledge, was present.

B-3 b) Stress Corrosion Cracking of an Aluminum–10% Magnesium Pump Cover

Cracking in a hold-down bolt hole rim of a pump casing was reported. Previously failures in this area on these sea water pumps had been considered the fair result of bolting down stresses, and the casings were simply replaced. The type of cracking which occurred at most of the dozen holes around the rim is shown in a close-up photograph, Figure 10-4.

Metallographic examination of the cracking adjacent to one of the holes showed intergranular branching cracks, typical of stress corrosion cracking, Figure 10-5. It may also be noted that porosity contributed to weakening in this area. The nature of the cracking and its relatively limited distribution (compare with 5B-3(d)) strongly suggested that stress was the predominant factor.

Chemical analysis of the aluminum alloy casting revealed that the magnesium content was 11.6%, higher than the 10.6% maximum allowed by

Figure 10-4 Hold-down bolt hole of Al-10% Mg cast pump cover showing stress corrosion crack.

the specification. The susceptibility of aluminum–magnesium alloys to stress corrosion cracking increases markedly once the magnesium content exceeds 10%.

Aluminum–magnesium alloys show resistance to stress-corrosion when in the solution heat treated condition; that is, when quenched from about

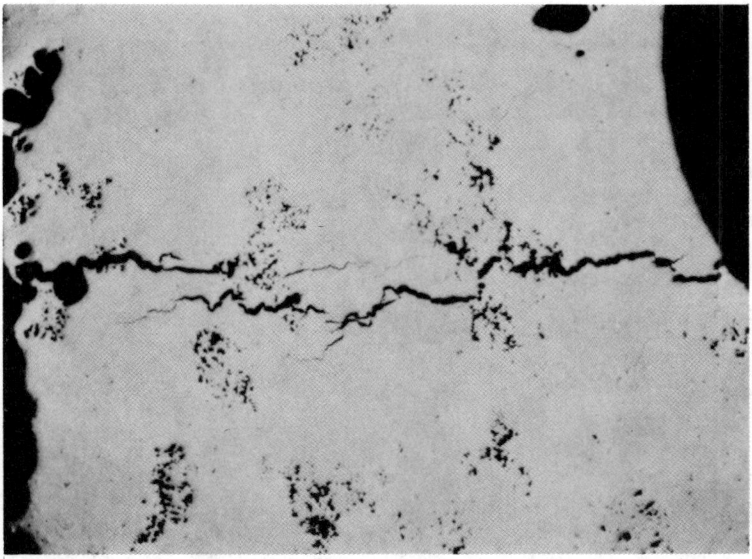

Figure 10-5 Section through bolt hole (Figure 10-4) showing stress corrosion cracks in the porous casting. (×5)

800° F. This heat treatment produces a one-phase microstructure. The precipitation of an anodic second phase at the grain boundaries and slip planes of the primary crystals results in the alloy becoming susceptible to stress corrosion. In aluminum 10% magnesium alloys, this precipitation will take place in short periods of time at high temperatures, or in long periods of time at room temperature.

In summary, the factors, in addition to stress, contributing to the failures were above-specified amounts of magnesium, improper heat treatment, and casting porosity. A wiser choice, even if it appears that one is getting a weaker alloy is the aluminum–6% silicon casting alloy. This alloy being more readily cast and therefore free of porosity and also not being susceptible to stress corrosion, would in final analysis prove to be stronger and more reliable.

B-7 Failures of Impellers and Sealing Rings by Impingement

Impellers and pump sealing rings from sea water pumps have been badly wasted in a relatively short period of service.

Examination of the surface of one of these impellers, shown in Figure 10-6, suggested the combined action of corrosion and erosion. Similarly, the sealing ring of Figure 10-7 shows the same destructive attack.

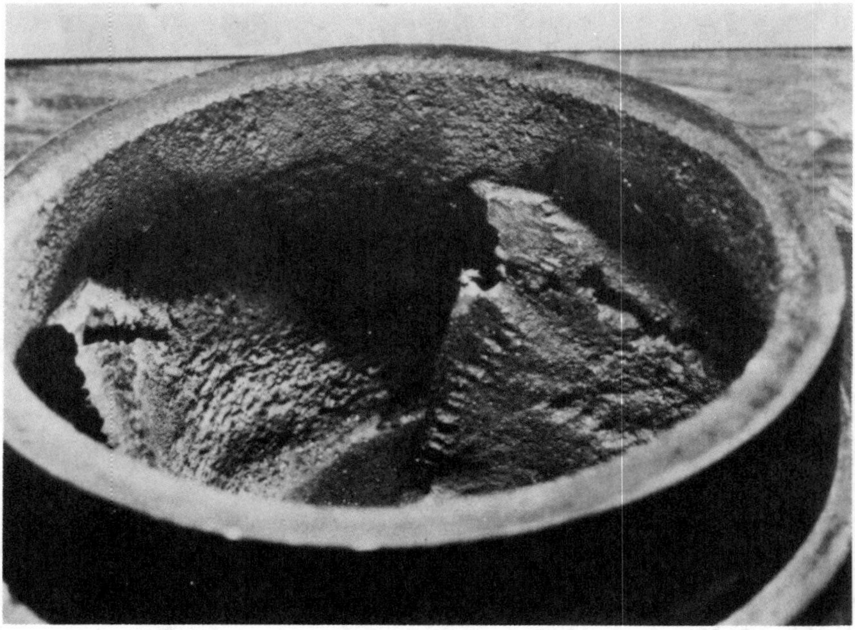

Figure 10-6 Failed sea water impeller. Note the ragged nature of the impingement attack on the tin bronze.

An erosive effect was indicated by the absence of corrosion products in most of the pits. A corrosive effect was indicated by the presence of corrosion products in some pits, as well as "skeletons" of metal which would not have been expected to survive if the metal was being destroyed by mechanical effects alone.

It was noted that the material of the ring was Ounce metal (85 copper, 5 tin, 5 zinc, 5 lead), not phosphor bronze as specified. Ounce metal cannot be expected to have as good corrosion resistance as an 8 or 10% tin bronze. It was also noted that these castings were very porous, (see Figure 10-8) which would reduce the corrosion resistance of this material even further.

Similarly, the impeller castings, although of a better material, an 8% tin

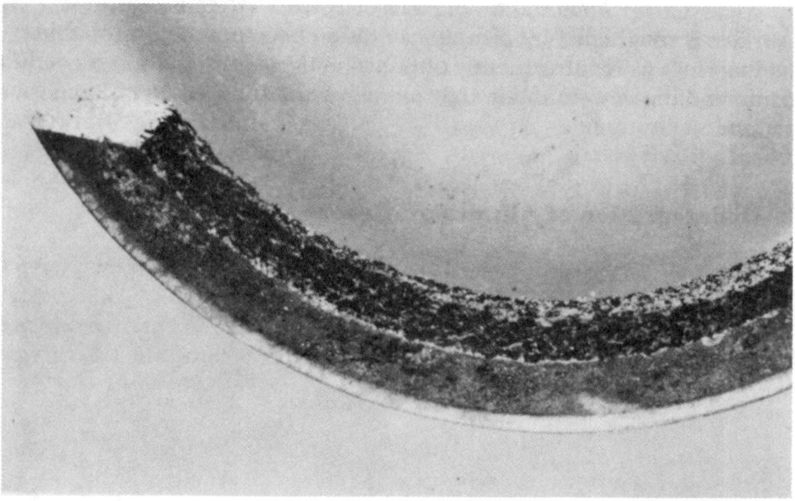

Figure 10-7 Sealing ring for impeller on pump showing impingement attack.

bronze, were quite porous and the porosity was particularly bad near the pitted areas. It is quite probable that this excessive porosity was at least in part responsible for the attack. Corrosion would start in the pores, because

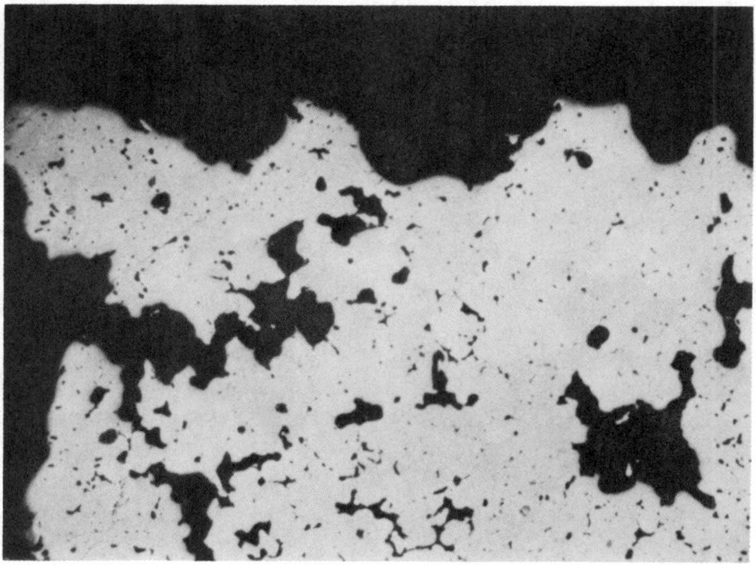

Figure 10-8 Microsection through impeller showing excessive shrinkage in attacked metal. (×35)

they would set up differential aeration cells with the outside surface. Once the surface is roughened by pitting, cavitation becomes more probable.

Sealing rings were subsequently obtained in the centrifugally cast condition. Recommendations were made that monel or titanium be considered for the impellers.

B-8 Disintegration of Aluminum Bronze Impeller

A quantity of "nuggets" retrieved from the casing of a sea water pump was received for assessment. The impeller was missing and it could only be surmised that it had disintegrated in service. It appeared that the fragments had been tumbled within the confines of the casing as shown by their rounded appearance in Figure 10-9.

Figure 10-9 "Nuggets" of sea water pump impeller which failed by dealuminification.

Spectrographic analysis showed that the material of the "nuggets" was a 10% aluminum bronze, the composition used for the other impellers.

Sectioning of the "nuggets", as shown in Figure 10-10, revealed that corrosion attack had all but penetrated from both surfaces. This attack, as shown in more detail in Figure 10-11, shows the dealuminification of the second phase. The weakening from this corrosion attack readily leads to cracking along the corroded phase, also shown in Figure 10-11.

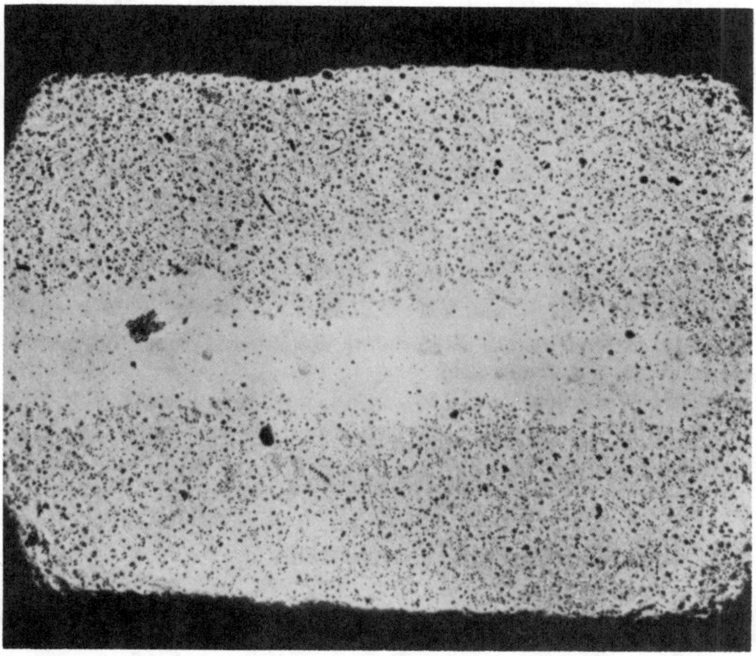

Figure 10-10 Sectioned "nugget" of aluminum bronze impeller showing depth of dealuminification. (×24)

These alloys, and indeed any aluminum bronze containing more than 8.5% aluminum, are of the two phase variety. Alloys containing less than 8.5% aluminum are single phase. For the alloy in question the two phases are a copper-rich alpha phase and an eutectoid. Difficulties with the material stem from the composition of the eutectoid which is an intimate mixture of alpha grains and an aluminum rich gamma phase. The gamma phase is anodic to the alpha and provides a ready-made preferential corrosion path in the eutectoid. This particular type of attack is sometimes referred to as dealuminization (or dealuminification) to liken it to the dezincification of some two phase brasses.

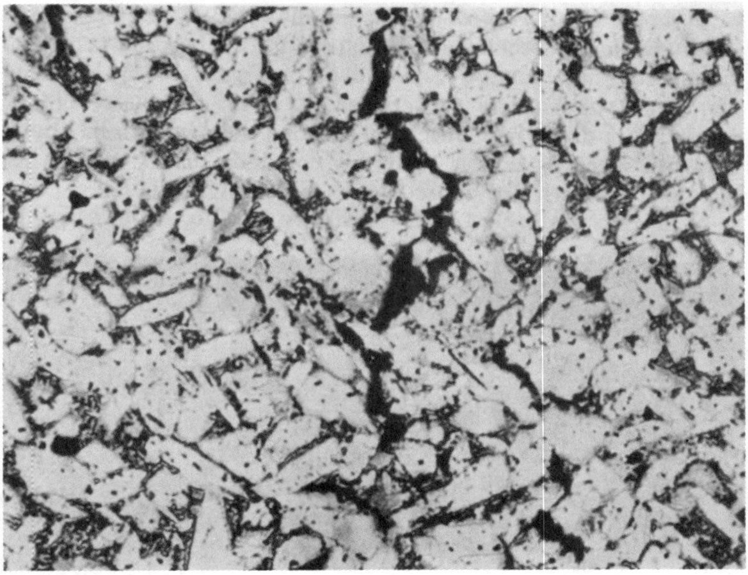

Figure 10-11 Microstructure of corroding aluminum bronze "nugget" showing preferential attack of dark eutectoid. (×100)

Dealuminification can be minimized by suitable heat treatment, i.e., solution heat treating. This particular impeller was in the as-cast condition.

C-1 a) Core Shift in Pump Piston Rod

A double-ended piston from a condensate pump was examined at a refit

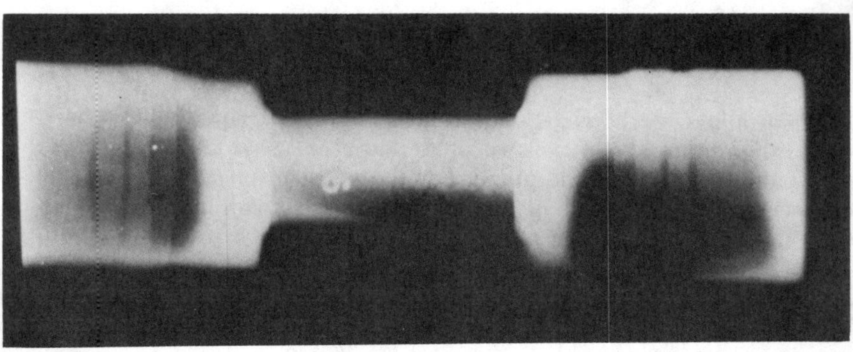

Figure 10-12 Radiograph of pump piston rod showing core shift.

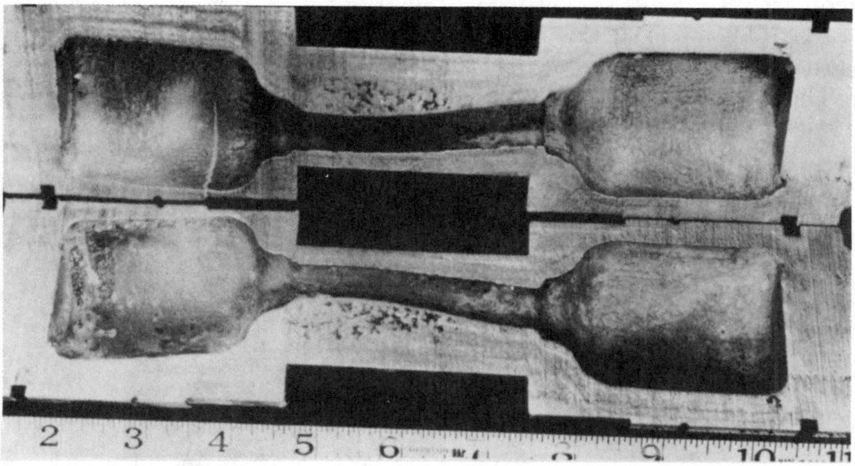

Figure 10-13 Sectioned piston rod (Figure 10-12) showing off-centre core and crack outlined by magnetic particle inspection.

by workmen who noted that water was bubbling from a crack in the wall of one of the pistons.

As a first step in investigating the source of water in an apparently solid piece, a radiograph was taken. The radiograph readily revealed the nature of the problem, as shown in Figure 10-12. Here it may be noted that a core shift was responsible for the unusually thin wall at the crack location. This was confirmed by sectioning as shown in Figure 10-13.

The value of radiography in revealing unknown or inaccessible interiors has often proven to be a most invaluable tool.

C-5 Heat Treating Quench Cracks in Crosshead Pin

A broken crosshead pin off a large reciprocating pump was brought in to determine the cause of failure. The unusual fracture surface of one end of the pin is shown in Figure 10-14.

As may be noted in the photograph the darker part of the fracture surface, adjacent to the transverse drilled hole, appears to have existed as two cracks for some time prior to final rupture. Examination of the other half of the crosshead pin showed a third crack, extending from the same hole at right angles to the fracture surface of Figure 10-14.

Spectrographic analysis showed that the pin was a low alloy steel (0.2% Cr, 0.2% Mo). The pin had been hardened to Rockwell C55.

The fracture by itself might have led to speculations about transverse forces producing the failure; however, the presence of 3 radial cracks almost equidistant strongly suggested stresses prior to service, resulting in quenching cracks. Other pins of this type were subsequently examined by magnetic particle inspection, and some had similar quench cracks.

The obvious remedy was to use a more alloyed steel, thereby permitting a gentler quench and reduced quenching stresses.

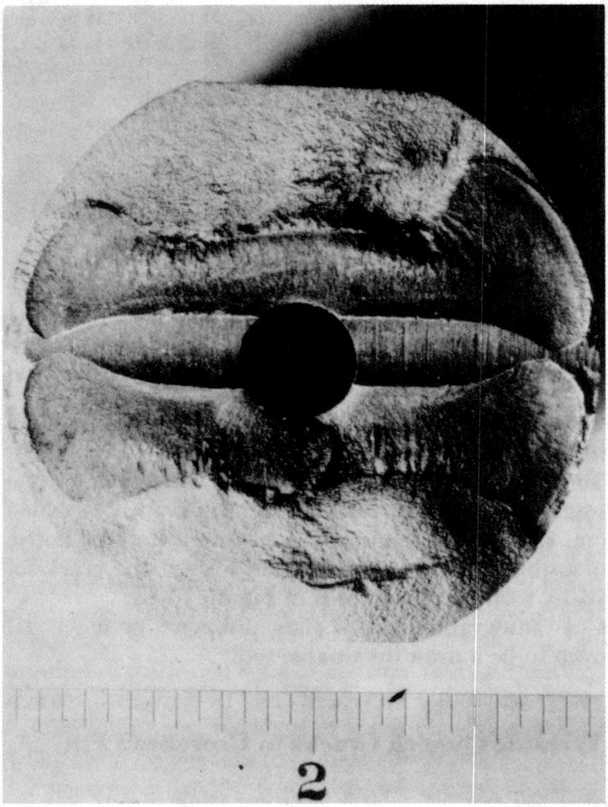

Figure 10-14 Fractured surface of cross head pin showing two quench cracks originating at the transverse drill hole.

XI

VEHICLES

TYPICAL FATIGUE cracking in bus steering assemblies, at failure and in the early stages of development are shown. Destruction of hardened steel surfaces in a complex drive unit are related to absence of lubricant. Corrosion of an aluminum brake support plate in a tanker truck is investigated. Leakage of water tanks in a fire truck are traced to troubles with welding of the type 430 stainless.

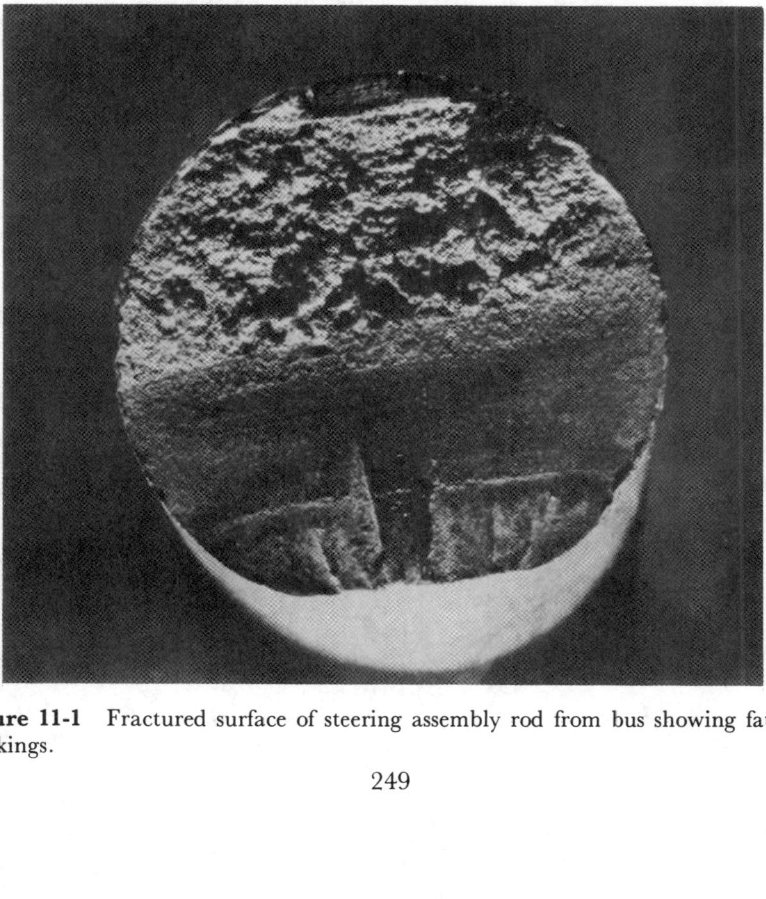

Figure 11-1 Fractured surface of steering assembly rod from bus showing fatigue markings.

B-4 Corrosion failure of tanker brake pedal support.

C-3 Weld failures in fire truck tank.

A-1 **Fatigue Failures of Bus Steering Assemblies**

Loss of steering control on a passenger bus, fortunately under favourable circumstances so that the bus was halted with negligible damage just off the roadside, resulted in an investigation. The investigators found that a fracture had occurred in the steering assembly. Both parts of the fractured linkage were brought in to the laboratory.

The fractured surface, shown in Figure 11-1, revealed the typical markings of a fatigue failure. In the course of bending this 1 inch steel rod, the manufacturer had introduced a surface indentation which apparently served as the initiation point for the fracture.

Although this failure occurred well into the vehicle's life, the hazard associated with loss of steering control is obvious. As a result, a wider use of magnetic particle inspection was introduced.

It was interesting that shortly after the first incident, a fine crack was

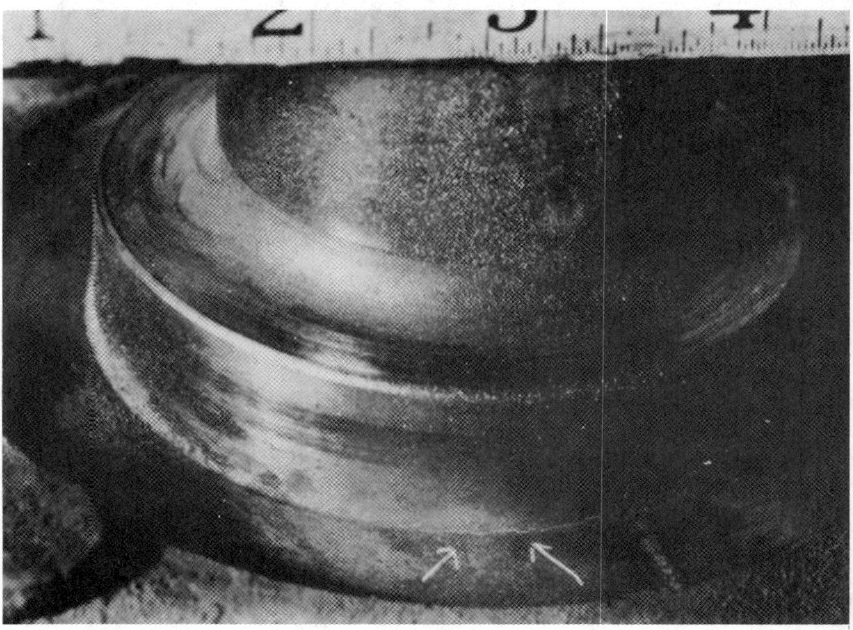

Figure 11-2 Magnetic particle inspection indication (arrow) outlining crack in axle of bus.

detected on the hub of a front axle on the same type of bus. This crack enhanced by magnetic particle powder is shown in Figure 11-2. A similar indication was also noted on the other axle of this bus. These cracks would undoubtedly have propagated and ultimately resulted in failure—with consequent hazard to life.

A-5 Crash Truck Drive Unit Seizures—A Lubrication Problem.

The drive units off a specially built crash truck, one part of which is shown in Figure 11-3, were submitted for examination because of seizure. Also submitted were samples of the oil used in this transmission.

Figure 11-3 Drive unit load transmitting surfaces showing galling caused by inadequate lubrication.

Visual examination indicated a gross destruction of mating surfaces of the drive unit. There was evidence of severe overheating and of metal welding and tearing. Micro-examination showed that the parts were of a good quality, properly heat treated steel (Figure 11-4) but the damaged surface showed

Figure 11-4 Fine tempered martensite microstructure of drive unit parts of Figure 11-3. (×400)

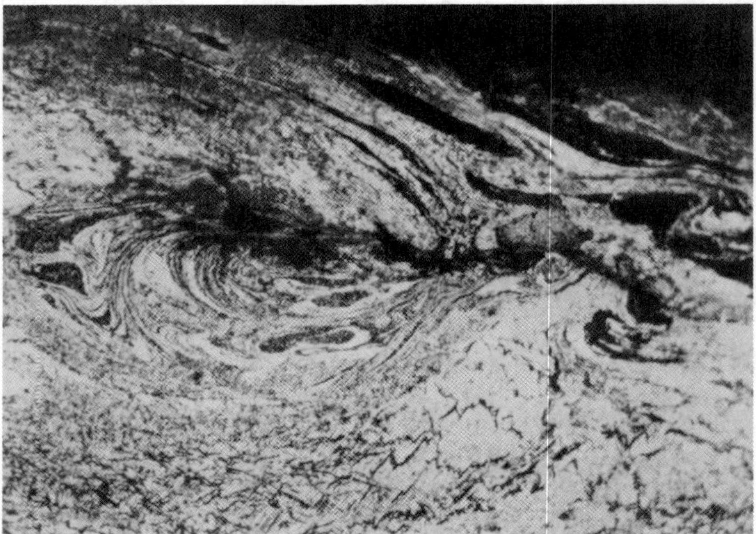

Figure 11-5 Highly deformed structure at galled surfaces of drive units. (×150)

severe plastic deformation and overheating. Temperatures attained were at least 1600° F as evidenced by some martensite just below the surface (Figure 11-5). This area also had a number of fine cracks.

The oil sample was checked and appeared to be in good condition. The tests performed showed that the oil met requirements of flash point, kinematic viscosity, and copper strip corrosion test. On the Alpha Lubricant Testing unit, designed primarily to evaluate film strength under extreme pressure conditions, this oil performed very well. A small dab of it touched on the steel rubbing parts ran for 29,000 revolutions without failure (the machine was stopped for other work), and also gave over 400,000 revolutions (without failure) when running in a bath of the same oil. Pressures exerted on the rubbing surfaces during these tests were of the order of 100,000 psi. The results lead us to the conclusion that, provided the oil was present, failure was most unlikely.

From discussions on these failures it became evident that these drive units were not necessarily always full of oil. The oil tended to drain out relatively rapidly because the oil seals leaked badly. There was also one filler hole on the casing and unless care was taken, it was possible to fill the casing incompletely.

Measures to ensure the presence of sufficient oil at all times seemed to be the most useful first step to avoid future troubles. The addition of another hole on top of the casing would be an obvious improvement as it would provide an air escape.

B-4 Corrosion Failure of Tanker Brake Pedal Support

The brake pedal support plate from an aircraft refuelling tanker truck fractured in service and was submitted for examination. The two pieces of the support plate shown in Figure 11-6 appeared to have fallen apart readily and corrosion was noted on the surface.

The support plate was found to be an aluminum alloy containing around 8% silicon and 3.5% copper. This high strength alloy has poor corrosion resistance because of its high copper content. It can be seen from Figure 11-7 that the material tended to suffer intergranular corrosion. From the orientation of the corrosion attack, it did not appear that galvanic attack between this aluminum part and the steel floor boards contributed much to the corrosion.

The corroded surface of the casting was found to be quite alkaline when tested with moistened pH paper. In addition to aluminum oxide, analysis of the corrosion products showed sodium, magnesium, and boron (negligible chloride). The liquid responsible for the corrosion attack may therefore have contained borax, a not uncommon alkaline ingredient of cleaning materials.

For future replacements of this support plate it was suggested that steel,

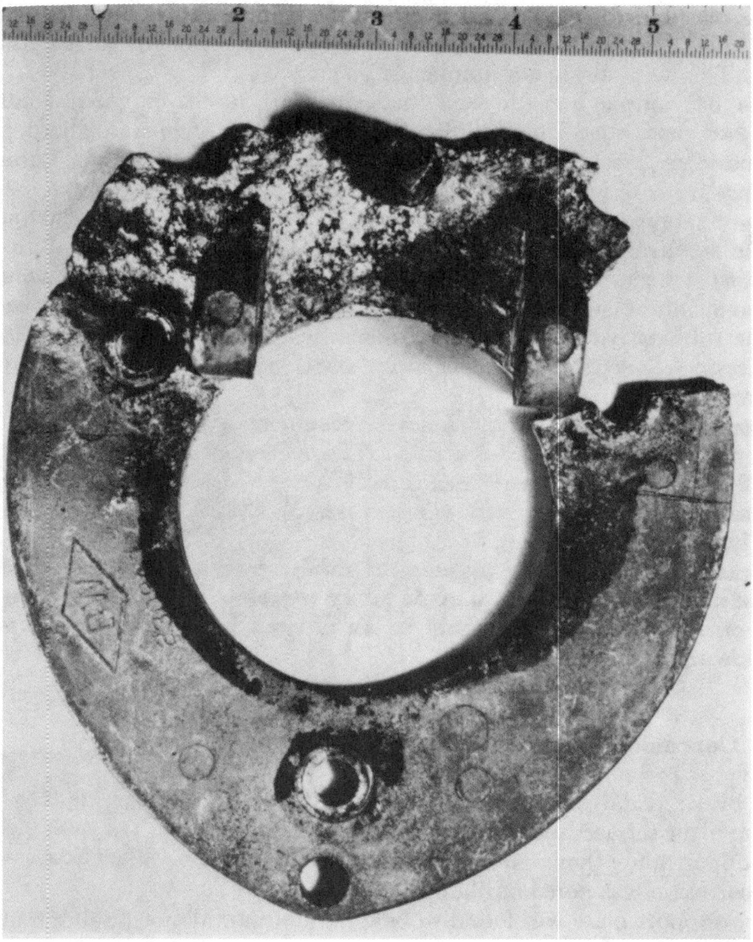

Figure 11-6 Corroded and fractured aluminum alloy support plate for air brake pedal of refuelling tanker.

or galvanized steel, would be a more satisfactory material. If an aluminum alloy had to be used for non sparking purposes, then it was recommended that the part be fabricated from AA 5154 or cast from a 6% silicon aluminum alloy. The thickness of the part should be increased for the latter cast material.

C-3 Weld Failures in Fire Truck Tank

Leaks in the water tanks of a fire truck were noted shortly after purchase. The leaks were traced to cracks adjacent to the welds.

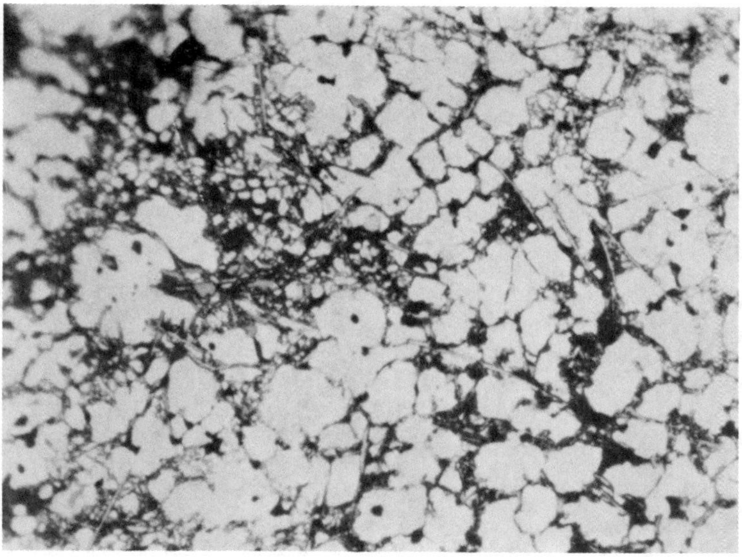

Figure 11-7 Intergranular attack along the needle-like copper-rich phase of the aluminum alloy casting. (×500)

Figure 11-8 Section through lap weld of tank of fire truck. Note the cracks originating in the crevice (which was present before the top piece was broken off). (×22)

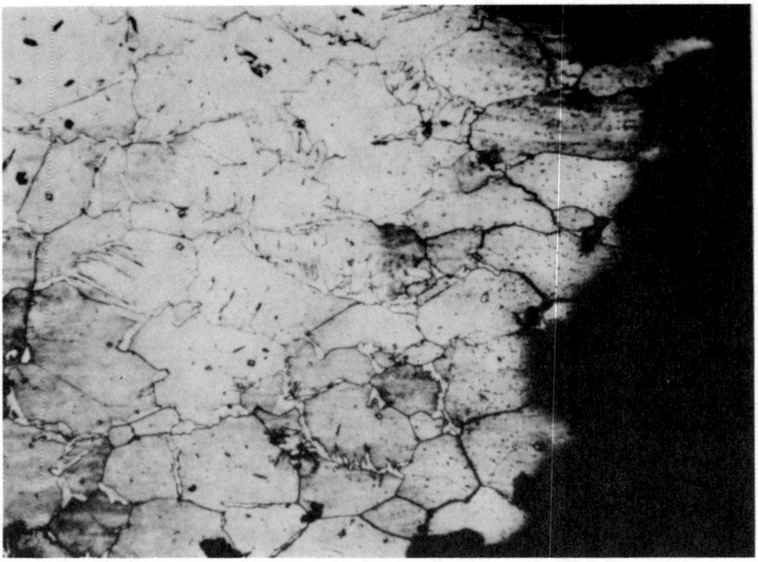

Figure 11-9 Edge of fracture of heat affected parent metal showing the many intergranular cracks, and the second phase. (× 145)

Radiographic examination showed that the cracks in the lap welds were even more extensive than those noted visually. Also, a number of the butt welds showed severe incomplete penetration, a defect that could be expected to propagate in service and ultimately result in additional cracking.

Spectrographic analysis showed that the tank had been fabricated from Type 430 Stainless Steel. The weld metal was high in nickel content, and considering the dilution, probably resulted from using an 18–8 welding rod.

Metallurgical examination showed that the cracking had initiated in the lap crevice and it was apparent that intergranular corrosion had contributed toward the cracking failure. Note Figures 11-8 and 11-9. The formation of the cracks was facilitated by the presence of the second phase which had formed in the heat affected zone of the parent metal.

A number of metallurgical problems can be encountered in the welding of ferritic stainless steels (Type 430). Embrittlement of the heat affected parent metal is expected if suitable post heat treatment is not carried out, and this treatment could not be adequately controlled in the case of the large tank. It was apparent that the tank in question needed to be completely rewelded. All joints should be butt welds and these should have complete penetration. Radiography was specified to ensure that the welds were, in fact, sound.

XII

INTERNAL COMBUSTION

ALTHOUGH IT was not possible to establish exactly the way an air compressor failed, the fatigue failure of a connecting rod was probably significant. An interesting failure of a piston in an opposed-piston diesel engine was traced to stress at an internal corner. Too small a fillet was responsible for a number of failures of diesel crankshafts. Potential rupture of a compressor crankshaft was averted by magnetic particle examination. Thermal cracking between valve ports in a cylinder head is discussed.

Corrosion fatigue cracks were proven by strain gauge studies to be responsible for leaks in aluminum cylinder heads. Stress corrosion cracking caused failure of several high strength brass fuel injectors. Cavitation corrosion to the point of perforation of numerous diesel cylinder liners is described together with some aspects of the mechanism involved.

Two problems tracing back to faulty casting procedures are outlined. Fatigue failure in a connecting rod was related to original flaws and to loose bottom end nuts.

A-1 a) Explosion of air compressor.
b) Piston failure in diesel engine.
c) Diesel crankshaft failure.

A-3 Scored crankshaft.

A-5 Thermal cracking of cylinder heads.

B-2 Aluminum cylinder head failures by corrosion fatigue.

B-3 Diesel injector failures by stress corrosion cracking.

B-7 Cavitation corrosion on cylinder liners.

C-1 a) Aluminum cylinder head failure—perforation by a sand core wire.
b) Piston ring failures—material and porosity.

C-9 Connecting rod failure in truck engine.

A-1 a) Explosion of Air Compressor

An air compressor in use on a diving tender failed catastrophically by flying apart—the cylinders and associated gear were broken away from the crankcase, Figure 12-1.

Visual examination of the main fractures in the body of the compressor

Figure 12-1 Explosive failure of cast iron compressor crankcase.

showed evidence that a portion of the surfaces had parted at some prior period. In addition, the fracture of one of the connecting rods also showed some prior parting in the nature of a fatigue crack. There was evidence that the broken connecting rod had pounded and chipped the piston pin, and had embedded the resulting loose bits into the top end bearing surface of the connecting rod.

The fracture in Figure 12-2 shows that failure was partly by fatigue—the rounded part of the break had advanced prior to the failure. This crack would serve as a severe stress raiser. The break at the opposite side was at an area of a severe notch ground in to accommodate the bolt, as shown in Figure 12-3. These two weaknesses would then make the final separation possible at a relatively low load.

Although not too well shown in Figure 12-2, the bolt head had suffered some pounding on its underside as well, suggesting that it was loose. Whether

Figure 12-2 Connecting rod fragment showing the fatigue failure (right) which initiated the break.

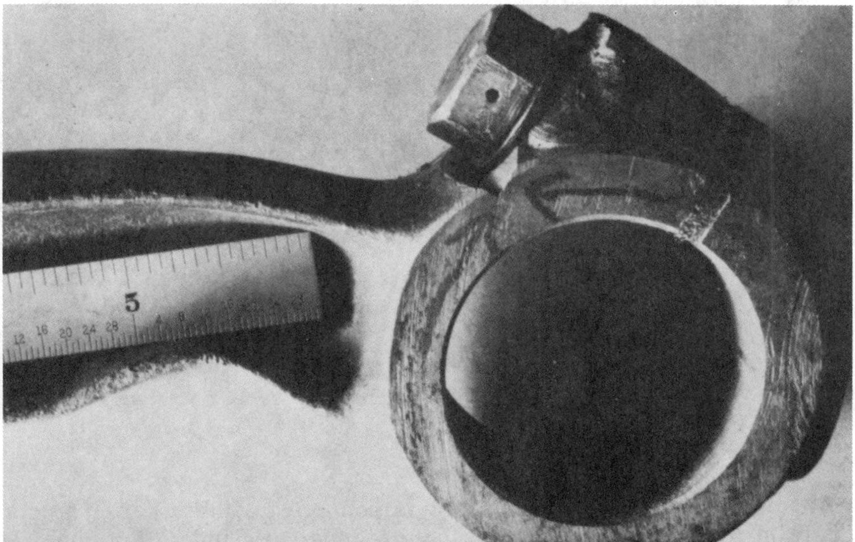

Figure 12-3 Other connecting rod from compressor showing the notch (arrow) ground into one side to provide space for the bolt head.

this was a factor contributing to the fatigue crack or resulted from it, was not established.

The other piston was badly broken up—probably the result of pounding after the compressor flew apart.

Regarding the possible mechanisms, certain main features required further consideration:

a) The broken connecting rod had several cycles of free motion during which it pounded itself, the pin and other areas.

b) The relatively uniform upward push which caused the main fracture and lifted the upward half of the compressor and threw it several feet away.

c) The cause of fracture of the connecting rod, which must have parted sometime before the main force lifted the top half of the compressor.

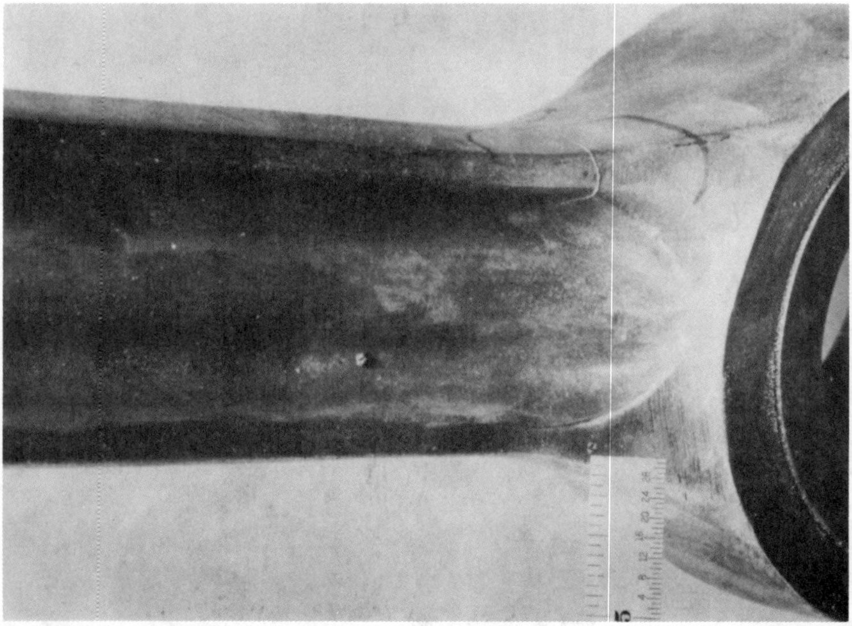

Figure 12-4 Diesel connecting rod showing crack outlined by a magnetic particle indication.

Any possible sequence of events must be connected with the partial fatigue failure of the connecting rod. It's faílure apparently initiated a chain of events which culminated in the compressor flying apart.

For the upward thrust which both ruptured the top of the crankcase and lifted the top half of the compressor, there were at least two possibilities:

a) The connecting rod with a well placed blow transferred its energy totally to the compressor body. The evidence does not tend to support this hypothesis unless water accumulated (from the broken cylinder wall) on top of the piston. Otherwise blows to the wrist pin would have no lifting effect and none of the other blows appeared to have been placed so that the desired lifting effect would result.

b) That the flaying connecting rod struck a spark and caused an explosion of oil vapours in the crankcase.

Although the exact failure mechanism was not positively established, the failure of the connecting rod was of significance. Admittedly this was a particularly poor example of a connecting rod but as shown in Figure 12-4, well-designed and well-machined connecting rods also develop fatigue cracks. These cracks can usually be detected as was the one in Figure 12-4 by rigid adherence to a periodic magnetic particle inspection.

Figure 12-5 Fractured surface of underside of piston top showing piston pin stud bosses where fracture initiated.

A-1 b) **Piston Failure in Diesel Engine**

The failure of a lower piston in an opposed-piston diesel engine was brought to the attention of the laboratory. The piston skirt was badly broken up below the second piston ring groove. The top end of this piston is shown in Figure 12-5.

Visual examination of several other pistons from this engine showed that cracks were present in the second piston ring groove of one of these, Figure 12-6. It was particularly surprising that this crack occurred in the center

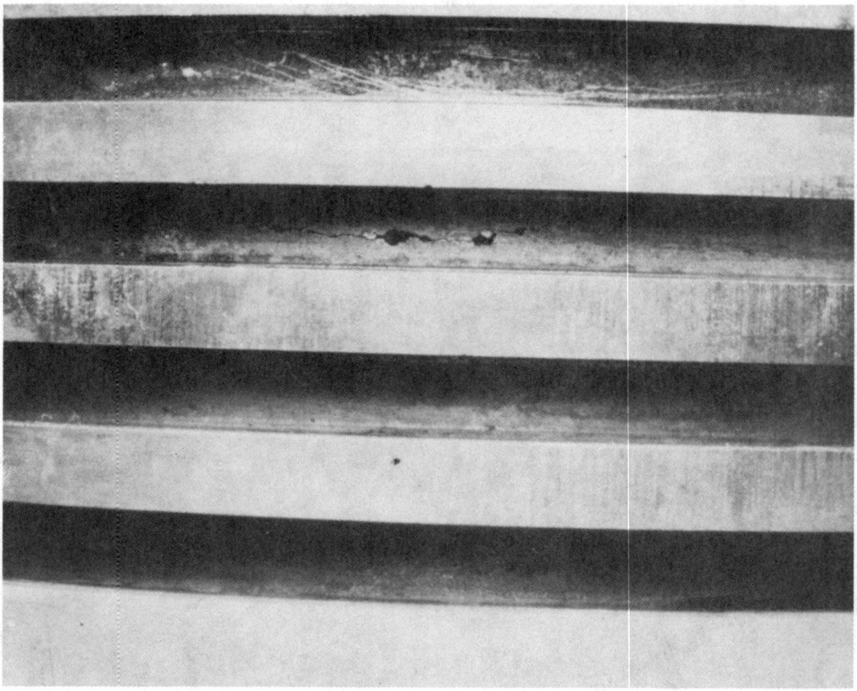

Figure 12-6 Cracks noted in second piston ring groove of piston. These cracks initiated inside at stud boss–skirt junctions Figure 12-7.

of the groove rather than as might be expected, in the corner. A section cut through this area, Figure 12-7, showed that this crack had originated inside the piston, at the junction of the skirt and the piston pin stud bosses. This location, in fact, was where rupture of the first piston, Figure 12-5, took place.

The engine from which these pistons came had been in service for a considerable period. The pistons in other similar engines, in the same service

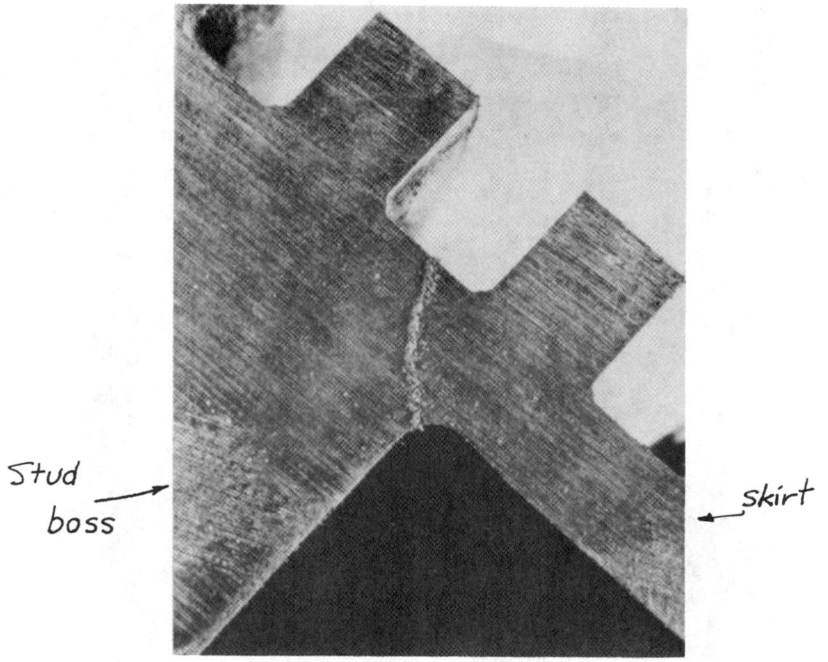

Figure 12-7 Section through cracked piston showing junction of stud boss and skirt where crack originated. (×3)

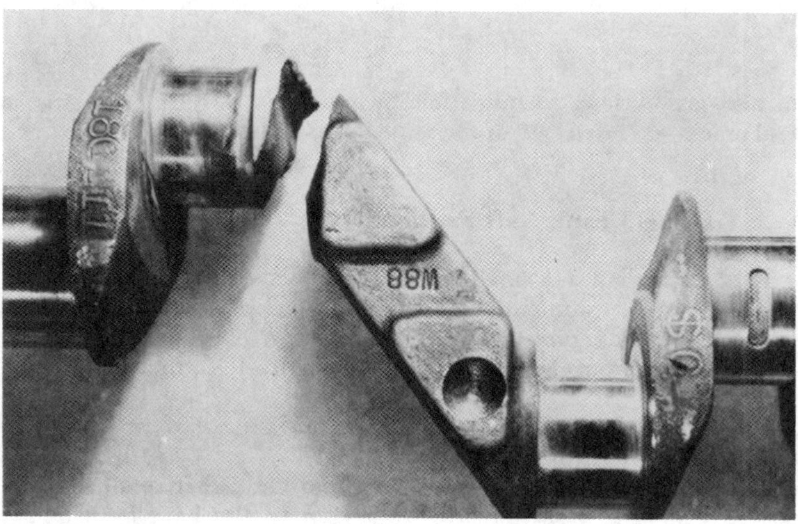

Figure 12-8 Fractured four-cylinder diesel crankshaft.

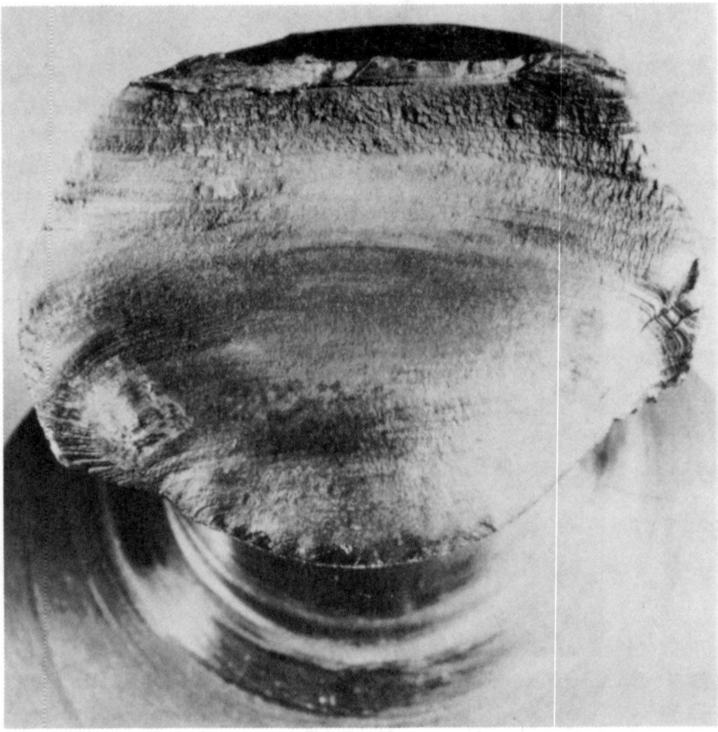

Figure 12-9 Fractured surface of diesel crankshaft showing classical fatigue markings.

were also given closer examination by magnetic particle inspection and several cracks were detected at the same location described above.

A-1 c) **Diesel Crankshaft Failure**

The failed crankshaft in a small diesel used to power a 25 foot motor cutter, was brought in for examination. The failure took place at a piston bearing surface as shown in Figure 12-8.

It was obvious from the fracture surface, Figure 12-9, that failure was by fatigue.

Cracking initiated at the fillet, and the initiating flaw was probably a machining mark in the corner.

It was suspected that if the machining and the rather small fillets were the cause of failure, other potential failures may also have been initiated. A check of the whole crankshaft by magnetic particle inspection indicated that this was the case. Three additional cracks were detected. One of these is shown in Figure 12-10.

Fatigue failures are not unusual when a machine has served for a very long period, or where abuse or an accident has overloaded a component so that a crack develops and propagates. In this instance, however, it was suggested that the manufacturer was liable for an inherently weak crankshaft. The manufacturer was so notified. Some time later, during overhaul of newer engines of this type, it was noted that the manufacturer had in fact, introduced a more generous fillet.

As a result of these failures and cracking of several engine components, a rigorous program of periodic magnetic particle inspection was instituted. A number of interesting potential fatigue failures were detected. One of these is described below.

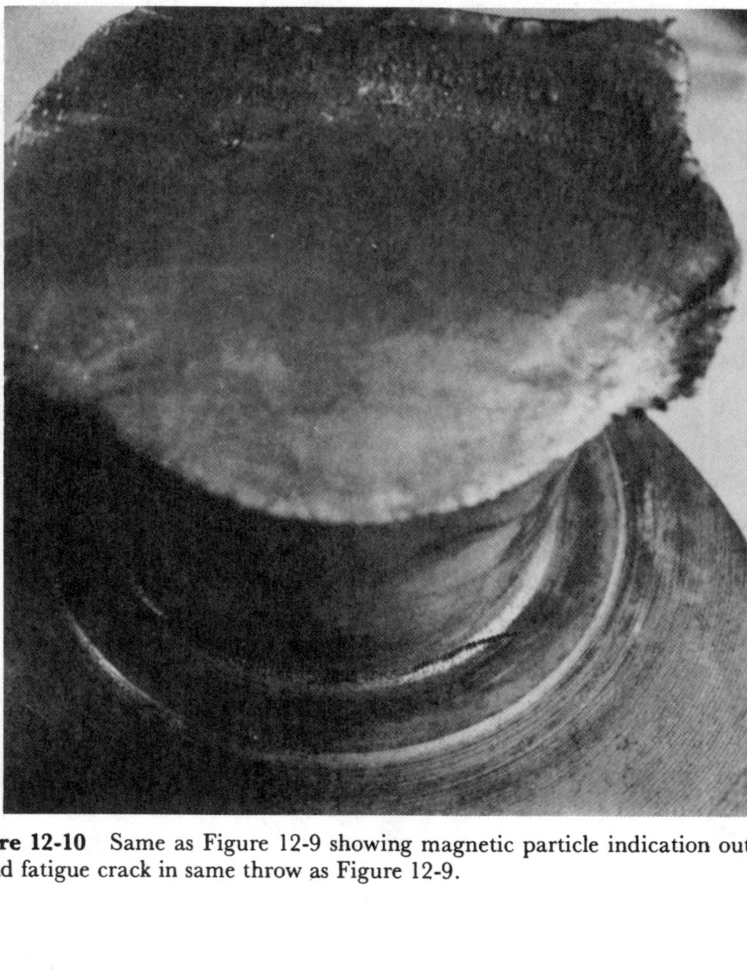

Figure 12-10 Same as Figure 12-9 showing magnetic particle indication outlining second fatigue crack in same throw as Figure 12-9.

An indication of cracking was picked up as shown in Figure 12-11 adjacent to the wrist pin bush on a large diesel connecting rod. To determine the extent of this crack, the bushing was removed, and an amazing network of additional small cracks was revealed as in Figure 12-12. While time did not permit detailed examination, these multitudinous cracks strongly suggested original forging flaws. The propagation of the larger, at least, represented a potential hazard.

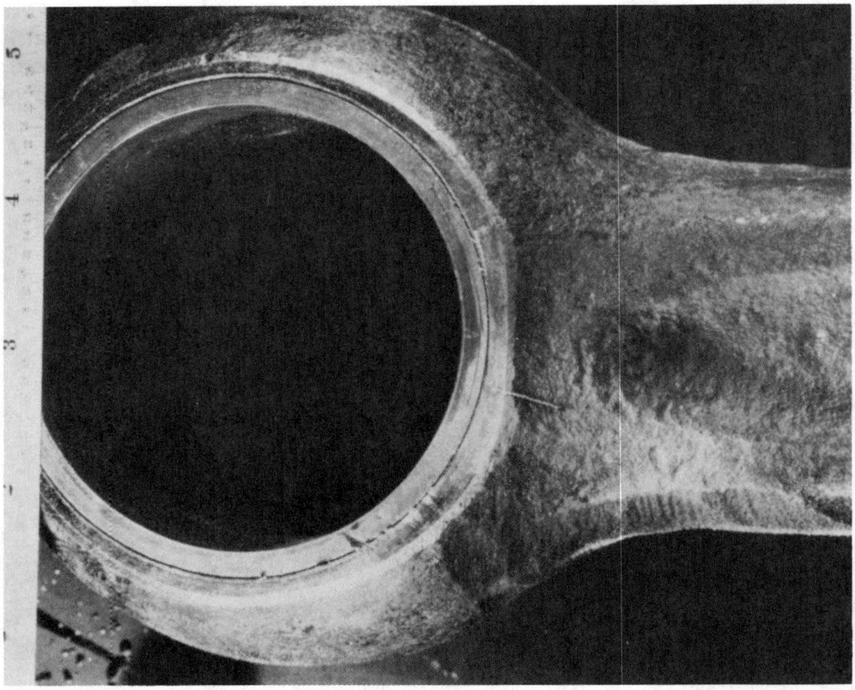

Figure 12-11 Diesel connecting rod showing crack (magnetic particle at pin end).

A-3 **Scored Crankshaft**

A badly worn air compressor crankshaft was brought in for examination prior to machining and building-up.

Magnetic particle inspection soon revealed, as shown in Figure 12-13, a rather serious display of cracking. These cracks resulted from service stresses acting in conjunction with the score marks as stress raisers. Light grinding established that these extended to depths greater than could be considered for the rebuilt crankshaft.

A-5 Thermal Cracking of Cylinder Heads

A problem which has occurred on numerous heavy duty engines is one of cracking between, or adjacent to, valve ports in cylinder heads. One of these is shown in Figure 12-14.

This cracking occurs in a very hot area and is undoubtedly the result of

Figure 12-12 Same as Figure 12-11 with bushing removed showing network of small cracks.

thermal expansion followed by contraction stresses on cooling. Many such cycles lead to a form of cracking called thermal fatigue cracking.

An interesting method of alleviating this cracking was described by the chief metallurgist of GMC Diesel.[38] This involved an aluminum coated steel insert which takes up the expansion and contraction.

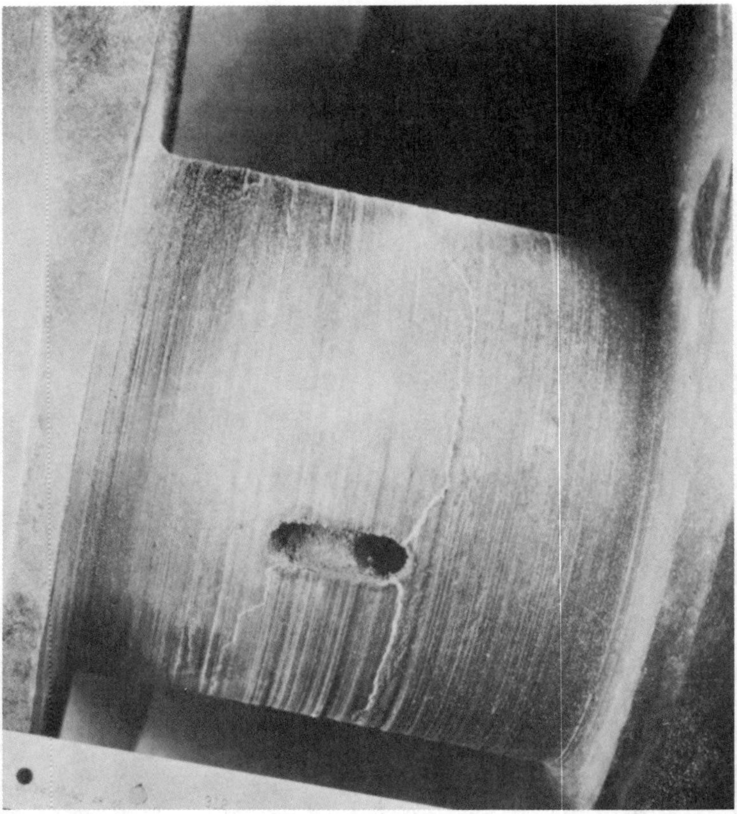

Figure 12-13 Crank pin bearing surface of connecting rod showing scored and cracked surface. Cracks are outlined by a magnetic particle inspection.

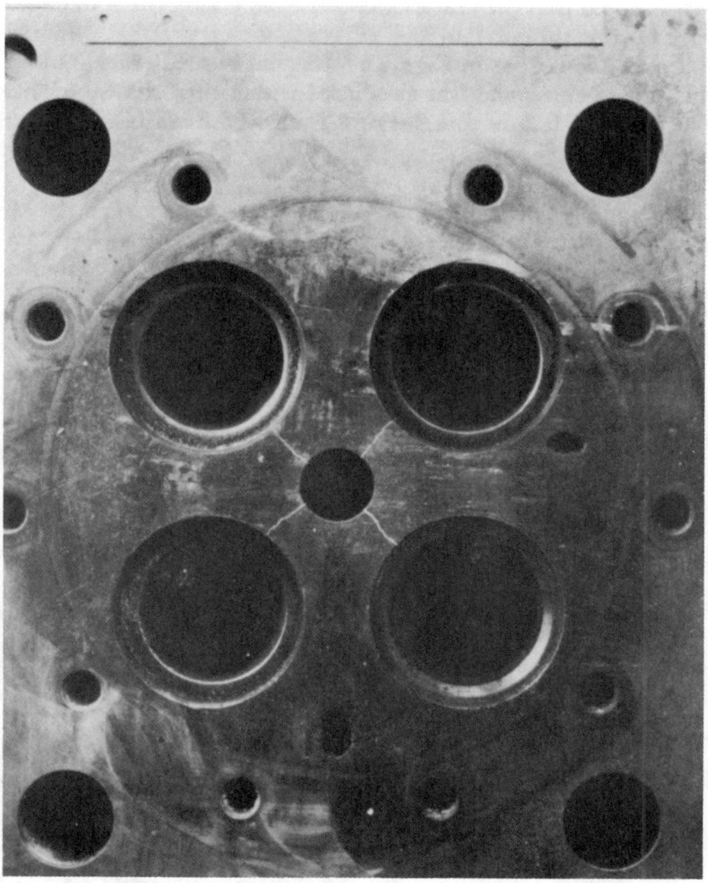

Figure 12-14 Cast iron diesel cylinder head showing cracks from injector opening to valves. The cracks are shown by magnetic particle inspection.

B-2 Aluminum Cylinder Head Failures by Corrosion Fatigue

Leaking from a crack at the side of a large aluminum cylinder head was noted by the engine room personnel. This head was then made available for examination. The crack is shown in Figure 12-15.

It was noted that the crack extended from a plug (used to close an opening left by a core support) to the adjacent water passage. Microexamination of this area showed, as in Figure 12-16, that related cracking had originated at the plug opening and was associated with a corrosive attack in this area.

The mode of failure that suggests itself where corrosion and cracking are linked is that of corrosion fatigue. The fact that none of these cracking failures took place in less than 8000 hours and that numerous failures took place at slightly more than 8000 hours supports a fatigue-type failure mechanism. The fatigue strength of a material is defined in terms of the level of stress

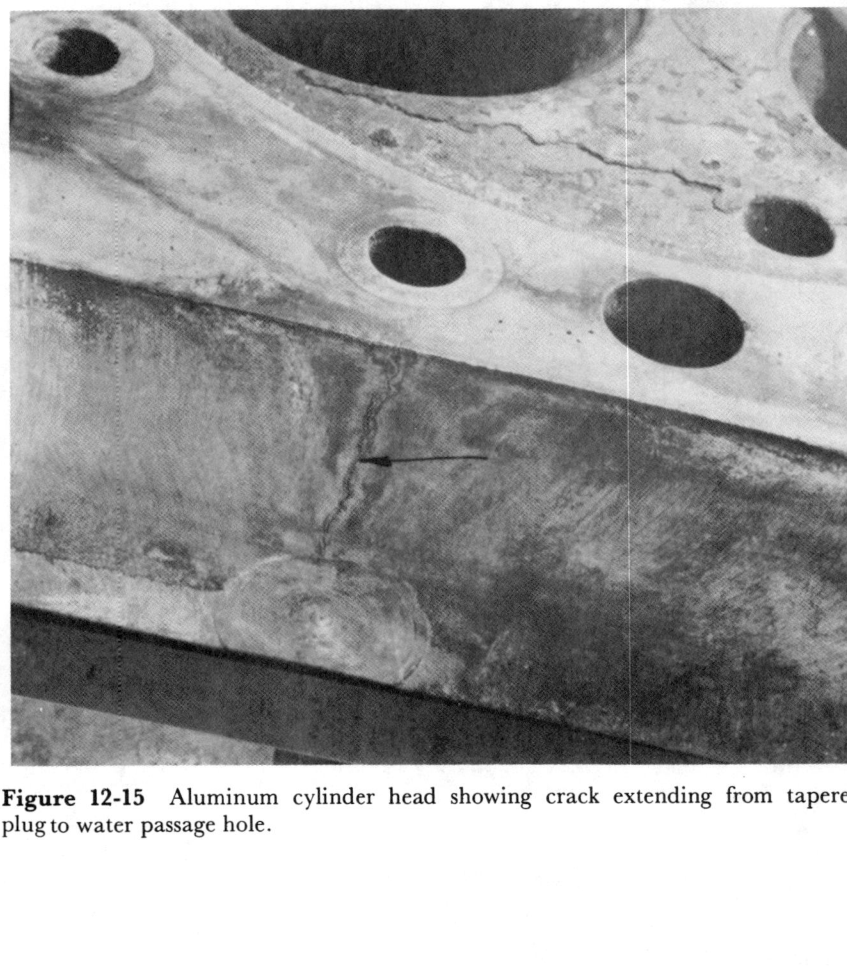

Figure 12-15 Aluminum cylinder head showing crack extending from tapered plug to water passage hole.

which will produce failure after a given number of repetitive stressing cycles. For the aluminum 5% silicon alloy the stress is 6500 psi for 500×10^6 cycles.* The effect of corrosion would be to decrease significantly the stress required for failure to take place.

In the case of the aluminum cylinder heads, cyclic stresses are associated with the firing of the cylinder. The magnitude of the stresses imposed during firing are not easily calculated in a complex casting, however, an approximate frequency may be obtained which suggests about 250×10^6 cycles will have taken place in 8000 hours.

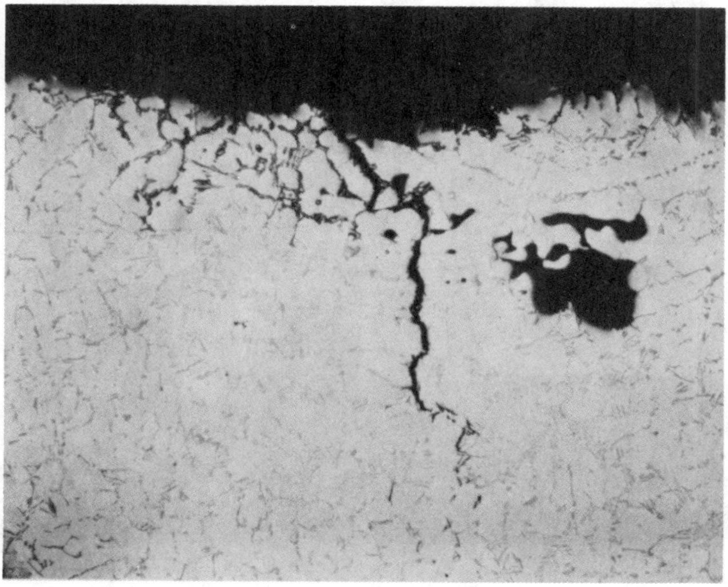

Figure 12-16 Initiation of corrosion fatigue crack in aluminum casting at the location of the tapered plug. $(\times 80)$

In addition to the cyclic stresses a number of static stresses were superimposed on the cyclic stresses. Some of these are: residual stress due to uneven cooling and shrinkage at the time of casting, thermal stress caused by uneven expansion of the casting because of temperature gradients through various portions of the head at operating temperature, torquing stress caused by the action of the holding down bolts modified by their different coefficients of expansion at operating temperature, and hoop stresses caused by insertion of tapered, threaded plugs. Pitting of course has the undesirable effect of concentrating the stresses.

* *Metals Handbook*, American Society for Metals, 1948, p. 806.

The effects of some of these stresses were evaluated using strain gauges. Type A-7 SR4 strain gauges were placed on a number of aluminum heads, as shown in Figure 12-17. Tensile stresses of as high as 6000 psi were measured at the location of the crack on uncracked heads. This value was partly due to residual casting stresses but mostly due to the hoop stresses produced when the tapered plug was inserted. It should also be noted that the evaluation of stresses at the tapered plug was made on a head which had not cracked. Heads which did crack may have had even higher residual stresses.

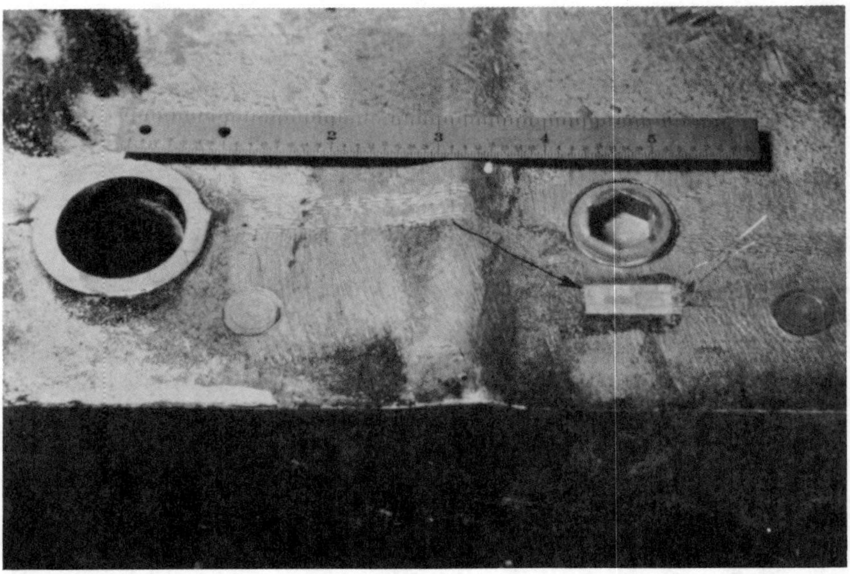

Figure 12-17 Strain gauge attached to cylinder head between the tapered plug and the water passage hole.

There had been some discussion concerning the severity of stresses produced at the location of the crack from torquing down of the cylinder head. To resolve this problem, strain gauges were mounted on several engine heads on a ship, the bolts untorqued and stresses determined. These torquing stresses ranged from 500 to 1000 psi—much smaller than those resulting from the tapered plug.

One remedy suggested was the elimination of these tapered plugs and the use of ordinary plugs having a gasketed shoulder.

B-3 Diesel Injector Failures by Stress Corrosion Cracking

The fracture of the spray tip of diesel fuel injectors was reported. Three

Figure 12-18 Three failed diesel fuel injectors. Fractures took place at the spray tips (arrows).

Figure 12-19 Sectioned injector spray nut showing location of fracture "A". A sharp corner also is noted at "B".

injectors, shown in Figure 12-18, were examined: the end was missing on one, the other two were cracked.

Spectrographic analysis identified the material as a high strength brass containing 6% aluminum, 4% manganese.

One of the cracked units was sectioned as shown in Figure 12-19. Here it was noted that the crack originated at the stress-raising corner at the change of section.

Microexamination showed transgranular corrosion cracking in the beta brass, as in Figure 12-20.

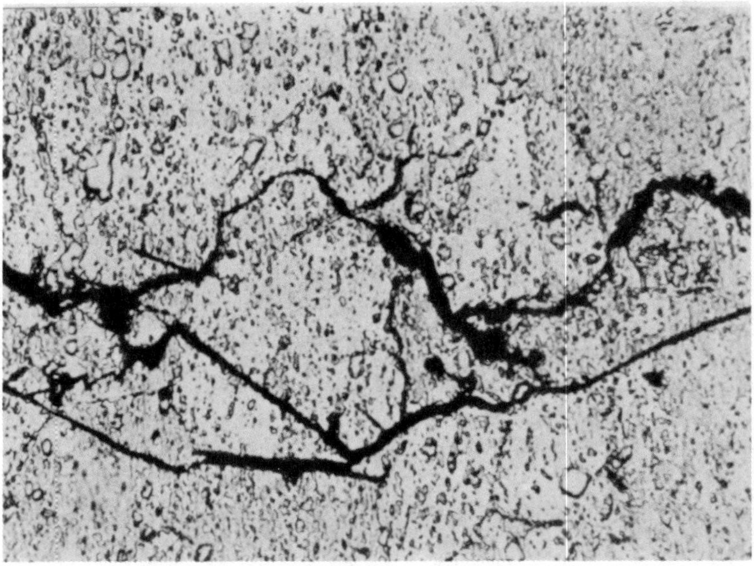

Figure 12-20 Transgranular stress corrosion cracks in high strength beta brass injector nut. (×150)

The high tensile brasses are known to be particularly susceptible to stress corrosion cracking in marine environments. In the case of the spray valve nuts exposure to raw sea water is unlikely, however, entrainment of fuel with salt water or other aggressive contaminant is certainly a possibility. The fact that three injectors failed within a brief period, and there was no previous history of failures of this component, strongly suggested the presence of a fuel contaminant. The stress component of the stress-corrosion mechanism appeared to be dominant. The presence of cyclic stresses and corrosion also suggested a corrosion fatigue mechanism as a distinct possibility in this failure.

Since these failures were the first of their type reported, in a component which has presumably given several thousand hours of satisfactory service, no change of material was recommended at that time. The obvious beneficial effect of a design change to provide a more gentle change of section at the spray tip bearing shoulder was drawn to the attention of the manufacturer.

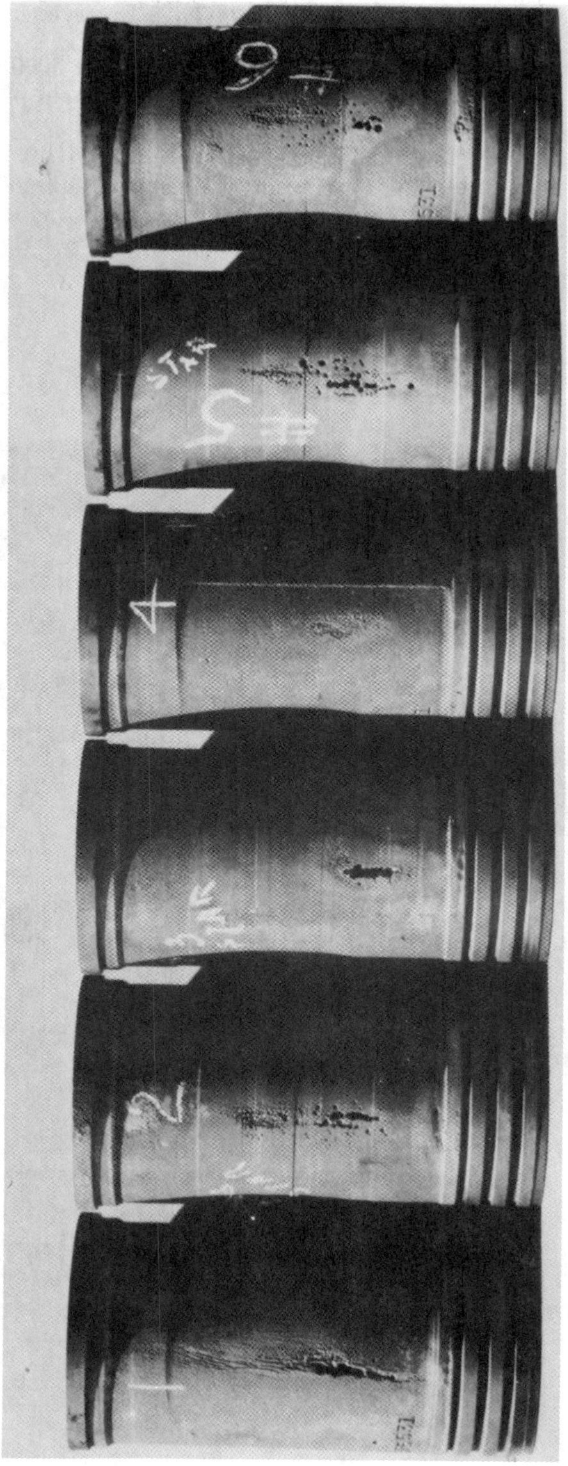

Figure 12-21 Cavitation pitting on six diesel cylinder liners.

B-7 Cavitation Corrosion on Cylinder Liners

Perforation of a diesel cylinder liner on a ship occurred after 3000 hours of running. The engine was fresh water cooled. Both of these main propulsion engines were dismantled for investigation.

Visual examination showed that severe external pitting had occurred on the waterside of all twelve liners. Six of these are shown in Figure 12-21. The pitting tended to be concentrated in a relatively small area and sometimes the pits coalesced to form a trench. The location of the pitting was the same in every instance, at right angles to the line of the crankshaft and in the power slap side of the liner. Similar pitting occurred directly opposite on the block. The preferred location and "honeycomb" form of the attack is typical of cavitation-corrosion.*

Figure 12-22 Cavitation pits on one liner. Note the ring of corrosion products around pits. (×4)

* Speller F. N. and LaQue F. L., "Water side deterioration of diesel engine cylinder liners", *Corrosion*, Vol. **6**, No. 7, July 1950, p. 209.

Figure 12-22 is an enlarged photograph showing the typical external appearance of the pitting. A ring of corrosion products was noted around each pit.

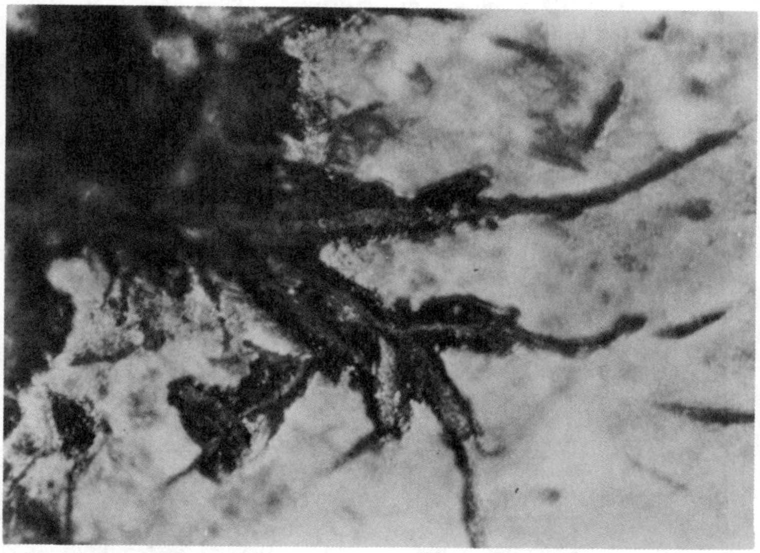

Figure 12-23 Section through corroded cast iron cylinder liner showing corrosion penetration along graphite flakes. (×400)

Microexamination of sections cut through a typical pit showed, as in Figure 12-23, an interesting effect. The graphite flakes from the original cast iron were embedded in the corrosion products and, because these relatively weak flakes have retained their original shape and size it was quite evident that mechanical breakdown was not the primary cause of the pitting. In fact, it could be concluded that corrosion was by far the predominant effect. Nevertheless, the preferred location and grouping of the pits substantiated a mechanical component.

Investigation soon showed that inhibitor control had been at best sporadic. Also evidence of sea water contamination (trace of chloride) was detected. The obvious remedial action was to tighten up the control of inhibitor dosage.

In those instances where the mechanical factor is predominant, the most promising means of damping out the power slap consists of a design change in the piston rings to permit an oil layer to be retained between piston and wall at all times.*

* Collins, H. H., "*Pitting of Diesel Cylinder Liners*", Diesel Engineering and Users Association, London, S 279, July 1961.

C-1 a) **Aluminum Cylinder Head Failure—Perforation by a Sand Core Wire**

A leak in the injector well of a large aluminum cylinder head was brought to our attention. These heads had an earlier history of leaks caused by corrosion, and it was expected that there might be some connection.

Radiographic examination showed no evidence of the corrosion pitting previously associated with the leaks and detected readily by radiography.

Figure 12-24 Sectioned aluminum cylinder head showing steel wire (part of sand core) which had perforated the casting at the injector well causing water leakage.

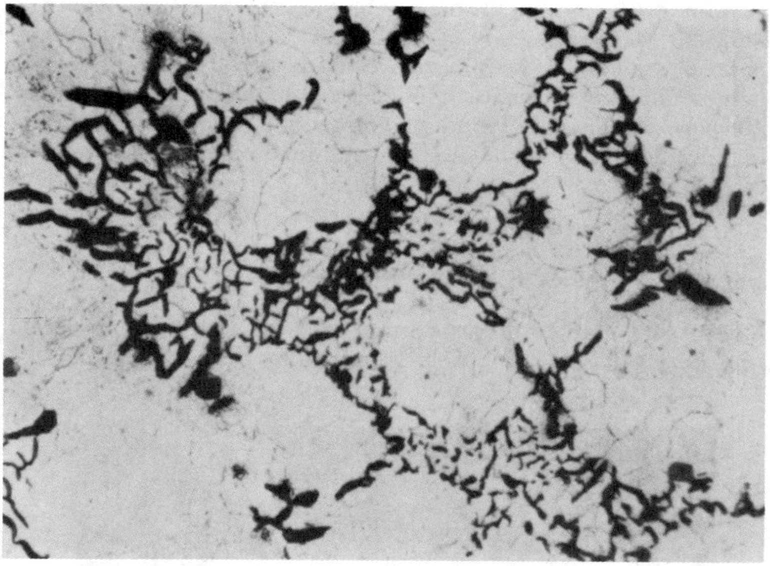

Figure 12-25 Microstructure of failed piston ring showing fine graphite flakes in matrix of ferrite. (× 100)

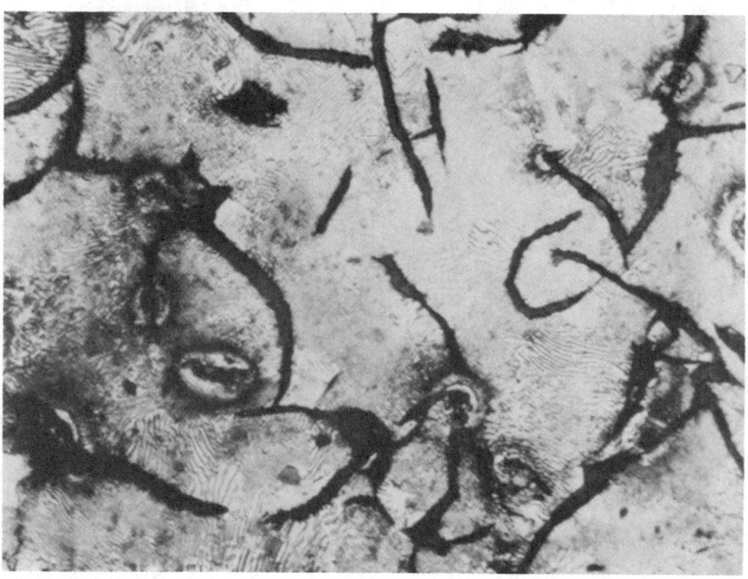

Figure 12-26 Structure of normal grey cast iron piston rings showing graphite flakes in a matrix of pearlite. (Note higher magnification than Figure 12-25.) (× 500)

Sectioning of the aluminum head, Figure 12-24, indicated that the leak in the injector well was due to the presence of a length of steel wire which had been part of a sand core used in the casting procedure. The presence of the steel wire resulted in the corrosion of the aluminum adjacent to the wire, and since the wire was loose, its vibration removed the corrosion products thereby accelerating the corrosion action. The wire finally perforated the casting at the well.

C-1 b) Piston Ring Failures—Material and Porosity

The failure of a number of piston rings on large reciprocating pumps was reported. Several of the pieces were submitted for examination.

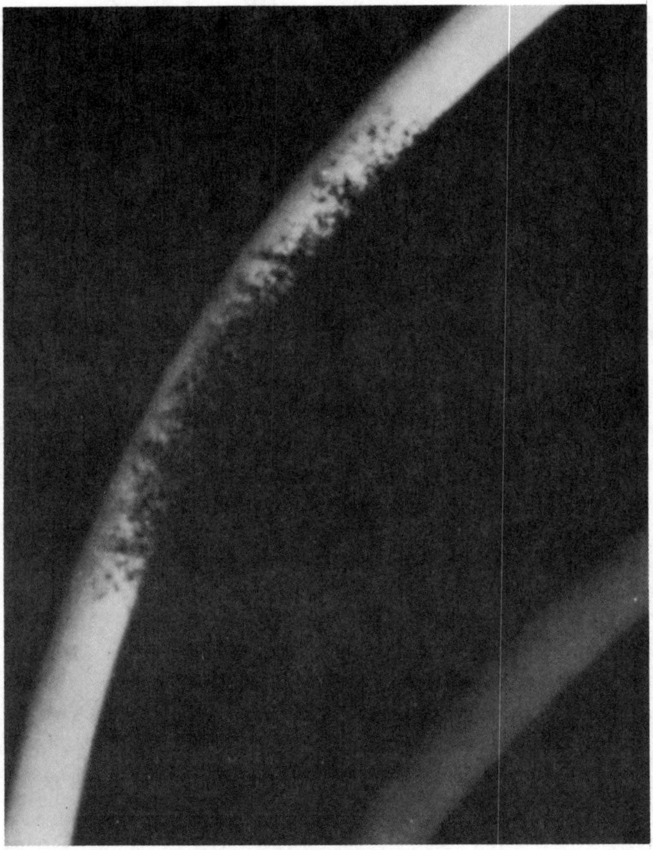

Figure 12-27 Radiograph of piston rings showing area of severe porosity in top ring.

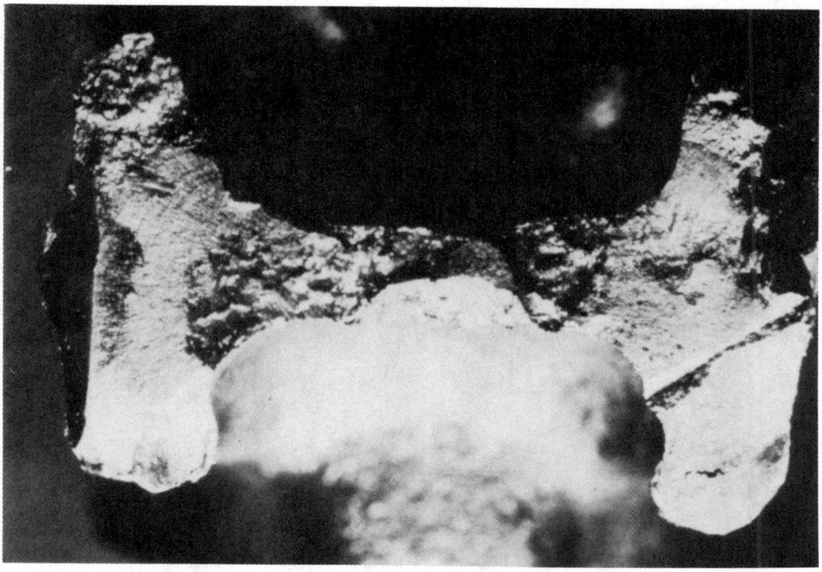

Figure 12-28 Fractured connecting rod showing fatigue markings originating in the web.

Radiographic examination showed that these were essentially sound.

Metallurgical examination indicated that the piston rings were fabricated from a cast iron with a ferrite plus graphite structure as shown in Figure 12-25. This material would not have the fatigue strength nor the spring properties of a good quality grey cast iron of the type usually used for piston rings shown in Figure 12-26.

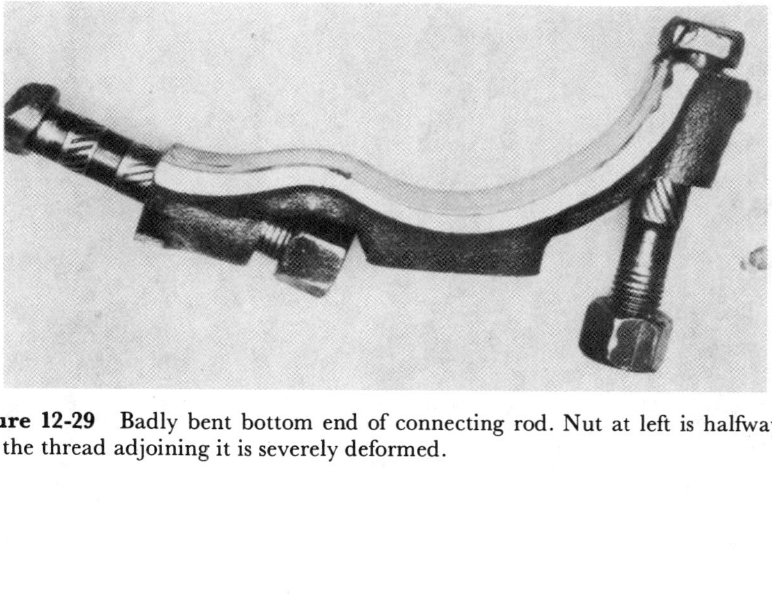

Figure 12-29 Badly bent bottom end of connecting rod. Nut at left is halfway off and the thread adjoining it is severely deformed.

Subsequently replacement rings were supplied and, prior to installation, were submitted for evaluation. In this instance, radiography, as shown in Figure 12-27, revealed gross unsoundness in a few of the rings.

The value of radiography prior to expensive machining was strongly demonstrated.

C-9 Connecting Rod Failure in Truck Engine

The laboratory was called on to examine broken pieces resulting from engine failure in a large truck. These pieces included the broken camshaft and parts of three pistons and connecting rods.

Careful examination of the broken surfaces of the camshaft did not indicate

Figure 12-30 Connecting rod showing small forging fold in web (arrow).

any flaws or other possible contributing factors. Examination of the remainder of the parts showed two features of interest:

a) One of the con rods had separated by a fatigue failure (Figure 12-28) approximately one inch from the wrist pin.

b) Most of the bottom end nuts were readily turned by hand—there apparently was no provision for locking them. One badly bashed bottom end (the one with the fatigue fracture further up) had apparently been free of nuts when the bashing took place. Another bottom end was also severely bent, both nuts were partly off and the thread adjoining one of these nuts was severely flattened (Figure 12-29), suggesting that they were loose prior to the accident.

In addition a small rupture (Figure 12-30) was noted near the big end of the con rod on which the fatigue fracture occurred. It seems possible that a similar weakness served as the initiating source for the fatigue fracture of Figure 12-28, which began in the centre portion of the rod.

In evaluating the overall picture it did seem that relying solely on torqued nuts to remain tight with no provision for locking, tended to leave a doubtful situation. Admittedly this is now the practice on countless vehicles on the road. There seems to be evidence that nuts were loose at the time of the catastrophe. The fatigue failure can most readily be explained from abnormal pounding related to loosening of the bottom end.

XIII

BEARINGS

MISALIGNMENT LED to overloading and failure of a compressor bearing. Another bearing from a main propulsion unit, failed by ingress of foreign material.

Severe corrosion of a tin-based babbitt resulted from water contamination and additives in the extreme pressure oil. Copper-lead bearings also suffered corrosion from possible sea water contamination of the oil. Corrosion primarily and some misalignment contributed to failure of a ball bearing in a sea water pump. An unusual corrosion attack of a thrust pad bearing involved selective removal of the matrix in the tin base babbitt.

Identification of a bearing metal as a lead solder resolved one diesel bearing failure. Ultrasonic examination proved valuable in determining adequacy of bond (babbitt to steel) in a number of main bearings.

A-4 Ball bearing failure.

B-1 a) Foreign material causes bearing failure.
b) Tin base babbitt bearing failures by corrosion.
c) Corrosion failure of copper-lead bearings.
d) Corrosion failure of ball bearings.

B-7 Thrust pad bearing failure.

C-7 Diesel bearing failure—a composition problem.

C-9 Poor babbitt-steel bonds cause bearing failures.

A-4 Ball Bearing Failure

A main ball bearing from an air compressor was removed for examination during an overhaul. The cage was broken, one ball was split in two, several other balls were heavily scored, and both inner and outer races showed signs of being badly overheated.

The outer and inner races were cut in half to facilitate further inspection. Examination of the inner race indicated that the bearing had been subjected to misalignment, putting an end thrust on the bearing, and to overloading. Figure 13-1 shows both halves of the inner race. The failed surface is to one side of the central track and extends about half way around the race.

Examination of the casting into which the bearing fitted showed that, unless the bearing was hard up against a locating shoulder, misalignment was feasible.

Figure 13-1 Both halves of inner ball race showing damage on one side only due to misalignment.

B-1 a) **Foreign Material Causes Bearing Failure**

The failure of a crankshaft bearing on a ship's main propulsion engine was reported and the bearing was made available for examination.

One half of the failed bearing is shown in Figure 13-2. Spalling could be noted at one end and numerous small dark pits were widespread. The failed bearing was sectioned and the dark material was examined under the microscope. As shown in Figure 13-3 the dark spot resulted from the presence of foreign material imbedded in the babbitt.

Spectrographic analysis confirmed that the bearing metal was a lead-base babbitt. X-ray diffraction indicated that the material imbedded in the babbitt was magnetic iron oxide (Fe_3O_4).

The dark spots shown in Figure 13-2 appeared to be corrosion pits. The presence of the magnetic iron oxide in the lead-base babbitt apparently resulted in some galvanic corrosion of the babbitt, the Fe_3O_4 being more cathodic.

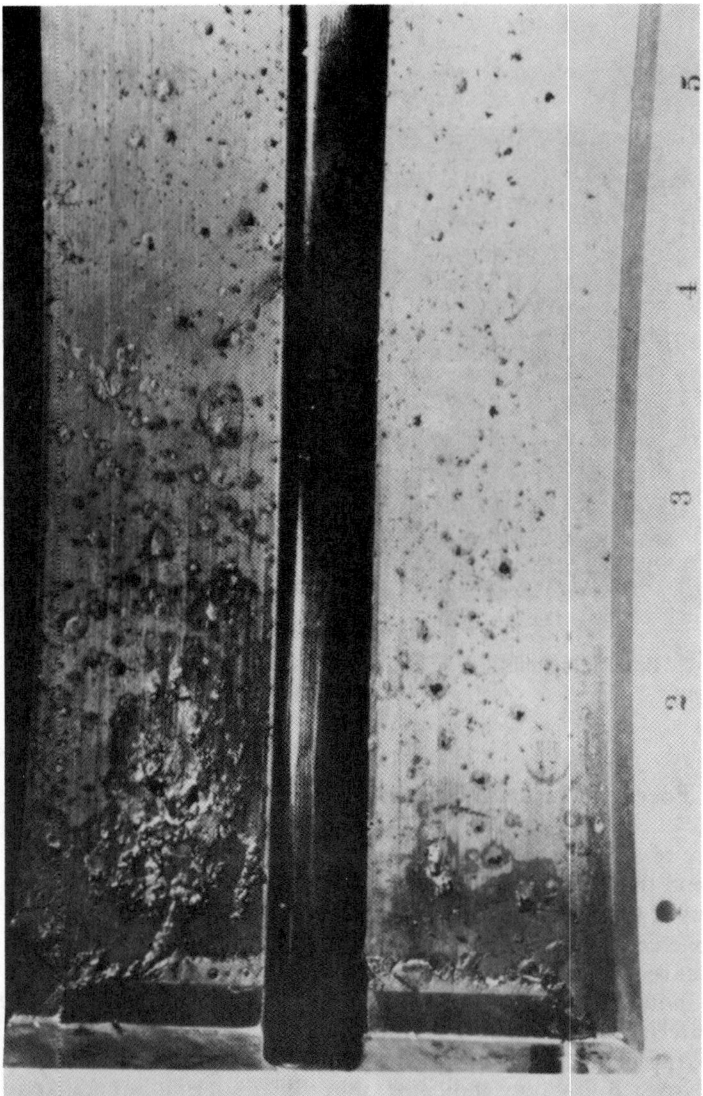

Figure 13-2 Failed bearing showing spalling and corrosion pits resulting from the ingress of foreign material.

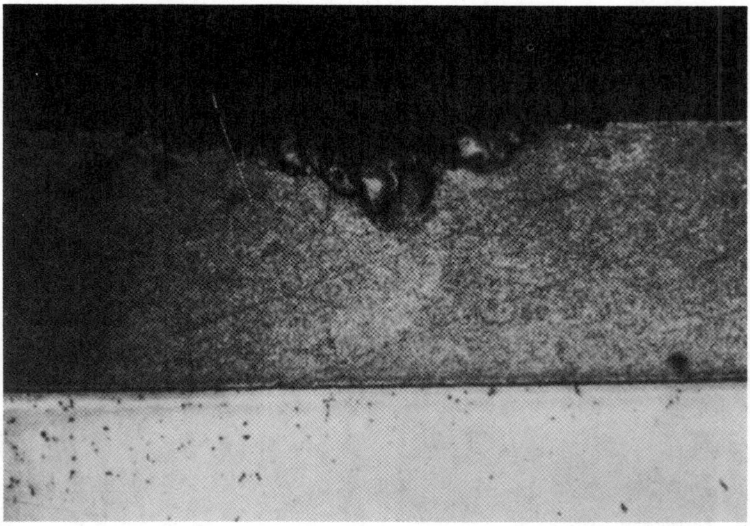

Figure 13-3 Section through pit in Figure 13-2 showing particles of magnetic iron oxide imbedded in the babbitt. (× 50)

The failure of the crankshaft bearing resulted from the presence of the non-metallic foreign material. The source of the magnetic iron oxide was not established.

B-1 b) Tin Base Babbitt Bearing Failures by Corrosion

Indications of overheating resulted in an inspection of journal bearings on a large pump. The bearing shown in Figure 13-4, failed after approximately 2000 hours in service.

Analysis of the babbitt metal indicated that it was a good quality tin-based material.

Metallurgical examination of the blackened babbitt (Figure 13-5) indicated that the bearing metal had been attacked by corrosion to a depth of approximately 0.002 inches prior to overheating and failure. This corrosion attack is attributed to the presence of moisture in the lubricating oil. This lubricating oil containing extreme pressure additives was known to have a greater affinity for water so that separation (water from oil) was more difficult.

The corrosion attack noted above probably initiated the breakdown. The corrosion resulted in a brittle bearing surface which has been known to crack and spall. Also, the black material has a greater volume than the original babbitt; that is, clearances would decrease as the corrosion attack progressed. It was recommended:

a) that greater effort be made to keep water content to a minimum level;
b) that, if permissible, a non-additive oil be used.

Figure 13-4 Large tin base bearing showing black tin oxide on worn and spalled bearing face.

Impending damage of this type has since been detected from the presence of tin in samples of oil analysed periodically to guard against early failures.

B-1 c) Corrosion Failure of Copper-Lead Bearings

Failed main bearings from a marine gasoline engine were removed and sent up for examination. One of these bearing surfaces is shown in Figure 13-6. Large areas of the bearing surface had spalled off.

As determined by a spectrographic analysis, the bearing metal was basically a copper–lead alloy. This alloy had been sintered onto a steel backing and flashed with a soft solder.

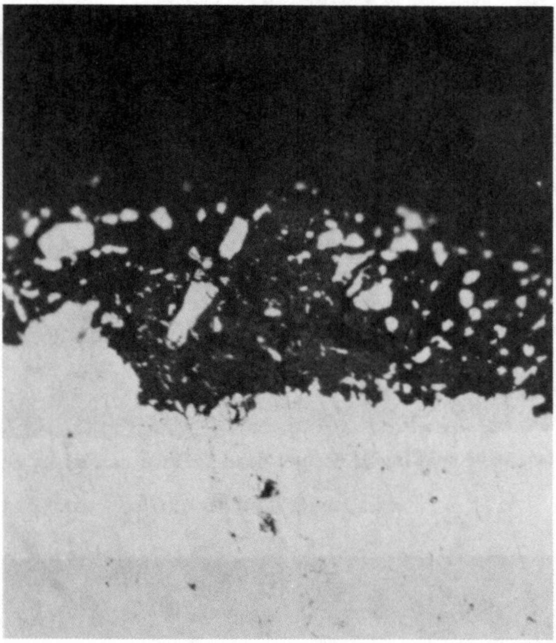

Figure 13-5 Section through babbitt (Figure 13-4) showing corrosion attack on tin-rich matrix. Note that the black hard corrosion product is cracked. (×400)

Metallurgical sectioning indicated that the bearing metal had been attacked by corrosion as shown in Figure 13-7. This section through one of the pits indicated that the lead had corroded preferentially between the copper globules.

It is known that copper–lead bearings show poor resistance to corrosion in certain oils. Although no samples of oil were available for analysis, it was noted that a trace of chloride was present on the corroded bearing surface, suggesting that the oil may have been contaminated with sea water.

Although the failure mechanism appeared to be corrosion, there was also evidence that the copper was not uniformly distributed in the bearing metal as shown in Figure 13-8. A non-uniform bearing metal could contribute to a decreased fatigue resistance.

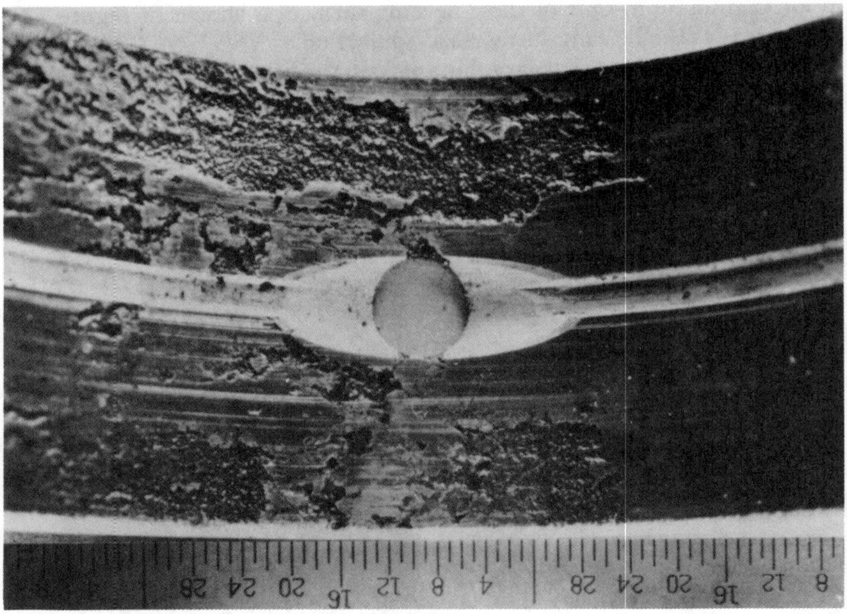

Figure 13-6 Extensive spalling of copper–lead babbitt caused by corrosion attack.

Figure 13-7 Section through pit in copper–lead babbitt showing preferential corrosion of lead. Globules show that the babbitt was a powder metallurgy product. (×150)

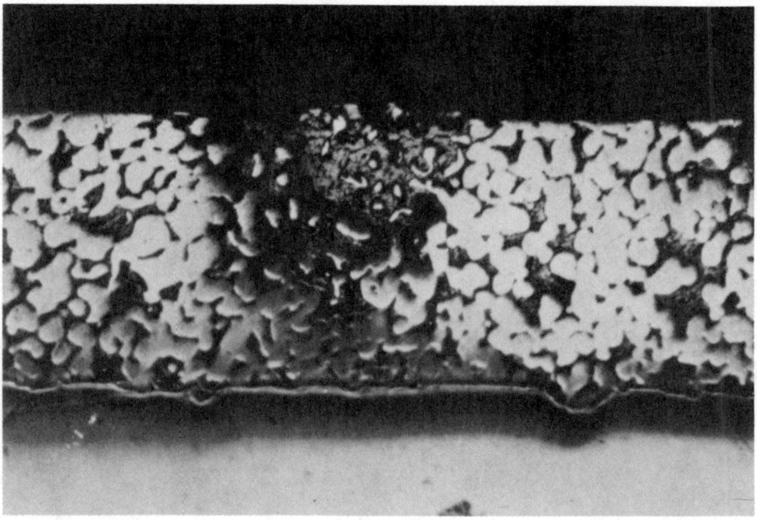

Figure 13-8 Section through copper–lead babbitt showing an area low in copper content. (×80)

B-1 d) **Corrosion Failure of Ball Bearings**

Frequent failures of ball thrust bearings from a sea water pump were reported. Several of these thrust bearings were made available for examination and one of the bearing races is shown in Figure 13-9.

Examination of the ball races indicated that all of the races showed wear. The outer edges of an inner race shown in Figure 13-9 were the most severely damaged areas. There was evidence that misalignment was a factor in the bearing failure: the inner races were worn on the outer edges whereas the outer races were worn on the inner edges.

Examination of the less severely worn races showed that corrosion was probably a major factor in the initiation of the bearing failure. Figure 13-9 shows several areas of corrosion attack extending down into the race. Once attack of this type has extended into the raceways, a breakdown of the bearing could quickly take place.

The presence of a considerable amount of corrosion on various parts of these ball bearings suggested that the oil seals allowed water to enter the bearing area.

B-7 **Thrust Pad Bearing Failure**

Wastage of the babbitt on thrust pad bearings of a turbine pump was noted

Figure 13-9 Hull and Fire pump thrust bearing inner race showing corrosion attack extending over spalled area.

during a refit. Two of these pads, shown in Figure 13-10, were brought in for examination.

Two Michell bearings, one on each side of a collar on the pump shaft, restrict the lateral movement of the shaft. The thrust bearings consist of a number of babbitted pads which are designed to tilt slightly in a backing plate. Once the shaft is rotating, no direct contact should exist between the bearing and collar because the rotation should force an oil wedge between the tilted pads and the collar. The pump is so designed that one of the bearings is always slightly loaded while the other acts only as a guide.

Visual examination suggested a surface porosity or surface attack. Examination of a section under the microscope showed that the metal had been severely corroded. The matrix material had apparently been preferentially

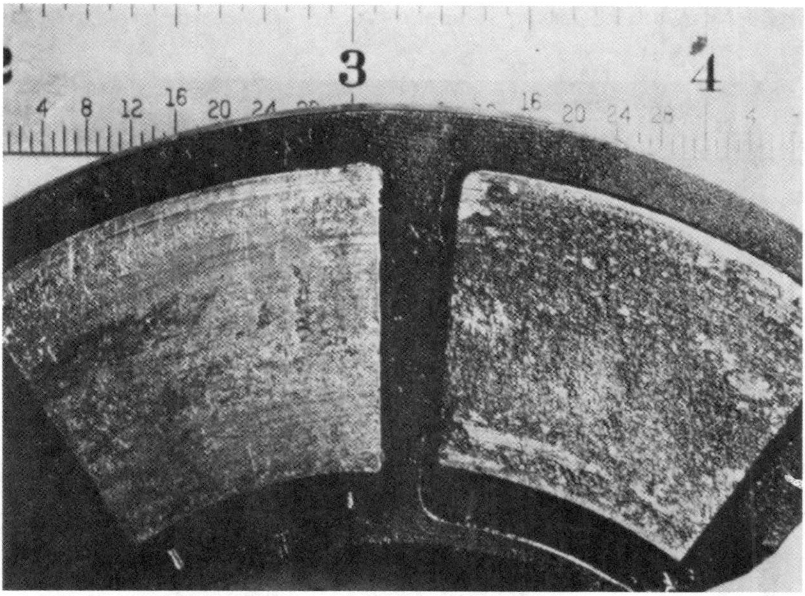

Figure 13-10 Thrust pad bearings showing damage on tin babbitt surface.

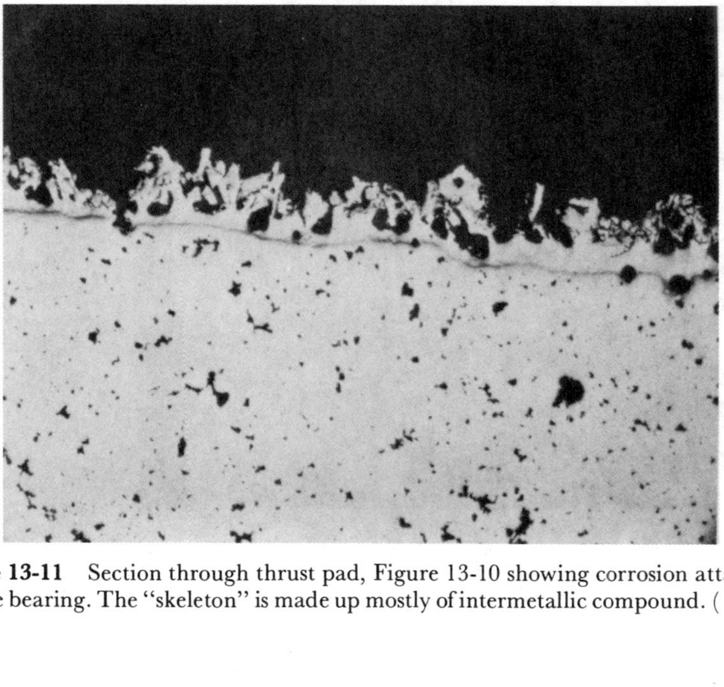

Figure 13-11 Section through thrust pad, Figure 13-10 showing corrosion attack on tin base bearing. The "skeleton" is made up mostly of intermetallic compound. (× 150)

Figure 13-12 Spalling of bearing metal on connecting rod bearings of a diesel generator.

attacked leaving essentially the intermetallic phase, as shown in Figure 13-11.

Spectrographic analysis showed that the babbitt metal was the specified good quality tin base, containing copper and antimony as hardeners.

Past experience with tin base bearing metal of this type has indicated that corrosion often occurs in oil contaminated with water. We learned later that the oil from this pump had, in fact, been contaminated with water and other impurities.

The corrosion attack usually blackens the bearings, the hard black material being the tin oxide corrosion product. However, in the above instance, the attack had been associated with a mechanism which has removed the corrosion products, even from well down in the babbitt. One possibility is that some ingredient in the contaminated oil resulted in the preferential dissolution of the matrix leaving behind the more resistant phases.

The exact mechanism or factors responsible for this specialized attack were not established.

C-7 **Diesel Bearing Failure—A Composition Problem**

Con rod bearings in a ship's diesel generator failed by spalling as shown in Figure 13-12.

Figure 13-13 Section through lead–tin alloy bearing showing failure initiation. Note absence of intermetallic compounds in bearing metal.

Visual examination suggested a poor bond because of the relatively clean separation at the bronze backing. Metallurgical examination of a piece cut near the failure area showed (as in Figure 13-13):

a) absence of intermetallic compounds normally expected in a babbitt metal;

b) indication of lack of bond at the interface, and

c) evidence of separation or weakness in the body of the bearing metal.

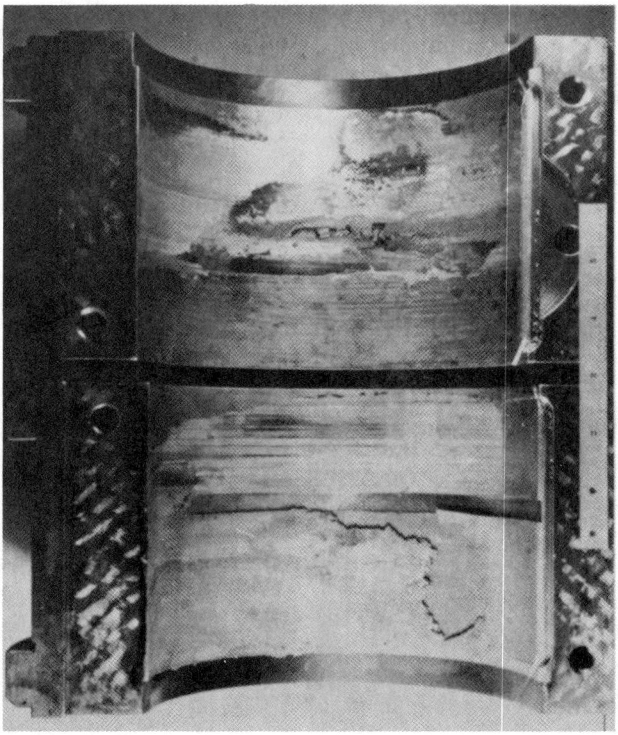

Figure 13-14 Failed main pinion bearing. Note that a large part of the babbitt has peeled away from the lower half of the bearing.

Spectrographic analysis showed this bearing to be essentially lead with a minor amount of tin. This composition might more properly be considered a solder, rather than a babbitt. Most lead-base babbitts contain from 10 to 15% antimony to provide the wear resisting intermetallic phases.

C-9 **Poor Babbitt-Steel Bonds Cause Bearing Failures**

The early failure of a number of critical main gearing bearings was a matter of great concern. These bearing failures, shown in Figure 13-14, resulted in premature and expensive maintenance. Large areas of the babbitt metal had spalled away from the backing and other areas could be readily lifted with a knife blade.

Analysis showed that the bearing metal was a good quality tin-base babbitt.

Because there were no replacements available, detailed examination of bond areas could not be made, and it was highly urgent that quality of bond of other bearings of this type in service, and of future replacements, be known.

Accordingly experiment work was undertaken and an ultrasonic method developed* to enable the quality of the bonds to be evaluated. Using this method a number of these bearings were examined, and it was shown that a few were poorly bonded over a large part of the interface. The problem appeared to be the difficulty in getting a tin-base babbitt to bond directly to a steel backing, even with prior tinning. A much better approach was the use of bronze backing to ensure proper bonding.

* Peters B. F. "An ultrasonic method for evaluating bonds in large metallic bearings", *Materials Evaluation,* Vol. **22**, No. 6, June 1964, p. 257–259.

XIV

SPRINGS

FATIGUE CRACKING of an unusually repetitive nature was the result of stressing bronze springs excessively. Inadequate operation of scroll springs by gumming-up with corrosion products was relieved by drainage and corrosion prevention. The value of shot peening springs to prevent corrosion action is discussed in two case histories.

Defects such as a seam incorporated in the original steel stock and those introduced by grinding are cautioned against.

A-4 Phosphor bronze hatch spring failure.

B-1 Scroll spring failure by corrosion.

B-4 Failure of non-shot-peened springs.

C-2 Longitudinal defect in steel spring.

C-4 a) Stress corrosion cracking of small springs.
b) Failure of spring at ground surface.

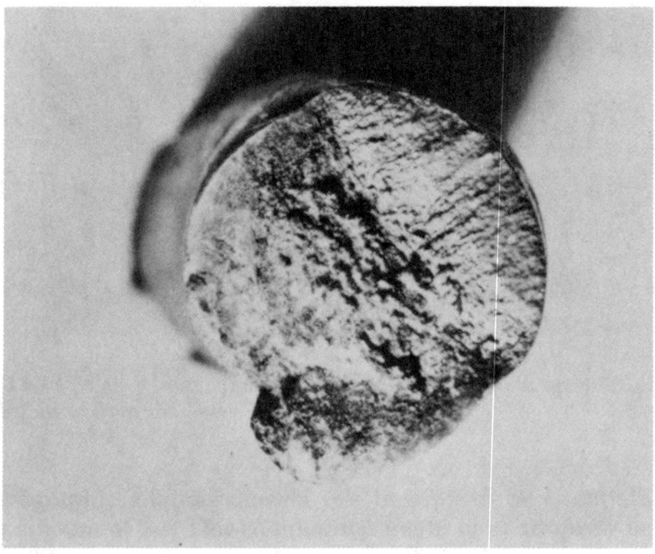

Figure 14-1 Fractured surface of phosphor bronze spring showing fatigue markings.

A-4 Phosphor Bronze Hatch Spring Failure

Early failure of hatch cover springs was brought to laboratory attention. Fracture of these springs shown in Figure 14-1 was occurring as early as 8 months.

The markings on this fracture surface indicated that failure took place at a fatigue crack which had propagated nearly half-way through the cross-section. A careful examination of the remainder of the spring, using dye penetrant methods, revealed a whole series of surprisingly uniformly spaced cracks on the inner surface as shown in Figure 14-2.

Figure 14-2 Failed phosphor bronze hatch spring showing numerous cracks on the inside surface outlined by dye penetrant.

Micro-examination confirmed that these were fatigue cracks propagating through the highly worked spring material.

Spectrochemical analysis showed that the spring was a good quality phosphor bronze (5% tin).

In the absence of evidence of corrosion or deleterious flaws, it was assumed that overloading was a likely mechanism. Accordingly, calculations using standard formulae were made and it was shown that with a 60 degree turn of the hatch, a stress of 76,900 psi was applied to the bronze spring. Since the maximum tensile stress should be limited to about 50,000 psi for the phosphor bronze it appears that the fatigue cracks in the springs were primarily due to the high stresses which resulted from normal service.

Because changing the over-all size of the spring was undesirable, a suitable steel spring was recommended.

B-1 Scroll Spring Failure by Corrosion

Failures and inadequate operation of scroll springs—used to achieve a quick shut off for main stop valves—was reported by several ships. These springs often broke up into two or three pieces. One of these springs is shown in Figure 14-3. The dark greasy deposit built up on the surface of these springs may be noted.

Figure 14-3 Scroll spring failure showing deposit on spring.

It was apparent that this deposit would hamper the spring in its operation by greatly increasing the friction between the metal surfaces when they slide against one another. There had been some discussion covering the possibility that this material might be a carbonaceous residue from overheating of the lubricant applied. Chemical analysis showed that it was almost wholly iron oxides.

As shown in Figure 14-4 these springs became rather badly corroded while in service, and the corrosion products (iron oxides) combined with the lubricant to form these unwanted deposits.

Study of a $\frac{3}{4}$ inch long crack in one of these springs suggested that corrosion also played a part in its initiation and propagation. There was no doubt that corrosion pits served as starting points for the fatigue failures.

The metal of the springs was analyzed and it proved to be a plain carbon steel. Examination under the microscope, as well as hardness measurements,

Figure 14-4 Cleaned surface of scroll spring showing corrosion attack. (×4)

showed that the steel had been heat treated to a reasonable spring temper. The actual hardness, Rockwell C40-43 is a bit on the low side for optimum resistance to permanent set. A more desirable hardness recommended was Rockwell C44-46.

It was recommended that steps be taken to prevent accumulation of moisture by appropriate drainage and that a grease which preferentially wetted the steel surface in the presence of water be used. Both of these measures were taken and proved successful.

B-4 **Failure of Non-Shot-Peened Springs**

Failure of springs on the explosive doors of a diesel engine were brought to our attention. These springs had been in service for approximately 15 years and, after being subjected to an alkaline cleaning, had fallen into several pieces. A few of the springs did not fail in this manner.

Examination at low magnification showed that those springs which fell apart had many additional fine cracks on the inner surface, such as those shown in Figure 14-5. The other springs had a lightly mottled surface (evidently shot-peened) and were free of any defects.

A chemical check of the alkaline cleaner failed to reveal any ingredient which might have acted in a deleterious manner, e.g. caustic; also many other parts had been subjected to the same cleaner with no bad effects.

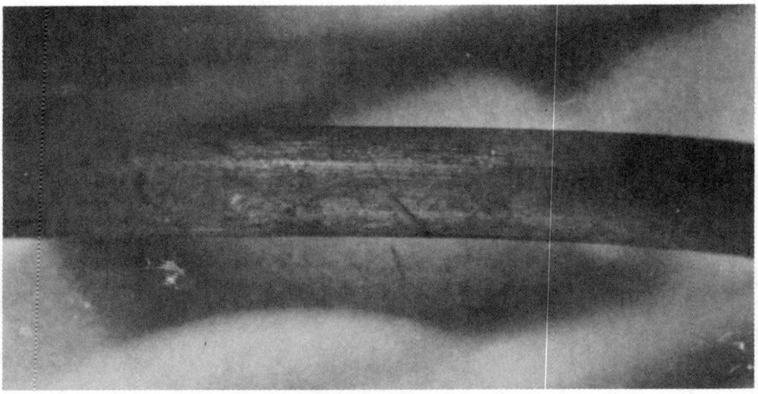

Figure 14-5 Inner surface of failed spring showing a fine diagonal crack. (×10)

Micro-examination of one of the areas containing cracks indicated that the cracking was intergranular. This is shown in Figures 14-6 and 14-7. The intergranular character of the cracking indicated that corrosion played a part in this failure.

The evidence available suggested that the long service period coupled with exposure to combustion products, i.e. stress and a corrosive (acid) environment, served to produce cracks to the point of failure in some instances. The shot peened springs, on the other hand, had a surface which had been

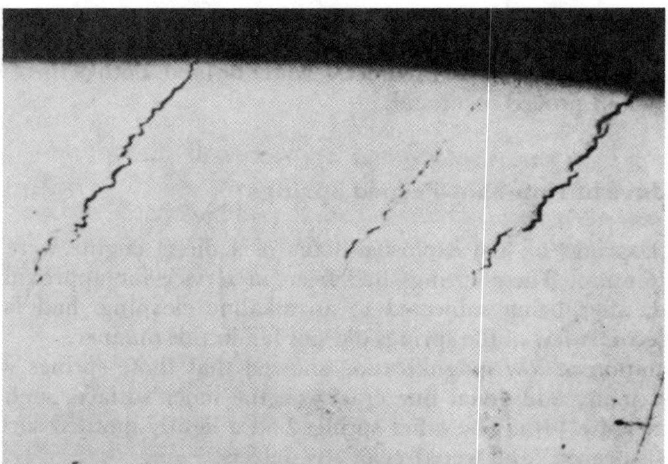

Figure 14-6 Section through cracked spring showing three diagonal cracks. (×80)

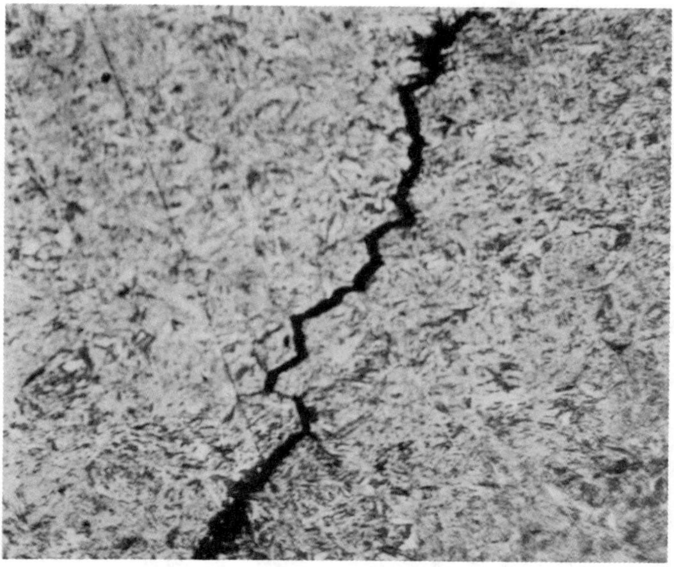

Figure 14-7 Intergranular type cracking in slightly tempered martensitic structure of spring. (×430)

pre-stressed in compression. The surface of a shot-peened spring was then only subject to tensile stresses which exceeded the compressive stresses imposed by peening. This resulted in a greater fatigue life under normal conditions, and certainly a better resistance to a combination of stress and corrosion.

Although the service life of the non-shot-peened springs had been satisfactory, it was considered good practice in future replacements to specify shot-peened springs.

C-2 **Longitudinal Defect in Steel Spring**

Failure of a breech spring on a large gun was brought to the attention of the Laboratory. The fracture surface is shown in Figure 14-8.

Analysis showed that the spring was an A.I.S.I. 1080 spring steel.

Visual examination of the spring fracture, Figure 14–8, showed that the failure had initiated at a longitudinal seam. Sectioning of the spring near the fracture, Figure 14-9, showed that a crack extended deep into the spring from the surface. A close look at the crack, shown in Figure 14-10 revealed

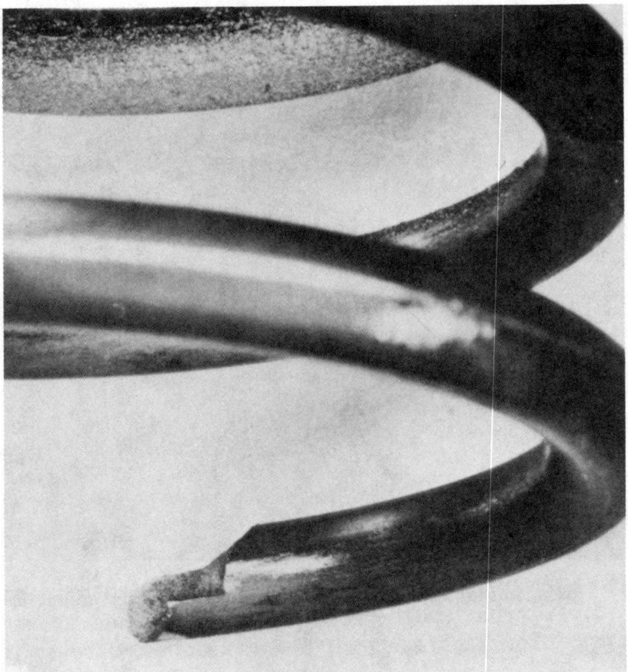

Figure 14-8 Breech closing spring showing longitudinal seam at fracture.

that about half of the separation was associated with an inhomogeneity in the metal structure; that is, the failure started at a manufacturing flaw. The extension of the crack likely resulted from the propagation of the original flaw under service stresses.

In summary, the failure of this spring was attributed to a metallurgical flaw which originated in the steel making process.

C-4 a) **Stress Corrosion Cracking of Small Springs**

A number of failed springs from a small pump were submitted to the laboratory. Figure 14-11 shows two of these springs: one fractured at each end in a manner typical of most of the failures, and one broke at the centre.

Spectrographic analysis indicated that the springs were manufactured from carbon steel.

Microscopic examination showed that the failure mechanism was that of stress corrosion cracking. A number of fine cracks extended into the spring from a fractured surface. It was likely that the action of residual chlorides

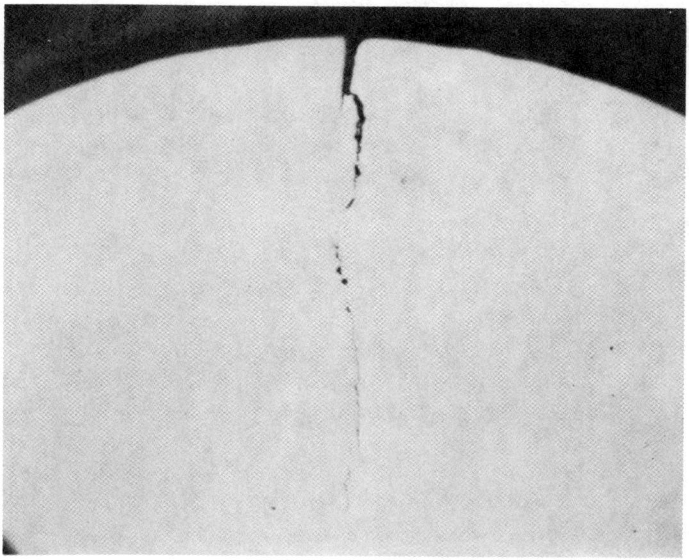

Figure 14-9 Section through spring near failure showing longitudinal crack. (× 30)

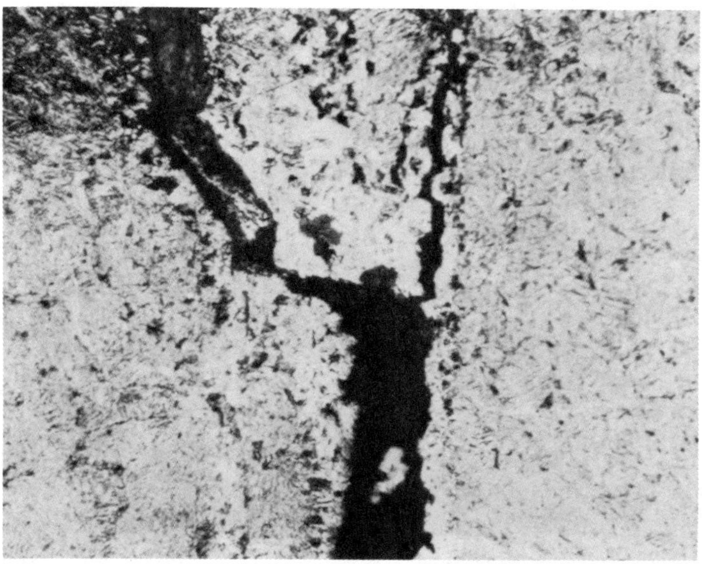

Figure 14-10 Microstructure near crack showing inhomogeneity of steel. (× 400)

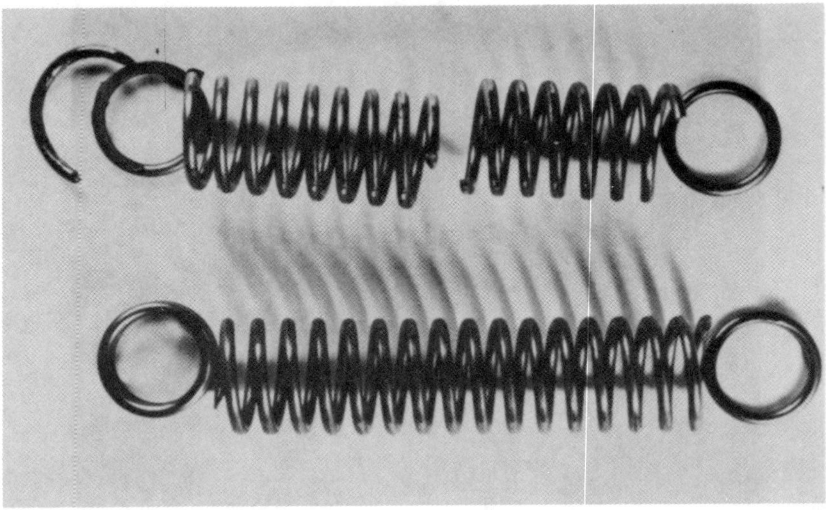

Figure 14-11 Two springs which failed by stress corrosion cracking.

in the fuel combined with the stresses present accounted for this type of cracking.

Springs of the type examined are usually shot peened to make them more resistant to failure by stress corrosion cracking (and fatigue). It is noteworthy that each spring examined showed grinding (or possibly wear or polishing) on both the inside and outside. The removal of the shot peened metal by any one of the above methods would result in an effect opposite to shot

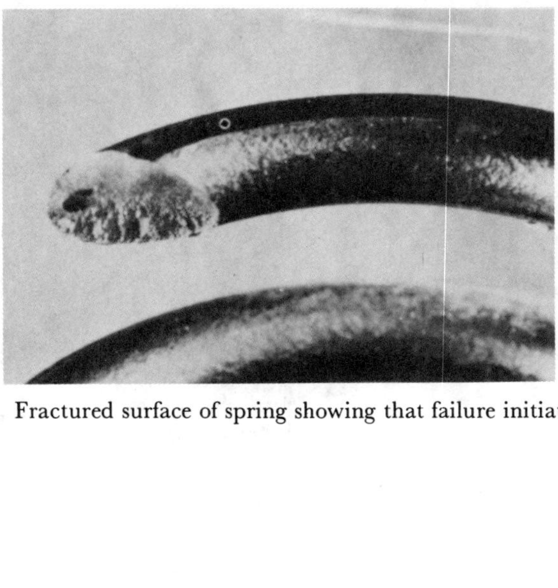

Figure 14-12 Fractured surface of spring showing that failure initiated at a ground surface.

Figure 14-13 Fractured surface of steel spring. Note that failure initiated at the ground area (arrow). (×4)

peening; that is, the susceptibility of the metal to stress corrosion cracking would increase. An examination of the fractured surfaces (Figure 14-12) indicated that the fractures initiated at a surface which had been ground. It is likely that the springs could have withstood the combined action of the stresses and corrosion had the resistant shot peened surface metal been left intact.

C-4 b) **Failure of Spring at Ground Surface**

The failure of a large spring on a steam engine was reported and the fractured spring was submitted for examination.

Spectrographic analysis showed that the spring was a good quality low alloy steel. The metal had been suitably heat treated to Rockwell C42.

Examination of the fracture (Figure 14-13) showed that its origin was associated with a surface mark which had apparently affected the structure of the metal to a slight depth (note the dark spot at the initiation of the fracture). Figure 14-14 shows that the mark that initiated the failure was probably produced when the spring was ground flat on the end.

Metallurgical examination showed that the spring had been heat treated after grinding and this might have been expected to alleviate adverse metallurgical changes. However, heat treatment could not eliminate a crack which developed from severe grinding.

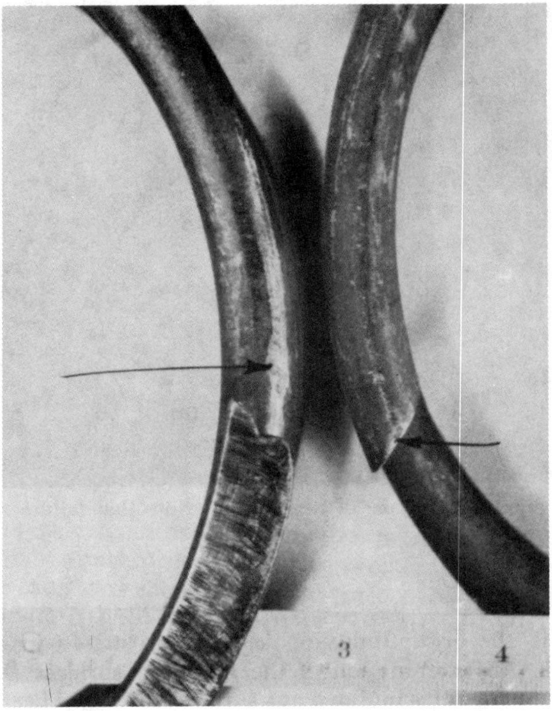

Figure 14-14 Fractured spring (right) alongside identical spring. Note that the fracture (arrow at right) occurred where grinding of the end of the spring extended onto the lower coil. Arrow at left also shows grinding marks.

XV

PROPELLERS AND ANCHORS

DIAGNOSIS OF A mysterious failure of a brass propeller proved it could only be by an impact blow. Other brass propellers can fail readily by stress corrosion cracking.

Anchor shanks need to be heat treated to provide useful toughness. Metallurgical examination showed why torpedo propellers were inadequate.

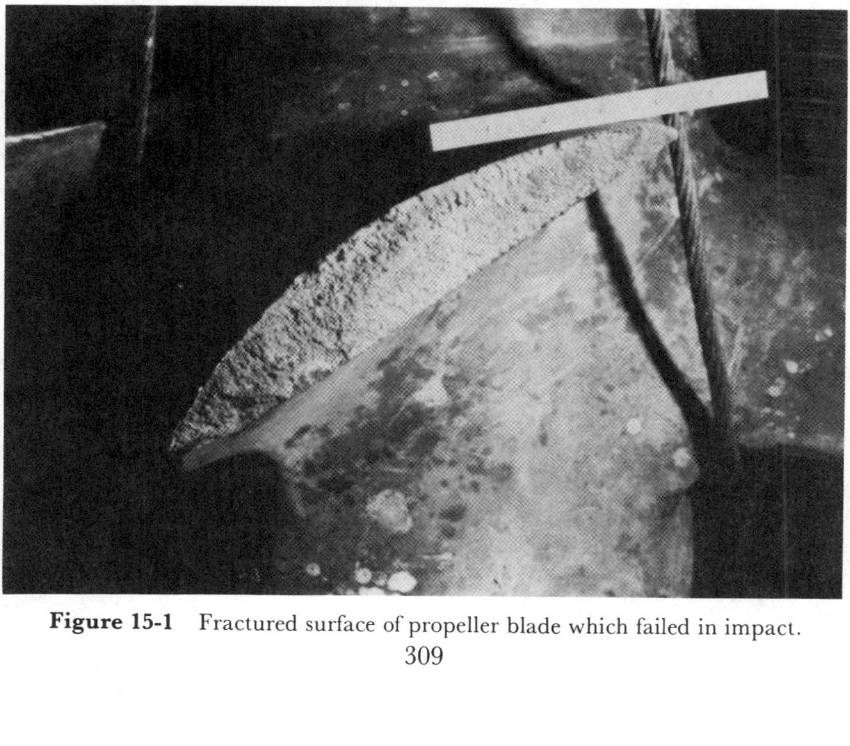

Figure 15-1 Fractured surface of propeller blade which failed in impact.

A-2 A Sudden Propeller Failure at Sea

That one blade of a four-bladed ship's propeller had "fallen off" was reported to the laboratory, and the remainder of the propeller itself was made available. The fracture surface is shown in Figure 15-1.

Visual examination showed little evidence of unsoundness and indicated that fracture initiated near the leading edge of the blade. The coarse character and markings on the fracture surface suggested that fracture was the result of a single blow.

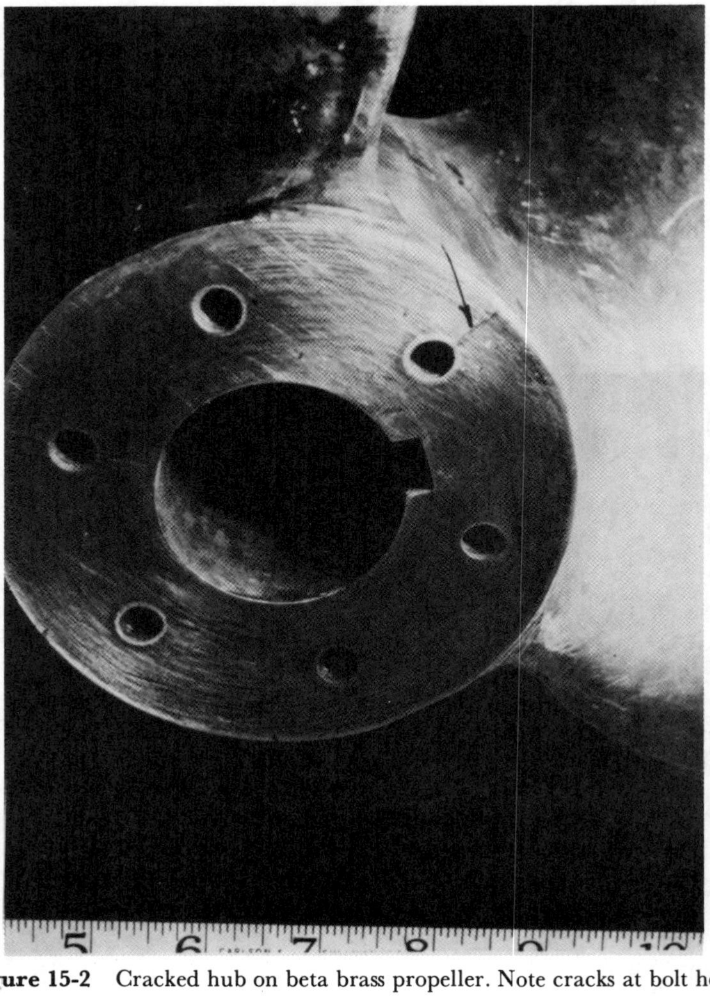

Figure 15-2 Cracked hub on beta brass propeller. Note cracks at bolt holes.

Spectrographic and metallurgical analysis confirmed that the propeller was a two-phase brass (a manganese bronze conforming to ASTM B147 alloy 8A).

Radiographic examination showed that the stub of the blade, near the fracture, was relatively free from casting imperfections. If anything, the blade stub was of better quality than the adjacent blades.

Impact tests performed on samples machined from the metal near the leading edge of the stub showed that the propeller metal had good impact strength. Charpy V-notch values of 38 and 41 foot pounds were recorded.

There was little doubt that failure could be attributed to a specific experience such as hitting a submerged object.

B-3 Cracking of Beta Brass Propellers by Stress Corrosion

Several propellers on small yard craft were cracking severely at the bolt holes as shown in Figure 15-2, and also in the blades, next to the hub. Of

Figure 15-3 Photograph of one of the fractured surfaces on beta brass propeller showing granular separation. (×8)

seven such propellers examined, the distribution of cracking varied as follows:

1) one propeller had 6 cracks (mostly small) on one of its blades and no cracks in the boss;

2) three of the propellers had two or three cracks at the bolt holes in the boss.

3) all propellers had one or more cracks in at least two of the three blades.

Radiographic examination failed to show any evidence of shrinkage or other weakness which might have contributed to the cracking.

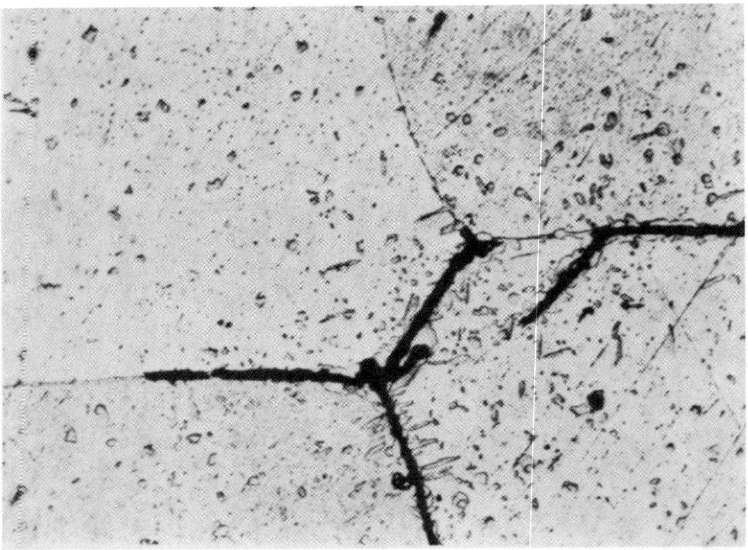

Figure 15-4 Intergranular stress corrosion cracking of beta brass propeller. Note that cracks follow precipitated alpha phase. (×600)

Chemical analysis confirmed that the composition of the propellers was essentially high strength manganese brass, as specified. Physical tests on representative samples cut from sound areas exceeded physical requirements of 75,000 psi and 15% elongation. Impact tests also provided evidence of good properties, i.e. 27 to 32 ft. lbs. in Charpy V-notch. Incidentally, fractures in the tensile specimens showed separation at the grain boundaries, giving a granular effect.

One of the cracks in the boss of this propeller was opened and the cracked surface observed. It was at once apparent that this was a rather unusual type of separation. As shown in Figure 15-3, the metal had parted in a granular fashion, i.e. there were practically no ruptures through grains.

Microexamination of areas near the cracks and within the body of the propeller provided confirmation of a weakness at the grain boundaries. As shown in Figure 15-4, a network of alpha outlined the beta grain boundaries. This structure, particularly evident near some of the cracks, would be expected to constitute a weakening or embrittling factor and it is known to be susceptible to stress corrosion cracking in marine service. Additionally, possible exposure to bird droppings while at dockside would provide an ammonia-type corrodent (p. 177).

C-5 Anchor Shank Failure

The loss of an anchor, a serious matter under some sea conditions, occurred most unexpectedly. The fracture on the recovered end of the shank, shown in Figure 15-5, had an unusually "blotchy" surface. It was reported that the

Figure 15-5 Failed anchor shank. Note the intergranular separation at top left.

failed surface of the shank was bright and metallic when the shank was removed from the water. Had a crack existed part way into the shank prior to the failure, a severely oxidized area would have been noted on the failed surface.

A number of gouges and indentations on the surface of the shank near the failure were noted. These defects probably contributed toward the failure in that they acted as initiating sites for the fracture.

Analysis showed that the metal was a medium carbon (0.5%) steel. Metallurgical examination of the steel near the break showed a ferrite network outlining original austenite grains, Figure 15-6. The grains noted in Figure 15-5 had parted at these ferrite bands.

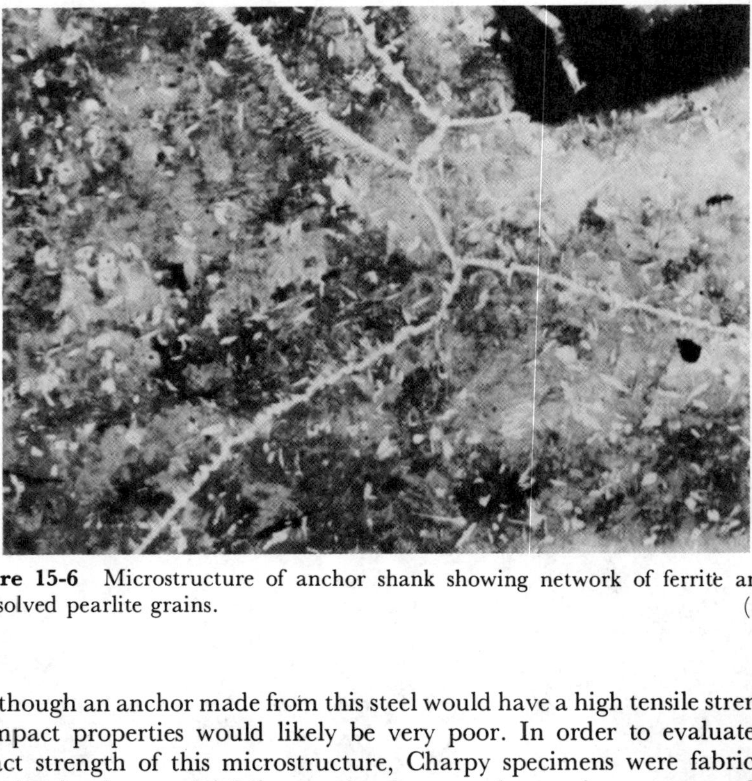

Figure 15-6 Microstructure of anchor shank showing network of ferrite around unresolved pearlite grains. (×80)

Although an anchor made from this steel would have a high tensile strength, its impact properties would likely be very poor. In order to evaluate the impact strength of this microstructure, Charpy specimens were fabricated from the shank near the failure and, when tested, gave impact strengths of only 2 ft. lbs. at room temperature. This indicated a very brittle material: impact strengths in excess of 30 ft. lbs. can be expected for forged mild carbon steels.

It was recommended that measures be taken to improve the impact properties of the steel. This could be accomplished by appropriately heat treating these castings: preliminary tests indicated that a homogenizing treatment would likely be most satisfactory. The castings were accordingly heated to 1800° F and held for about eight hours and then air cooled (normalized). The impact strength of the shank steel heat treated in this way was about 18 foot pounds. The appearance of the fractured surface of an impact specimen of the heat treated steel is compared with that of the as-received anchor shank in Figure 15-7. Note that the heat treated steel is much finer grained and shows some ductility.

Figure 15-7 Comparison between fractured surfaces of Charpy impact specimens The as-received anchor shank steel is at right. The homogenized steel at left is finer grained and shows some deformation, indicating ductility.

C-7 The Bending of Blades of Torpedo Propellers

Propellers on small experimental torpedoes were bending in an unexpected manner, and caused concern to the people involved. Up until that time similar propellers had been giving satisfactory service. Accordingly one of each, new and old, were made available for examination.

As a ready means of comparing physical properties, hardness readings were taken on each propeller. It soon became apparent that there was a considerable difference:

The new was Brinell 63–64,
the old was Brinell 85–89.

Because destructive tests were not permitted at the time, a light macro-etch on the surface of a blade of each propeller served to confirm, as shown in Figure 15-8, that there was in fact a metallurgical reason for the difference in hardness: a considerable difference in grain structure.

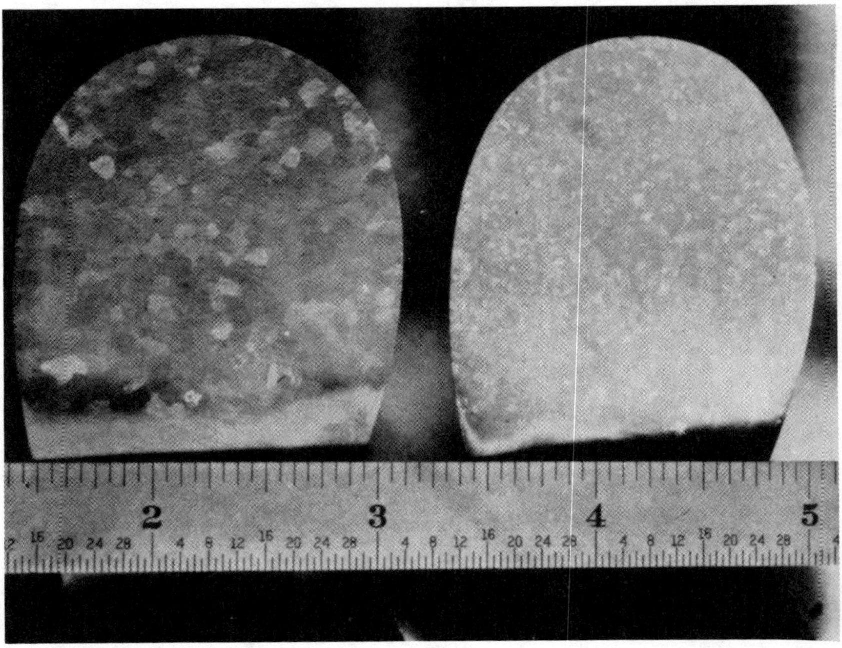

Figure 15-8 Two aluminum alloy torpedo propellers showing difference in grain size (old on right).

Spectrographic analysis showed that both were of a similar aluminum–silicon alloy except that the old one had 0.75% magnesium compared to 0.15% magnesium for the new one. The difference in grain size might be related to the difference in magnesium content—additional magnesium resulting in more nucleating particles and therefore fine grain size.

XVI

MISCELLANEOUS FAILURES

MECHANICAL ASPECTS contributing to failure of bolts in a helicopter engine are evaluated. An interesting difference between aluminum brass and aluminum bronze drive gears is brought out in a failure in which the temperature rose through lack of oil.

High temperature cracking of stainless steel by molten solder occurred during welding. Ultrasonic examination for laminations also revealed weld defects. A bronze valve arm on a submarine fell apart because of defective welding on an unsound casting.

Crescent-type wrenches gave inferior service because of inadequate heat treatment. Hardening of teeth in bandsaw blades must be controlled to limit hardness to the teeth alone. A Naval gun part failed through inadequate heat treatment. Hydrogen "flakes" in steel shafting were attributed to faulty steel mill practice.

A-3 Helicopter fan and hub assembly failure.

A-5 Brass and aluminum bronze gear failure.

B-8 Stainless steel sink weld failure.

C-2 Laminations in steel support plate.

C-3 Separation of submarine's valve disc arm.

C-5 a) Crescent-type wrench failures.
b) Band saw blade failures.
c) Brittle failure of cartridge support.

C-9 Hydrogen cracking in low alloy steel.

A-3 Helicopter Fan and Hub Assembly Failure

The laboratory was asked to investigate the failed fan and hub assembly, illustrated in Figure 16-1, from a helicopter engine. Attachment of hub to fan is accomplished by means of one fitted bolt and seven non-fitted bolts which have appreciable clearance in the receiving holes of both fan and hub. The fitted bolt and four of the non-fitted bolts had failed in service.

The bolts were spectrographically analysed and were found to be 0.3% Mo steel, cadmium plated. It is believed that the quality of bolt, which had been

Figure 16-1 Components from failed helicopter fan and hub assembly. Note elongated bolt holes.

hardened to about Rockwell C20, was adequate for the service intended. The fan and hub were of light alloy composition.

Examination of the fractured fitted bolt, the only bolt that had failed at the hub-fan interface, indicated that it had parted by fatigue. The fatigue fracture had apparently initiated where the bolt had rubbed against the hub. Fretting was noted on the surfaces of the bolt and hub over a small area.

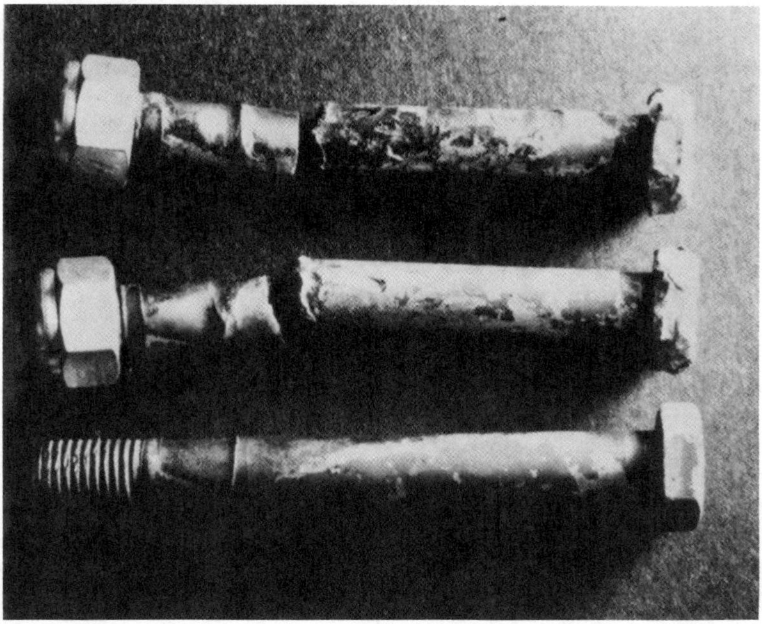

Figure 16-2 Non-fitted helicopter fan and hub assembly bolts showing deformation and failure.

It would appear, then, that in the initial stages, the vibrations set up in service resulted in slight movement between the hub and fan; the vibrations would have resulted in motion around the fitted bolt, causing some deformation to the non-fitted bolts and fretting on the fitted bolt. As the bolt holes enlarged the relative motions increased and wear took place between the hub lands and fan bore.

Examination of the non-fitted bolts (shown in Figure 16-2) indicated two distinct stages of deformation, the second overlapping the first but with a change in position. It is believed that the first stage of deformation took place when the vibrations pivoted around the fitted bolt. With the failure of the fitted bolt the direction of the vibration would change becoming completely circumferential. Also, at this stage, the hub and fan would be held only by the loosely fitted securing bolts, and considerable movement could be expected. The severe pounding of the bolts took place in this latter stage along with the circumferential deformation in the bolt holes (Figure 16-1).

The possibility that the bolts had not been adequately tightened when the fan was installed was considered; however, a much more rapid failure would likely have taken place under such conditions.

With all other factors apparently accounted for, the question of dynamic balance arose. A replacement hub-fan unit assembled in similar fashion to the previous unit, was subsequently tested for dynamic balance. This test confirmed that the replacement unit, although it had successfully been statically balanced, was very unbalanced dynamically.

The importance of dynamic balance, in assemblies otherwise satisfactory, was dramatically illustrated by this failure.

A-5 **Brass and Aluminum Bronze Gear Failure**

Failure of accessory drive gears on a non-magnetic minesweeper was brought to the attention of the laboratory.

The failure consisted of the stripping of the main drive gear and the two pump drive gears, one of which is shown in Figure 16-3; the idler gears, while badly bruised, were whole. The white metal of bearings associated with the gears was partly melted out and white metal was sprayed over the broken teeth of the stripped gears as well as on the idler gears.

In addition, it was reported that end cover bolts had fallen out. Failure appeared, therefore, to be associated with a displacement of the large drive gear and the subsequent cutting off of the oil supply. The mechanism responsible for loosening the bolts so that the large drive gear could be displaced was not apparent.

Spectrographic analysis showed that the large drive gear and the pump drive gears were aluminum manganese brass whereas the idler gears were an aluminum bronze (no zinc).

The compositions were approximately as follows:

	Copper	Aluminum	Iron	Manganese	Zinc
Al Mn Brass	55–60	10	2–3	2–3	rem.
Al Bronze	92	7–8	—	—	—

These compositions were probably chosen so as to have dissimilar alloys mating at the teeth, and because each is a high strength metal. It was puzzling, however, that the brass gear teeth failed but the bronze gear teeth did not.

Figure 16-3 Aluminum manganese brass pump drive gear showing stripped teeth.

Close examination of the fractured surfaces suggested an intergranular type of separation. This was confirmed by examination under the binocular microscope and is shown in Figure 16-4. It was suspected that exposure to temperatures above normal as well as sudden loading may have been responsible for this intergranular separation, and, accordingly, impact tests at several temperatures were run.

The results of these impact tests on Charpy V-notch specimens were as follows:

	70° F	400° F	600° F
Al Mn Brass	13·5 ft. lb.	8·0	3·5
Al Bronze	5·0	5·5	6

These results showed that the brass became weaker as the temperature increased. In addition, the fracture appearance of the brass became more nearly that of the actual failures only as the temperature went over 400 °F.

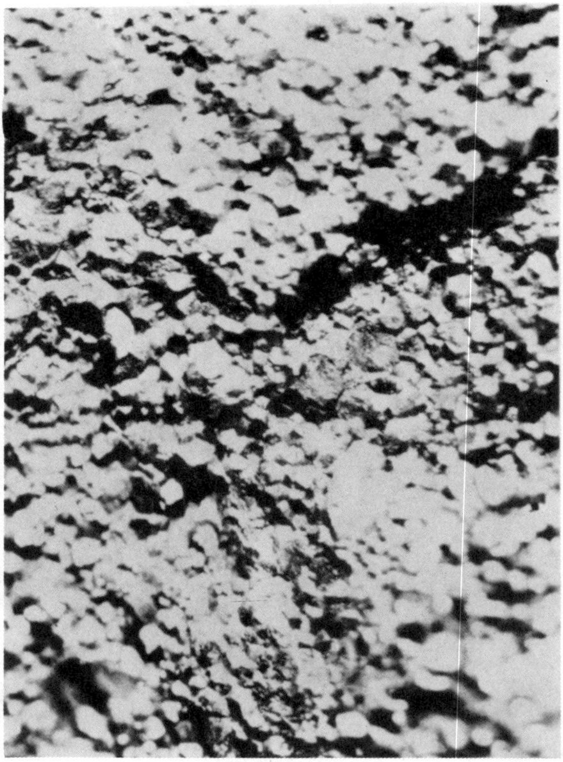

Figure 16-4 Surface of brass gear, Figure 16-3, showing granular separation.

This evidence indicated that failure of a brass tooth at ordinary operating temperatures was unlikely. The fracture of the bronze under test was at all times fine grained and suggestive of a ductile metal.

The failure was attributed to heating resulting from a cut-off oil supply. The mechanism responsible for loosening the end cover bolts so that the large drive gear could be displaced and thereby cut off the oil supply was not, however, determined. Some means to ensure that end cover bolts be locked in place was required.

B-8 Stainless Steel Sink Weld Failure

Samples of cracked stainless steel welds were submitted by the Welding Shop for examination. These welds were made as a sealing edge on stainless steel sinks. The seal had previously been obtained with a solder filler between spot welds.

Figure 16-5 Crack associated with weld of stainless steel sink.

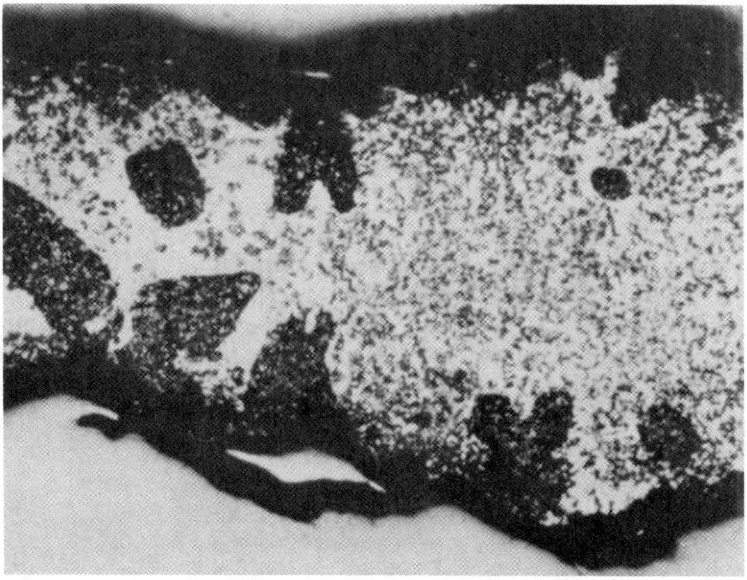

Figure 16-6 Lead tin solder in one of the cracks shown in the stainless steel sink, Figure 16-5. (×300)

The cracks which resulted when the edge on existing sinks was welded by either ordinary electric welding or by argon shielded welding are shown in Figure 16-5.

It was noted that the cracks were adjacent to the spot weld area. This was reasonable because the metal adjacent to the spot weld is in tension and would be subjected to an additional tensile stress from the heating and subsequent cooling of the edge weld. These additive tensile stresses, while considerable, may not be sufficient to rupture the stainless steel. That these stresses were somewhat below the rupture strength of the stainless steel was proven by making up an identical sample, including spot welds and edge weld.

It is known that steel under tensile stresses can be cracked in the presence of molten metals such as copper or lead. In this instance, molten solder appears to have been responsible for the penetrating action which produced the cracks. In Figure 16-6 is shown a typical microstructure of a lead–tin solder as it occurred in one of the cracks. The presence of solder was confirmed in other cracks, whether coarse or fine.

Stainless steel should be kept scrupulously free of solder if welding is called for.

C-2 **Laminations in Steel Support Plate**

An unusual separation at the edge of a $1\frac{1}{4}$ inch plate, used to support a marine turbine, was noted after welding. The separation and weld are shown in Figure 16-7.

Because laminations in the steel were suspected, all the support plates were ultrasonically examined. This confirmed that laminations extended over large areas of several of the plates. Ultrasonic examination also revealed an incomplete penetration in the support bracket welds.

Subsequently one of these was made available for destructive examination. A section cut through a typical area is shown in Figure 16-8. Both the lamination and the incomplete penetration were confirmed.

Metallurgical examination showed the exact nature of these laminations, Figure 16-9, in this AISI 1020 steel. Although the presence of the lamination was not structurally significant in this application, there was concern over the quality of the welding revealed. In service, the welds are subject to severe vibration, and therefore propagation to failure was a possibility.

C-3 **Separation of Submarine's Valve Disc Arm**

The falling apart during dismantling of an exhaust valve disc arm on a submarine in refit was a matter of considerable concern. Failure of this unit underway could result in destruction of a main diesel. The fractured casting

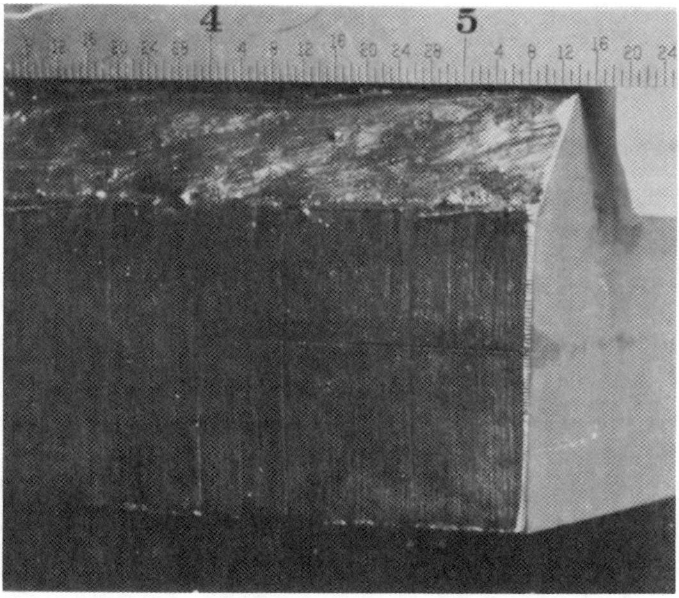

Figure 16-7 Lamination in steel plate opened by stresses from weld (top).

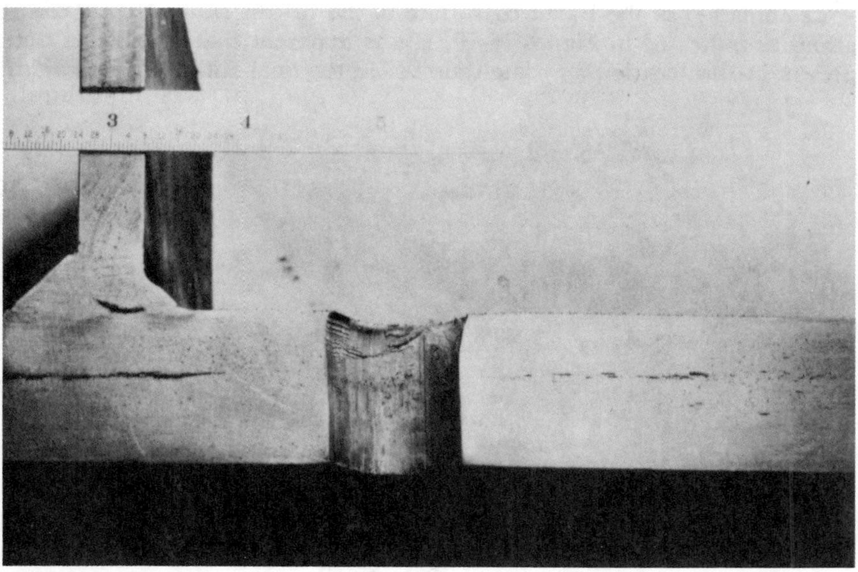

Figure 16-8 Lamination in plate outlined by magnetic particle indications. Note also the incomplete penetration of the weld.

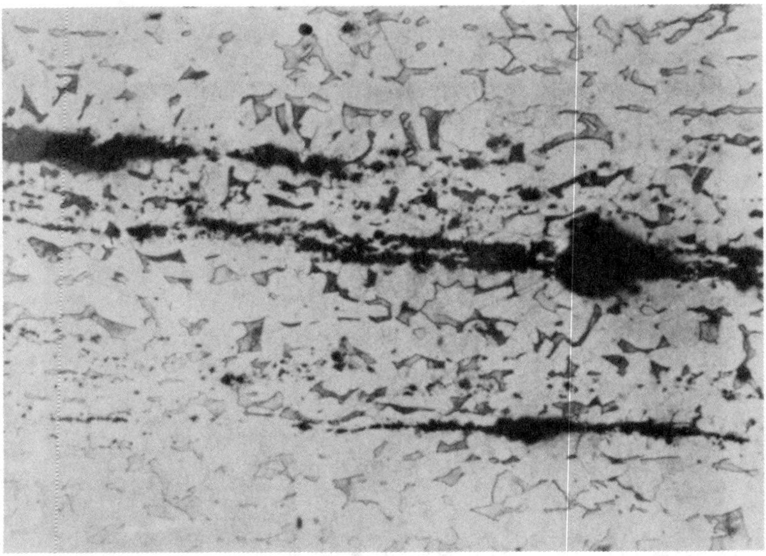

Figure 16-9 Microstructure of laminated area of plate showing voids in steel. (× 80)

was made available. Failure was in the area of a square hole into which the hinge pin fitted.

Examination of the fractured surface of the casting showed large coarse grains as indicated in Figure 16-10. It was apparent that cracks had been present at this location for some time before the final failure. Metallurgical

Figure 16-10 Fractured surface of valve disc arm. Note the large grains at the separation. (× 2)

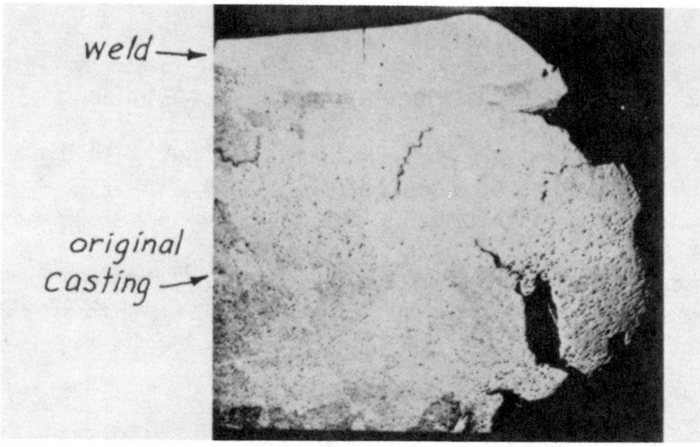

Figure 16-11 Section at fracture showing weld repair and subsurface cracks below weld.

examination showed that the inside surface of the square hole in the casting had been rebuilt by welding. The weld appeared to have been at least partially responsible for cracks forming in the parent metal of the original casting. Cracks, just below the weld surface, are shown in Figure 16-11. This type of

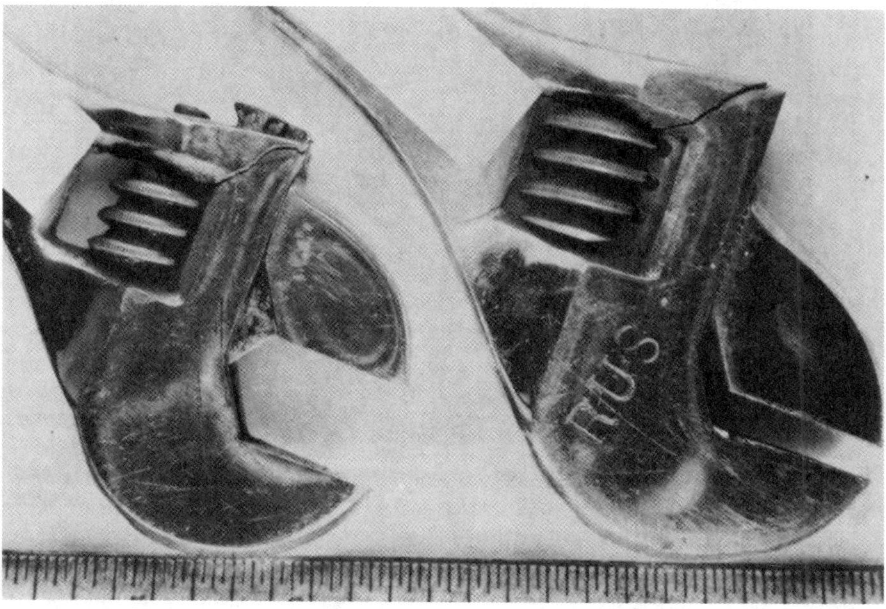

Figure 16-12 Photograph of 12 inch and 10 inch wrenches showing region of failure.

cracking was not present in the parent metal away from the weld area. A crack was also noted in the weld metal.

Spectrographic analysis indicated that the casting was a gunmetal conforming to ASTM Specification B143, alloy 1A. The weld metal was a 2–3 percent silicon bronze.

Radiographic examination of the failed disc arm indicated that it was a reasonably good casting which could be repaired by welding. The disc arm was subsequently repaired by brazing a square wrought tin bronze insert into the casting.

Examination of three other disc arms showed that they also had been rebuilt by welding. It was recommended that they be repaired to ensure reliability.

C-5 a) **Crescent-Type Wrench Failures**

A sudden rash of failures of 10 inch and 12 inch Crescent-type wrenches was reported by a works foreman. In view of a prior total absence of failure of this type of tool an investigation was requested.

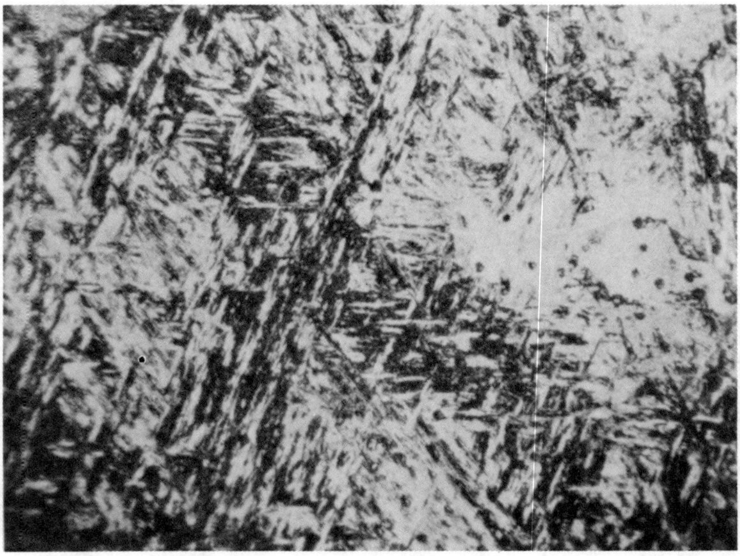

Figure 16-13 Structure of cracked wrenches showing coarse incompletely transformed martensitic structure. (×430)

It soon became apparent that these were from a new supply. Accordingly samples of each type were made available for examination.

The failed wrenches are shown in Figure 16-12. The type of cracking

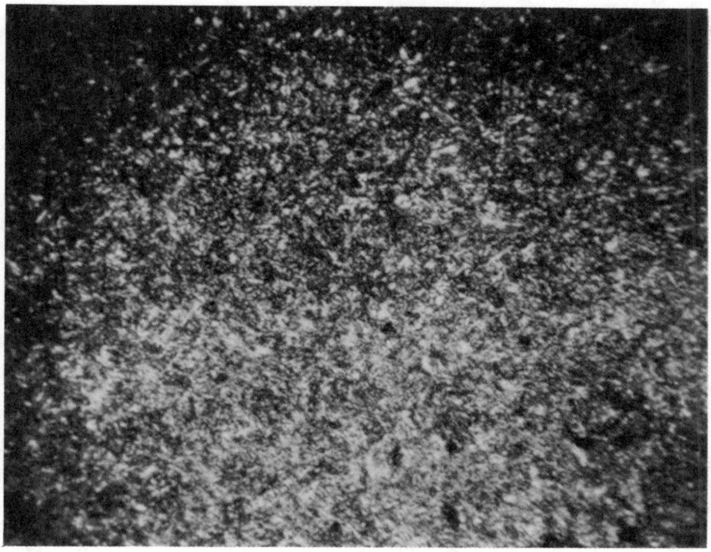

Figure 16-14 Structure of satisfactory Crescent type wrenches. (×430)

suggested the possibility of abuse but then all wrenches would likely have been exposed to similar abuse. Incidentally, the practice of stamping identification (RUS in Figure 16-12) can result in crack initiation.

Microscopic examination proved rather revealing as shown in Figures 16-13 and 16-14. The major difference was that the new stock was poorly heat treated; i.e. contained a coarse structure and was incompletely transformed. The old stock, on the other hand, had an appropriate fine grained tempered martensite structure.

C-5 b) **Band Saw Blade Failures**

Early failure of the first two bandsaw blades made up from a large shipment of rolls of bandsaw blading was reported. A sample of one of the blades which failed in 30 minutes use is shown in Figure 16-15. The cracks are periodic with the wave set of the blade and appear only on alternate crests; i.e. the crests on the same side of the blade.

Metallurgical examination showed that in hardening the teeth of the saw the manufacturer also extended the hardened zone into the blade. The extent of the hardening is shown in Figure 16-16. This photograph also shows a crack which started near the root of a tooth in the brittle, hardened area. Figure 16-17 shows a properly heat treated unused blade from another supply in which the root of the tooth is not hardened, thereby giving the blade greater toughness and flexibility.

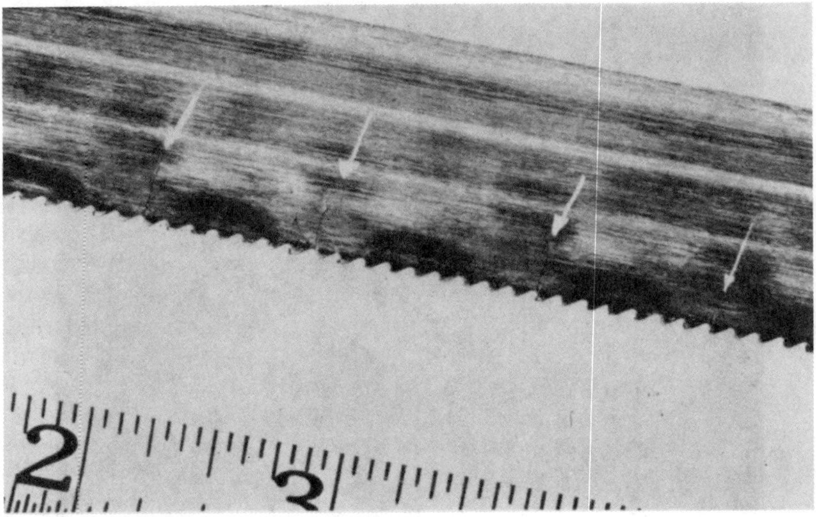

Figure 16-15 Band saw blade cracked after 30 minutes.

Flexibility is obviously required in a band saw since the blade is continually being bent around the radius of the pulley. The poor blade cracked because, being improperly hardened, it lacked this flexibility. The cracks formed at the region of highest stress; i.e. the valleys of the waves in the blade which are opposite in curvature to that of the band saw pulley.

Why the improperly heat treated blade should fail by fatigue was not immediately apparent because the heat treatment should not decrease the strength of the blades. It is noteworthy, however, that in the properly heat treated blades, the parent metal at the root of the teeth, being suitably

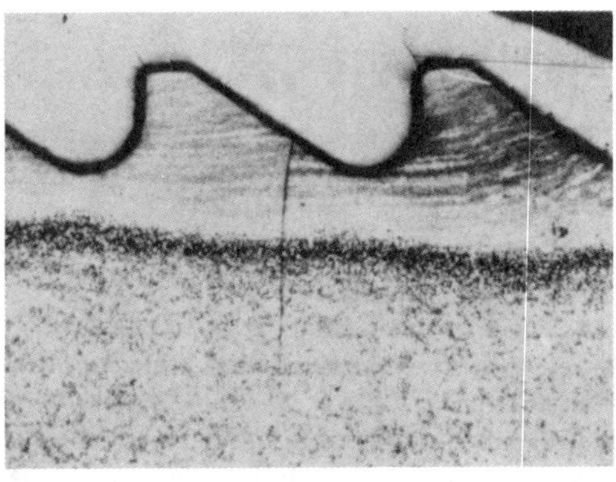

Figure 16-16 Cracked blade showing that teeth were hardened well past the teeth roots. (×12)

tempered, would have low or negligible residual stresses. However in the fully hardened steel in Figure 16-16, high residual stresses would be present. It appears that the additive effects of the residual stresses to the service stresses have, in the case of the poorly heat treated blades, resulted in the fatigue cracking. In addition, the fully hardened brittle steel would have been more notch sensitive.

The poor quality band saw blades were returned to the manufacturer. Subsequent shipments of saw blades served satisfactorily.

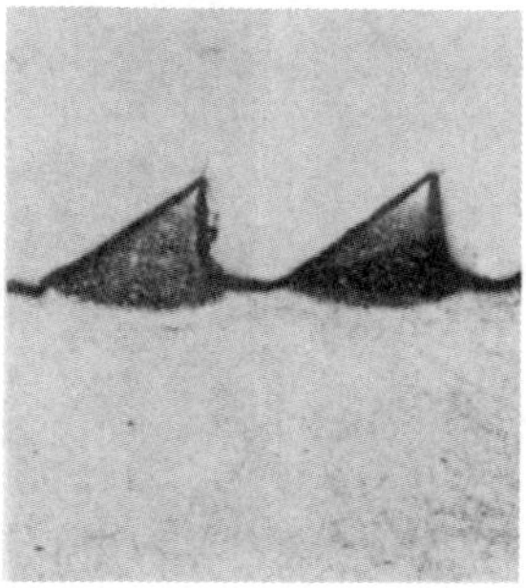

Figure 16-17 Properly hardened blade. Only teeth have been hardened. (×12)

C-5 c) **Brittle Failure of Cartridge Support**

The brittle failure of a cartridge support on a large gun was of concern to the armament people. This part of the loading mechanism is shown in Figure 16-18.

Spectrographic examination showed that the shell support was a low alloy steel containing 0.2% chromium and 0.3% nickel. The steel had been carburized to a depth of about 0.030 inch and had been hardened to Rockwell C52.

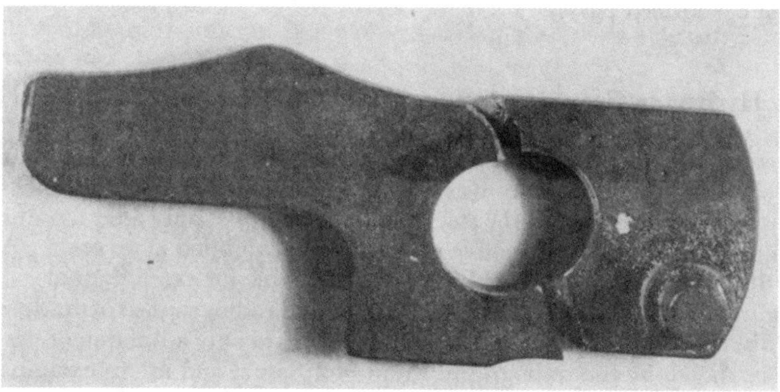

Figure 16-18 Fractured carburized steel shell support part. (×2)

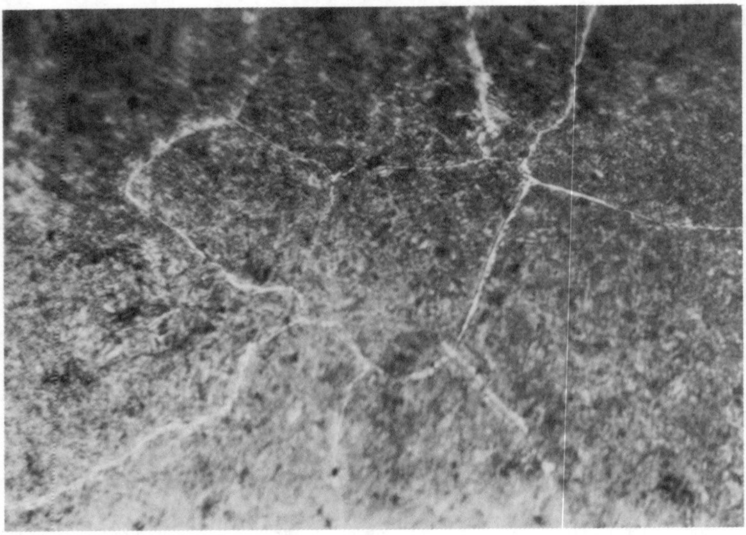

Figure 16-19 Microstructure of carburized area of shell support showing deleterious carbide network around tempered martensite. (× 400)

Metallurgical examination of the failed part indicated that the steel had been inadequately heat treated following the carburization. Figure 16-19 shows a carbide network at the grain-boundaries which formed during carburization and should have been redissolved before subsequent heat treatment. A grain boundary carbide network is a frequent cause of failure from brittleness in carburized parts.

C-9 **Hydrogen Cracking in Low Alloy Steel**

Following the discovery by the Machine Shop of cracks in 6 inch diameter steel shafting, the laboratory was asked to examine and assess another piece of this stock. The cracks in the steel had been observed only after a number of forgings, made from the defective material, had exhibited large cracks. A half inch disc, representing the steel shaft was available for examination.

Figures 16-20 and 16-21 are photographed and radiographed of the defective steel disc. The cracks shown in the photograph are only a fraction of the total number shown by the radiograph. From both visual and micro-examination it was determined that these cracks were "shatter cracks" or "flakes".

These "flakes" in alloy steel forgings are permanent defects caused by a

Figure 16-20 End of defective shafting showing hydrogen cracks (white marks).

faulty steel-making process. For alloyed steels, "flakes" can be attributed to the following:

a) high hydrogen content,

b) segregation—producing regions of high alloy content,

c) rapid cooling.

No equipment was available to determine whether or not the steel contained an unusually high hydrogen concentration. However, the microstructure indicated that homogenization of the original cast structure had not been attained, and there had been a relatively rapid cool from forging temperature. The rapid cool coupled with the segregation could induce "flakes" in this alloy steel even with normal hydrogen content.

Comparison with the microstructure of another sample of the same type

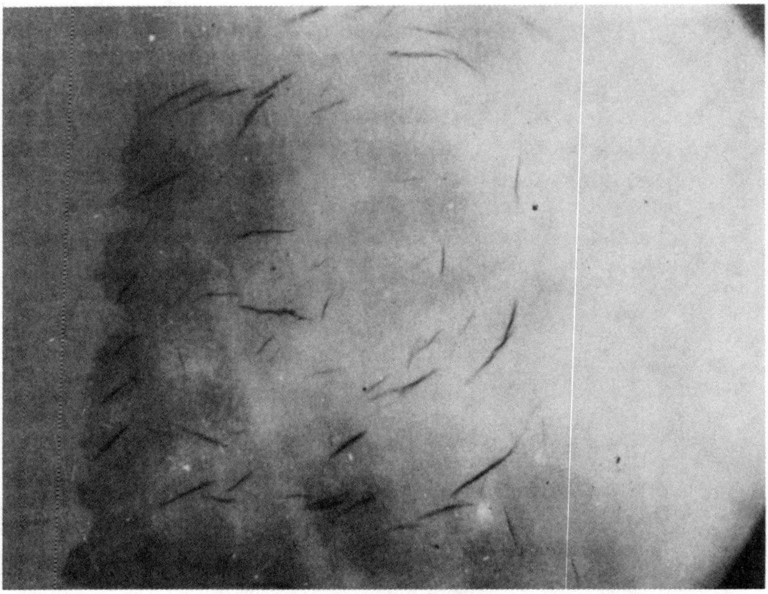

Figure 16-21 Radiograph of disc taken from defective shafting showing many "flakes" (cracks).

of steel shafting, which showed no visible cracks, confirmed the above indications. The structure of the crack-free shafting showed that it had been cooled very slowly after forging. Radiographic and magnetic particle inspection also proved that the steel was free from cracks. The manufacturer was notified that this shipment of steel was defective.

APPENDIX

REFERENCES AND SUPPLEMENTARY SOURCES

THIS LISTING is obviously not at all complete, but should prove useful within the objectives of the text. The list of texts and other sources is provided for those who wish to extend their knowledge in specific fields. Many of the texts listed have extensive references, further extending the opportunity to dig deeper.

Important articles in the desired field in the current technical literature also have extensive reference lists.

General Metallurgy

1. *Metals Handbook,* published by the American Society for Metals, 1948.
2. *Metals Handbook,* 8th edition, Vol. 1, 1961, published by American Society for Metals.
3. Camp, J. M. and Francis, C. B., *The Making, Shaping and Treating of Steel,* by United States Steel Co., 1951, Pittsburgh, Pa.
4. Wulff, J., Taylor, H. F., and Shaler, A. J., *Metallurgy for Engineers; Casting, Welding, and Working,* John Wiley and Sons Inc., N.Y. 1952.
5. Samans, C. H., *Engineering Metals and Their Alloys,* The MacMillan Co., N.Y., 1950.
6. Brick, R. M., Gordon, R. B., and Phillips, A., *Structures and Properties of Alloys,* Third Edition, 1965, McGraw-Hill Book Co. Inc.
7. Gillett, H. W., *The Behaviour of Engineering Metals,* John Wiley and Sons, 1951.
8. Kent, R. van Horn, Editor, *Aluminum Properties Physical Metallurgy and Phase Diagrams,* American Society for Metals, Metals Park, Ohio, 1967.

Casting

9. Briggs, C. W., *The Metallurgy of Steel Castings,* McGraw-Hill Book Co. Inc., N.Y., or Pergamon Press Ltd., London, 1954.
10. Murphy, A. J., *Non-Ferrous Foundry Metallurgy,* McGraw-Hill Book Co. Inc., New York, 1946.

11. Wallace, John F., *Fundamentals of Risering Steel Castings,* Steel Founders Society, Cleveland, 1960.
12. Charles F. Walton, Editor, *The Gray Iron Castings Handbook,* Gray Iron Founders' Society Inc., Cleveland, Ohio, 1958.
13. St. John, Harry M., *Brass and Bronze Foundry Practice,* Penton Publishing Co., Cleveland, Ohio, 1958.

Welding

14. Phillips, A. L., Editor, *Welding Handbook,* Fourth Edition in 5 volumes, published by American Welding Society, New York, 1962.
15. Udin, H., Funk, E. R., and Wulff, J., *Welding for Engineers,* John Wiley and Sons Inc., N.Y., Chapman and Hall, London, 1954.
16. West, E. G., The *Welding of Non-Ferrous Metals,* Chapman and Hall, London, 1951.
17. Jefferson, T. B. and Woods, G., *Metals and How to Weld Them,* published by James F. Lincoln Arc Welding Foundation, Cleveland, Ohio, 1954.
18. Stout, R. D. and Doty, W. D., *Weldability of Steels,* Welding Research Council, New York, 1953.

Metallurgical Laboratory

19. Kehl, G. L., *Principles of Metallographic Laboratory Practice,* McGraw-Hill Book Co. Inc., 1949.
20. *Fifty Years of Progress in Metallographic Techniques,* ASTM Technical Publication No. 430, 1968.

Metal Failures (General)

21. Polushkin, E. P., *Defects and Failures of Metals,* Elsevier Publishing Co., Amsterdam, London, New York, Princeton, 1956.
22. *Technical Report,* New Series, Vol. 3, 1957, by British Engine, Boiler and Electrical Insurance Co. Ltd., Manchester, England.
23. Thielsch, H., *Defects and Failures in Pressure Vessels and Piping,* Reinhold, N.Y., Chapman and Hall, London, 1965.
24. Wulpi, Donald J., a series of 10 articles on "How components fail", *Metal Progress,* September 1965, Vol. **88**, No. 3, p. 73–77, and in succeeding issues of *Metal Progress* with completion in September 1966, Vol. **90**, No. 3.
25. Tittel, R. C., "Analyzing diesel engine failures", *Metal Progress,* Vol. **95**, No. 4, April 1969, p. 74–78.

26. Averbach, B. L., Felbeck, D. K., Hahn, G. T. and Thomas, D. A., Editors, *Fracture*, MIT Press and John Wiley and Sons, 1959.
27. *Analysis of Service Failures,* Republic Alloy Steels, 1961.
28. Osborn, C. J., Editor, *Fracture,* the proceedings of Tewksbury Symposium, Faculty of Engineering, University of Melbourne, 1965.
29. Read, Harold J., *Hydrogen Embrittlement in Metal Finishing,* Reinhold Publishing Corporation, N.Y., 1961.
30. Osgood, Wm. R., Editor, *Residual Stresses in Metals and Metal Construction*, Reinhold Publishing Corp., N.Y., 1954.

Fatigue

31. Battelle Memorial Institute Staff, *Prevention of Failure of Metals Under Repeated Stress,* John Wiley and Sons, 1941.
32. Cottell, G. A., "Fatigue failures, with special references to fracture characteristics", in *Technical Report, New Series,* Vol. **2**, British Engine, Boiler and Electrical Insurance Co. Ltd., Manchester, 1954, p. 221–256.
33. Murray, W. M., Editor, *Fatigue and Fracture of Metals,* a Symposium held at M.I.T., 1950, John Wiley and Sons Inc., N.Y., or Chapman and Hall, London.
34. Cazaud, R., *Fatigue of Metals,* Chapman and Hall Ltd., London, 1953.
35. Grover, H. J., Gordon, S. A., and Jackson, L. R., *Fatigue of Metals and Structures,* Bureau of Aeronautics, Department of the Navy (U.S.), 1954.
36. Burke, J. J., Reed, N. L., and Weiss, V., *Fatigue—An interdisciplinary Approach*, Syracuse University Press, 1964.
37. Stulen, F. B. and Schulte, W. C., Fatigue failure analysis and prevention, in *Metals Engineering Quarterly,* ASM, August 1965, Vol. **5**, No. 3, p. 30–39.
38. La Belle, Jack E., "Practical aspects of fatigue" in *Metal Progress,* May 1965, Vol. **87**, No. 5, p. 68–73.
39. Gohn, G. R., "The mechanism of fatigue" in *Materials Research and Standards,* ASTM, February 1963, Vol. **3**, No. 2.
40. Hardrath, H. F., "Crack propagation and final failure", *ibid.*
41. Peterson, R. E., "Engineering and design aspects", *ibid.*

Design (Metals)

42. Peterson, R. E., "Engineering and design aspects", *ibid.*
43. Horger, O. J., Editor, *Metals Engineering Design,* Second Edition, an ASME Handbook published by MacGraw-Hill Book Co. Inc., 1965.

Brittle Fracture

44. Parker, E. R., *Brittle Behaviour of Engineering Structures,* John Wiley and Sons, Inc., N.Y., 1957.
45. Pellini, W. S., and Puzak, P. P., *Practical Considerations in Applying Laboratory Fracture Test Criteria to the Fracture-Safe Design of Pressure Vessels,* Naval Research Laboratory Report 6030, November 5, 1963.
46. *Conference on Brittle Fracture in Steel* reprinted from the Journal of the West of Scotland Iron and Steel Institute, 1952–53, Vol. 60, published by the West of Scotland Iron and Steel Institute, Glasgow.
47. Shank, M. E., *Control of Steel Construction to Avoid Brittle Failure,* Welding Research Council, New York, 1957.
48. Brown, Jr., W. F., and Strawley, J. E., *Plane Strain Crack Toughness Testing of High Strength Metallic Materials,* ASTM S.T.P., No. **410**, published by ASTM, 1966.
49. Tetelman, A. S., and McEvily, Jr., A. J., *Fracture of Structural Materials,* John Wiley and Sons, Inc., 1967.
50. Kenny, P. and Campbell, J. D., *Fracture Toughness, An Examination of the Concept in Predicting the Failure of Metals,* Pergamon Press, 1967.

Corrosion (General)

51. Uhlig, H. H., Editor, *The Corrosion Handbook* of the Electrochemical Society, Inc., N.Y., published by John Wiley and Sons, Inc., N.Y., or Chapman and Hall Ltd., London, 1948.
52. Evans, U. R., *Metallic Corrosion Passivity and Protection,* Edward Arnold and Co., London, 1946.
53. Evans, U. R., *The Corrosion and Oxidation of Metals: Scientific Principles and Practical Applications,* Edward Arnold (Publishers) Ltd., 1960.
54. Uhlig, Herbert H., *Corrosion and Corrosion Control,* John Wiley and Co., New York, 1963.
55. T. Howard Rogers, *Marine Corrosion Handbook,* McGraw-Hill Company of Canada Ltd., Toronto, 1960.

Stress Corrosion

56. *Symposium on Stress-Corrosion Cracking of Metals,* 1944, jointly published by ASTM and AIME, 1945.
57. Robertson, W. D., Editor, *Stress Corrosion Cracking and Embrittlement,* a 1954 Symposium by the Electrochemical Society, John Wiley and Sons, New York, 1956.

58. Nielsen, N. A., "The role of corrosion products in crack propagation in austenitic stainless steel. Electron microscopic studies", in *Physical Metallurgy of Stress Corrosion Fracture,* edited by Rhodin (Thor N.), Interscience 1959, p. 121–154.
59. Pickering, H. W., Beck, F. H., and Fontana, M. G., "Wedging action of solid corrosion product during stress corrosion of austenitic stainless steels" in *Corrosion,* June 1962, Vol. **18**. No. 6, 230t–239t.
60. Ohtani, N. and Dodd, R. A., "Mechanisms of transgranular stress corrosion cracking of solid solution alloys", in *Corrosion,* May 1965, Vol. **21**, No. 5, p. 160–172.
61. Tromans, D. and Nutting, J., "Stress corrosion cracking of face-centred-cubic alloys", in *Corrosion,* May 1965, Vol. **21**, No. 5, p. 143–160.
62. Swann, Peter R., "Stress corrosion failure", in *Scientific American, February* 1966, Vol. **214**, No. 2, p. 72–81.
63. Peters, B. F., Carson, J. A. H. and Barer, R. D., "Stress corrosion cracking in marine service" in *Materials Protection,* May 1965, Vol. **4**, No. 5, p. 24–37.
64. *Stress corrosion cracking in aircraft structural materials,* AGARD Conference Proceedings Series, No. **18**, 1967.

Non-Destructive Testing and Stress Analysis

65. McMaster, R. C., Editor, *Non-Destructive Testing Handbook,* in two volumes, Ronald Press Co., New York, 1959.
66. Hetényi, M., Editor, *Handbook of Experimental Stress Analysis,* John Wiley and Sons, 1950.
67. Hinsley, J. F., *Non-Destructive Testing,* MacDonald and Evans, London, 1959.
68. *Radiography in Modern Industry,* Eastman Kodak Co., Rochester, N.Y., 1947.
69. *Proceedings of the Third International Conference on Non-Destructive Testing,* Pan-Pacific Press, Tokyo, 1961.
70. *Proceedings of the Fourth International Conference on Non-Destructive Testing,* Butterworths, London, 1964.

CURRENT TECHNICAL LITERATURE

Current technical literature which may be consulted for recent work in a particular field includes:

British Corrosion Journal, London
Corrosion, NACE, Houston, Texas
Corrosion Science, Pergamon Press, Oxford
International Journal of Nondestructive Testing, Gordon and Breach, London
Journal Institute of Metals, London
Journal of the Iron and Steel Institute, London
Metallurgia, London
Metals Engineering Quarterly, ASM, Metals Park, Ohio
Metal Progress, ASM, Metals Park, Ohio
Metals Review of Literature, ASM
Material Evaluation, ASNT, Evanston, Ill.
Materials Protection, NACE, Houston, Texas
Non-destructive Testing, Research and Practice, Iliffe Pub. Ltd., Guildford, England
Transactions Institute of Marine Engineers, London
Welding Journal, AWS, New York

INDEX